Principles and Techniques for an Integrated Chemistry Laboratory

David A. Aikens
Ronald A. Bailey
James A. Moore
Rensselaer Polytechnic Institute, Troy, New York
Gary G. Giachino
Baker University, Baldwin City, Kansas
Reginald P. T. Tomkins
New Jersey Institute of Technology, Trenton, New Jersey

WAVELAND

PRESS, INC.

Prospect Heights, Illinois

For information about this book, write or call:

Waveland Press, Inc.
P.O. Box 400
Prospect Heights, Illinois 60070
(708) 634-0081

Copyright © 1984, 1978 by David A. Aikens, Ronald A. Bailey, Gary G. Giachino, James A. Moore, and Reginald P. T. Tomkins
The 1978 version of this book was entitled *Integrated Experimental Chemistry: Volume 1, Principles and Techniques*.

ISBN 0-88133-102-3

Printed in the United States of America

9 8 7 6 5 4

Contents

Preface

A number of chemistry departments have concluded that the undergraduate chemistry laboratory program can be improved by replacing the traditional compartmentalized laboratories in analytical, organic, inorganic, and physical chemistry with an integrated laboratory program. In comparatively few of these approaches, however, has integration been complete, and usually one or more of the traditional areas has not been included. Other approaches have been based on specialized or costly experiments. The lack of a comprehensive textbook for a unified laboratory has certainly been a barrier towards progress in the adoption of this concept.

It has become apparent that the modern practice of chemistry demands that students become familiar with a variety of instrumentation. The use of techniques such as infrared, visible-ultraviolet and nuclear magnetic resonance spectroscopy, gas chromatography, and liquid chromatography are becoming routine in industrial, academic, and governmental laboratories.

In 1971, the Chemistry Department at Rensselaer Polytechnic Institute initiated a four semester unified laboratory program which combined all of the required chemistry laboratory courses taken by chemistry majors after the freshman year (analytical, instrumental, organic, and physical chemistry). The sequence is based on two 4-hr laboratory periods a week in each semester, and a discussion period each week in the first two semesters.

This text has been found useful in providing students with a description of basic techniques and procedures in wet chemistry and instrumental areas.

D.A. Aikens
R.A. Bailey
G.G. Giachino
J.A. Moore
R.P.T. Tomkins

Introductory Remarks to Students

The importance of a good laboratory program lies in the fact that chemistry is fundamentally an experimental science; theories are necessary to correlate experimental observations and to guide the development of new experiments, but they are valid only in so far as they can do this. Most students majoring in chemistry will spend a significant portion of their future careers involved with laboratory work; either as direct participants, or as those who must use and evaluate experimental results. Consequently, a good understanding of how these results are obtained, and a sensitivity for their probable validity and reliability, is important.

The basic concept of an integrated or unified laboratory program is the incorporation in a logical fashion of several techniques and procedures into one experiment. For example, an experiment may involve a particular synthetic procedure, use various methods of purification and characterization, include some physical property measurements on the product and, where appropriate, study the thermodynamics or kinetics of the reaction. This approach emphasizes that techniques should not be regarded as being specific to the various traditional areas of chemistry, i.e., organic, physical, analytical, or inorganic chemistry, but rather that they can be applied to study problems in any area of chemistry, under appropriate circumstances.

This book provides background information useful in performing many types of chemical experiments. This level of understanding of techniques as well as theory behind the techniques is essential if laboratory work is to be done correctly and efficiently. However, it is assumed that students will have taken theory courses appropriate to the assigned experiments and that they are familiar with textbooks which deal more thoroughly with theoretical principles.

The experimenter must be aware of the safety precautions necessary in the laboratory and must read the chapter on safety before beginning laboratory work. All safety regulations must be understood and obeyed. Organized work in which the procedures have been thought out carefully is the best safety precaution.

Safety

A chemistry laboratory is an inherently dangerous place. Fire, toxic vapors, exploding containers, poisonous substances, broken glass, and corrosive materials are among the more common dangers. All of these hazards* are aggravated by careless or thoughtless work.

The main goal of this section is to encourage an attitude that will prevent accidents. We will also discuss some of the problems that may arise in the laboratory, but the rules, suggestions, and examples given here are not all-inclusive (nor can they *ever* be).

The following guidelines must be read *and followed* to minimize the chances of accidents. They are supplemented by more detailed instructions dealing with specific instances which should be considered as examples of proper means of disposal, of how to prevent accidents, or of how to act to minimize danger when an accident occurs.

1. Approved eye protection (shatterproof goggles or safety glasses *with* side shields) *must be worn at all times in the laboratory*. Contact lenses must *never* be worn in the laboratory because corrosive materials can become entrapped between the lens and the cornea where the chemicals may not be reached, even during intensive washing. These residual amounts of substances, although small, can cause irreparable damage.

2. It is *most* strongly recommended that each student wear a lab coat, or at least a lab apron, during experimentation. Protective gloves (e.g., disposable plastic gloves) should be used when handling dangerous materials. Shorts and

---

* The hazards arising from using glass tubing and from the potential danger of broken glass will be treated in the section on glass tubing. For detailed procedures with specific chemicals see Refs. 1,2,3.

mini-skirts provide little protection from splashed chemicals and are *not* safe lab attire. Sandals are similarly unsafe, and bare feet are absolutely forbidden in the laboratory. Individuals wearing long hair must tie it back to minimize the possibility of contact with burners and other equipment. Similarly, necklaces, bracelets, and other loose articles of attire are not to be worn in the laboratory because entanglement with equipment or fixtures can cause accidents and could prevent emergency departure from a dangerous area.

3. Note the location of eye-wash fountains and eye-wash bottles and learn their proper use from the laboratory instructor. In the event that chemicals should enter the eye(s) (or for chemical burns) call for the assistance of the instructor and wash carefully and as quickly as possible with large amounts of water.

4. *All* reactions using (or evolving) volatile, noxious or highly combustible chemicals *must* be performed in the hoods.

5. Know the location of fireblankets and showers and ask the instructor for information about their proper use. Note emergency exit routes. In the event of a fire do *not* attempt to extinguish it yourself, but call the instructor.

6. Water on the floor can cause slips and falls, and should be mopped up immediately. Chemical spills must also be removed at once. Each work-area (this includes balances, instruments, desk tops) should be cleaned after each use to remove, as completely as possible, any chemical residues.

7. Eating, drinking, and smoking are not permitted in the laboratory. They invite ingestion of poisonous chemicals and a burning cigarette may be overlooked as a possible source of ignition of flammable vapors.

8. Be especially cautious in situations where special hazards may occur.

    a. Vessels under vacuum can implode violently. Dewar flasks and similar containers should be wrapped with tape to help contain flying fragments should they break.

    b. Vessels under pressure can explode. Never heat a sealed container unless it is specifically designed for that purpose. Be sure that gases entering a system from a compressed gas cylinder cannot generate dangerous pressures.

    c. Distillation of flammable, volatile liquids (e.g., ether) requires special caution. *No* flames may be in the vicinity. Condensers must be cooled with an adequate flow of water. Most organic vapors are heavier than air and can flow for considerable distances, at explosive concentrations, along a bench or the floor.

    d. Some chemicals (e.g., alkali metals, metal hydrides) react violently with water; be sure you know how to handle them and how to dispose of them safely.[1-3]

    e. Electrical equipment provides a possible cause of electrocution. Sparking from frayed or broken cords also may cause ignition of solvents. Report such hazards to the instructor and do *not* use equipment in this condition.

9   Under *no* circumstances is unsupervised experimentation permitted. Violation of this rule will result in permanent loss of access to the laboratory and its facilities.

10.   Disposal of chemicals.[3]

   a.   Organic wastes (both solid and liquid) should be placed in appropriately labeled containers.

   b.   *Small* amounts of water-soluble substances may be flushed into the drains *with large amounts of water*. Acids and bases must first be neutralized or highly diluted, and then properly discarded.

   c.   *Never* put acids, bases, or oxidizing agents in the organic waste containers. They may catalyze (possibly violent) chemical reactions resulting in the evolution of heat and fumes.

   d.   Solid waste must *not* be put in the sink. Use garbage cans for filter paper, boiling stones, towels, and similar debris. Clogged drains can cause flooding.

   e.   Check bottle labels for warnings specific to the particular chemicals in use (e.g., lachrymator, carcinogen, toxic substance) and for information about antidotes, or proper procedure(s) to be used in case of accidental exposure or to be followed for disposal. *Note*! The absence of a warning should *not* be misconstrued as an indication of the absence of dangerous properties. The physiological activity of many materials is unknown.

11.   Extraneous materials in the laboratory (e.g., books, coats) clutter the working area and create unsafe conditions. Clothing, books, and other nonessential materials, should be stored outside the laboratory, in the places provided.

12.   Excess noise, disruptive behavior, pranks, and socializing with classmates, have no place in the laboratory.

13.   The prime cause of accidents in laboratories is a lack of forethought. Think *before* acting! Be aware of your neighbor's activities, because *his* carelessness may be your undoing.

In recent years, the United States federal government has recognized the severity of hazards that arise in many professions, particularly the chemical profession. This recognition has taken the form of the "Occupational Safety and Health Act"[4a,b,c] (abbreviated OSHA; pronounced oh-shaa). This act has resulted in many regulations intended to promote safe working conditions. However detailed these regulations may be, *you* are the controlling factor. If you take the time to read the experiment and to prepare a checklist of procedures to be performed during the period *before* you come to the laboratory, questions should arise as to the *exact* way an experiment is to be performed. These questions (and any that arise *during* the experiment) must be clarified before you proceed with the work. Be aware of the toxicity of the chemicals used.[5,6] *Think* about the results of your actions in terms of common sense and not only in terms of a sophisticated chemical procedure, and 99% of the causes of potential laboratory accidents will have been eliminated.

Despite the best efforts of the most well-intentioned investigators, accidents will happen. The reality of this eventuality prompts the requirement to wear appropriate eye protection (at a minimum, shatterproof lenses in unbreakable frames, fitted with side shields) and the urgings above concerned with proper laboratory attire. In addition to goggles, full-face shields of transparent plastic (at least 4 mm thick) reaching to the chest (and thereby protecting the throat from sharp projectiles) are available when reason, experience, or instinct, indicates their use. Similarly, transparent, standing safety shields of shatterproof glass or thick, transparent plastic are available to interpose between you (and also your colleagues across the bench) and a reaction vessel. Leather gauntlets, heavy aprons, fireproof clothing and similar items are available when their use is indicated.[1-4]

In the event that you are struck in the eye by a projectile do *not* attempt to remove it yourself, but enlist the aid of classmates in calling the accident to the laboratory instructor's attention. Do *not* leave your station until told to do so by the instructor, who will arrange for emergency treatment. If liquid should splash in your eyes, respond to nature's reflex and keep your eyes *closed*. Call for help and ask your colleagues to move you to the nearest eye-wash fountain (eye-wash bottle) or sink and to notify the instructor of the situation. Do *not* open your eye(s) until a gentle stream of water has begun to flush the affected area. Only then should every effort be made to open your eyes as widely as possible to facilitate complete removal of the foreign material, particularly that trapped under the upper or lower eyelids. Washing should be continued until emergency transportation has been arranged. The affected area should be covered with cool, wet cloths during transportation, if washing cannot be continued.

Contact of small areas of the skin with corrosive (and/or readily absorbed) materials is best treated by washing the affected area with large amounts of water while asking a classmate to notify the instructor. If no damage is apparent, washing with a mild soap is helpful in removing the last traces of a contaminant. Any clothing that may be contaminated should be removed immediately and placed in a hood. Judgement about further treatment should be left to the instructor. However, if subsequent discomfort occurs, medical assistance should be sought immediately.

If corrosive (and/or readily absorbed) chemicals should come in contact with large areas of the body and/or clothing, the best course of action is to disrobe the injured individual under the emergency shower while a colleague activates the flow of water. If large areas of skin are contaminated, washing should continue while emergency transportation is arranged by the instructor or his designated agent. Modesty should not be allowed to interfere with the emergency procedure. When transporting the injured individual it is advisable to wrap the person in a blanket (fireblanket). Damaged areas of skin should be covered with clean, wet (ice-cold if possible) cloths during transportation. Do *not* use neutralizing agents or any agent (such as so-called "burn cream") that seals the damaged skin from contact with the air and also obscures the extent of the injury.

The exhaust hoods should be used whenever noxious and/or flammable chemicals must be used or when the possibility exists that such materials may be evolved during a reaction. Should noxious and/or flammable vapors be released into the main body of the laboratory, turn off all apparatus and, under the supervision of the laboratory instructor, leave the contaminated area as rapidly and in as orderly a fashion as possible.

Flammable vapors can be ignited in a variety of ways in addition to exposure to a direct flame. Combustible vapors can ignite in air when brought in contact with a hot surface (e.g., hot plates, heating mantles) at a temperature above the "flash point" of the vapor. The "flash point" is defined as the temperature at which the vapor pressure of the liquid is high enough to form a combustible mixture with air. All of the common solvents with the exception of carbon tetrachloride, chloroform, methylene chloride, and trichloroethylene have flash points below room temperature, e.g., diethyl ether, $-45°C$. Electrical sparks from faulty electrical connections, from static electricity, from brush-type electrical motors, and from electrical relays, among others, may serve as ignition sources. Hence the utmost caution is necessary to avoid all sources of ignition when using organic solvents.

If a fire occurs do *not* attempt to put it out yourself, but step away from it and call the instructor. If the fire is contained in a flask he will probably smother it with a watchglass or a beaker. If the fire is more extensive he will use the appropriate extinguisher, labeled according to the descriptions given in Fig. 1.1. All students not specifically designated by the instructor to help, should leave the immediate area of the laboratory.

Should an individuals' clothing begin burning, dousing the burning area with cold water is the best course of action. This not only smothers the fire, but cools burned skin areas, and also removes chemicals. The emergency showers are the most effective means of dousing. Smothering the fire with a fireblanket or a lab coat does not cool the affected area and does not remove any chemical contaminants. The clothing will continue to smolder, and may reignite. Get the injured person emergency medical attention immediately.

Spills of chemicals (including water) should be immediately cleaned up under the instructor's supervision, if no hazard from flammability or inhalation toxicity exists. Sand and/or sodium carbonate are effective absorbants for such spills, and rubber or plastic gloves should generally be worn when cleaning up such spills. If a flammability or toxicity hazard exists, notify colleagues and the instructor, and evacuate the affected area immediately.

In several of the above cited instances the possibility of the need for evacuation of the laboratory has been raised. It should therefore be clear that all entrances and exits and the aisles leading to them should be kept clear. Specifically, desk drawers and cabinet doors should be closed immediately after the item sought has been removed from the drawer or cabinet.

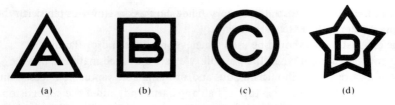

(a)   (b)   (c)   (d)

Figure 1.1. Symbols used to denote the applicability of a variety of fire extinguishers for particular types of fires.

(a) This symbol indicates the extinguisher is applicable for use on ordinary combustibles such as wood, cloth, paper, rubber, etc. The background of the symbol will be either metallic or green. It always will be found on water, multi-purpose dry chemical, and foam type extinguishers.

(b) This symbol indicates the extinguisher is applicable for use on flammable or combustible liquids, flammable gases, greases or similar materials. The background of the symbol will be either metallic or red. It always will be found on multi-purpose dry chemical, carbon dioxide, and foam type extinguishers.

(c) This symbol indicates the extinguisher is applicable for use on fires involving energized electrical equipment. The background of the symbol will be either metallic or blue. It always will be found on multi-purpose dry chemical and carbon dioxide extinguishers.

(d) This symbol indicates the extinguisher is applicable for use on certain combustible metals, e.g., sodium, magnesium, potassium. The background of the symbol will be either metallic or yellow. It always will be found on special dry powder extinguishers.

Many specific reagents and procedures exist for the proper disposal of waste chemicals. Information is available in a variety of places such as the references cited at the end of this section, and the labels of the bottles in which the chemicals are sold. At the appropriate points in the experiments to be described, such information will be supplied, but in a research environment the burden of seeking out similar material rests *entirely* on the researcher.

The question of chemical toxicity is receiving more and more concern as it is discovered that chemicals which have been used routinely with few precautions can produce harmful effects after many years (e.g., benzene). It must be realized that the long-term toxicity of many compounds, even relatively common ones, is often unknown. Therefore, all chemicals must be regarded as toxic, and ingestion (swallowing, inhalation, absorption through the skin) must be avoided by following procedures such as outlined in the preceding pages — especially use of hoods, gloves, washing of hands, and avoidance of eating, drinking, or smoking in the laboratory. Some of the common classes of organic compounds which are of most concern include alkylating agents, halogen-containing compounds, aromatic hydrocarbons, aromatic amines, aromatic nitro compounds, phenols, nitrosamines, hydrazines, aromatic azo compounds, and reactive olefins. Among inorganic compounds are heavy metals and their compounds (esp. Pb, Cd, Hg, Tl, Sb, Bi), carbonyls, cyanides, chromate salts, arsenic, selenium, beryllium and their compounds and asbestos. This list is not complete and does not imply that other classes of compounds are harmless.

Pregnant women may be at special risk with regard to chemical toxicity, and should be especially careful in the laboratory. They should consult with a physician concerning this.

REFERENCES TO CHAPTER 1

1. "Safety in Academic Chemistry Laboratories," American Chemical Society, Washington, D.C., March 1974. Single copies are free to ACS members. Multiple copies are available for 25¢ each.

2. "Safety in Handling Hazardous Chemicals," Matheson Coleman and Bell Co., Norwood, Ohio, 1968 (available as a separate reprint or as part of the MC/B Chemical Catalog).

3. "Laboratory Waste Disposal Manual," Manufacturing Chemists Association, 1973.

4. (a) "Occupational Safety and Health Act of 1970," Public Law 91−596, 84 Stat., 1590.

(b) "Code of Federal Regulations, 29CFR1901" (detailed specifications and regulations for (a)).

(c) "Federal Register" (a daily supplement to (b)).

5. "Toxic Substances List," National Institute for Occupational Safety and Health, Rockville, Maryland 20852 (revised annually).

6. N. I. Sax (ed.), "Dangerous Properties of Industrial Materials," 3d ed., Van Nostrand Reinhold, New York, 1968.

Information Sources

2.1. INTRODUCTION

It is frequently necessary in laboratory work to make use of the chemical literature. The information needed may be rather broad in nature, such as an analytical method or technique, or very specific, such as a physical constant or spectrum for a compound. It is therefore imperative that the student become familiar with the library and its resources as soon as possible.

A large variety of literature sources exists, and it is the purpose of this chapter to indicate some of the important data collections. In addition to the bibliography at the end of this chapter, several references are given in other chapters. It should be emphasized that while some of the collections are critical compilations, others are of a survey or review nature and often do not indicate a preferred or recommended value in those cases when several investigations have given different results. The reader should always be aware of the date of the publications and should determine whether supplemental and possibly more reliable information has been published subsequently.

The various sources of information can be broadly classified as follows:

chemical dictionaries and encyclopedia
handbooks (general and specific)
journals (primary data and review articles)
abstracting journals and current awareness sources

Those of the above classes that are of most general use to the student are discussed separately below. There are several textbooks devoted to the chemical literature and a selection of these books has also been included in the bibliography.[1-5]

For routine numerical data, handbooks are the most convenient sources, with general data compilations as sources of specialized data next. As indicated earlier, their limitations must be appreciated, and for critical purposes the original literature may have to be consulted. (This is also the case for data not found in handbooks or other compilations.) "Chemical Abstracts" is used to locate references to original journal articles. General review articles, appearing in a selection of review journals, are also useful for locating original publications. "The Science Citation Index" is a useful method for searching forward in the literature once a key article has been identified. Spectral data, or other specialized data, would normally first be sought in the appropriate specialized compilation or tables, with "Chemical Abstracts" being used as a source for original literature references and for data not available in any of the compilations.

In a search for information about a particular compound the Subject Index should be used if the compound is simple and the name is known, but for more complex compounds the name may not be obvious or may differ in each source. In this case the Empirical Formula Index should be consulted.

If the student experiences difficulty in locating information in the library he or she should consult with the reference librarian.

2.2. GENERAL HANDBOOKS AND COMPILATIONS

Most laboratories have general handbooks which are extensive compilations of the physical properties of chemical compounds. The most common handbooks are:

"Handbook of Chemistry and Physics"[6]
"Lange's Handbook of Chemistry"[7]
"International Critical Tables of Numerical Data . . ."[8]
"Tables of Physical and Chemical Constants"[9]
"Physico-Chemical Constants of Pure Organic Compounds"[10]
"Landolt-Börnstein Numerical Data. . . ."[11]
"Beilstein's Handbuch der Organischen Chemie"[12]
"The Chemist's Companion, A Handbook of Practical Data, Techniques and References"[13]

Although handbooks are secondary sources of information, citations to the original references are often omitted. The extent of coverage in the handbooks listed above varies but they all contain information on molecular weights, crystalline form, densities, melting and boiling points, refractive indices, viscosities, and common solubilities for both organic and inorganic compounds. A large selection of data covering thermodynamics, electrochemistry, spectroscopy, and various molecular properties is also included.

2.3. ABSTRACTING JOURNALS AND CURRENT AWARENESS SOURCES

The five most commonly used information sources for abstracts and titles are "Chemical Abstracts," "Index Chemicus," "Science Citation Index," "Chemical Titles," and "Current Contents." "Chemical Abstracts" and "Index Chemicus" give abstracts of the papers with references to the original articles, whereas "Chemical Titles" and "Current Contents" focus on very recent literature and only provide the title of the publication. However, information usually appears in "Chemical Abstracts" between 6 months and 1 year after the title has been cited in "Chemical Titles" or "Current Contents." Formula, Author, and Subject Indices are provided with "Chemical Abstracts" and use of these make it relatively simple to trace the abstract and hence the original literature citation. Collective indices (subject, formula, and author) are now published on a 5-year basis, and individual volume indices are published semiannually. The most recent individual indices use the "keyword" notation. The abstracts are identified by using either the empirical formula index or compound name index (Subject Index). Careful inspection of the abstract will often determine whether it is worthwhile to retrieve the original article, but abstracts often are not as complete as they should be. A search in "Chemical Abstracts" should be initiated with the most recent indices because this will often lead to the discovery of a comprehensive review article on the subject or system, thereby minimizing further searching. The Author Index is valuable if one is searching for work reported by a certain author.

The "Science Citation Index" is essentially a list of publications that use a given paper as a reference. That is, if a paper on a given topic is known, the "Science Citation Index" permits one to find more recent papers that use the known one as a reference, and that can be expected to provide new information about the same topic. It began in 1961, and covers approximately 2,700 journals. The index is arranged alphabetically by cited author and within this arrangement is chronological by cited year.

The overall Index is an integrated search system consisting of three separate but related indices:

Citation Index
Source Index (Author, Organization)
Permuterm Subject Index

Once some relevant articles are retrieved, the information they provide (title words, reference citations, authors, etc.) can be used to enter the other indices to continue the search.

A search in the Citation Index is started by looking up the name of an author known to have published material relevant to the subject of interest. If any of the previously published works of that author have been cited during the indexing period specified, the item will appear and those doing the citing will be listed. The names of the citing authors are then used to enter the Source Index for a complete description of the citing articles.

If no earlier relevant papers on the subject are known, the Permuterm Index may be consulted by matching key words that would appear in titles. Once authors have been established, the Citation search can be followed.

2.4. ORGANIC SYNTHESIS

The main sources for organic synthetic methods are summarized in Table 2.1. A few specific points on some of the common sources are indicated below.

1. "Beilstein's Handbuch der Organischen Chemie"[12] (in German)
   This is probably the most widely used source for information on organic compounds. It is encyclopedic in nature and attempts to cover all known organic compounds and includes the patent literature. The use of Beilstein is fairly complex, and a useful reference that serves as an introduction to the proper use of the Handbuch is "A Brief Introduction to the Use of Beilstein's Handbuch der Organischen Chemie" by E. H. Huntress, Wiley, 4th ed., 1938.
2. "Elsevier's Encyclopaedia of Organic Chemistry"[14] (in English)
   This source gives information on organic synthesis and reactions, and on physical properties, and includes references to the original literature.
3. "Reagents for Organic Synthesis"[15]
   This source is a comprehensive treatise of reagents useful in the synthesis of organic compounds. Physical properties, preparative routes, and methods of purification and application to the various reagents, with original references, are also given.
4. "Organic Functional Group Preparations"[16]
   This discusses representative chemical procedures for the introduction of functional groups into organic molecules, giving a general background and approach as well as specific examples. Literature references are given.
5. Houben-Weyl's "Die Methoden der Organischen Chemie"[17] (in German)
   An exhaustive treatise of the basic methods used in synthetic organic chemistry that includes a description of mechanisms, procedures, and apparatus used in all aspects of organic synthesis.
6. Theilheimer's "Synthetic Methods of Organic Chemistry"[18]
   Published annually, this series reviews the literature for each year. It contains an index based on functional group modification, and stresses modern methods and how they can be applied.
7. "Preparative Methods of Polymer Chemistry"[19] and "Macromolecular Syntheses"[20]
   Tested representative methods for polymer synthesis with specific examples and literature references.
8. "Organic Syntheses"[21]
   This series consists of annual volumes and collected volumes. It was originally

developed as a compilation of directions for the preparation of many specific organic chemicals. Recently the emphasis has changed towards model procedures that illustrate important types of reactions.

9. "Dictionary of Organic Compounds"[22]
    Reactions, physical properties, and references for some 25,000 compounds and their derivatives.

2.5. INORGANIC SYNTHESIS

Important sources are summarized in Table 2.1. The most widely used are:

1. "Inorganic Syntheses"[23]
    A series of Volumes similar to "Organic Syntheses" and a reliable source for the preparations of inorganic compounds reported in the literature.
2. "Handbook of Preparative Inorganic Chemistry"[24]
    This work includes information on inorganic synthesis as well as discussions of preparative methods such as high vacuum, high temperature, low temperature, and purification methods.
3. "Gmelin's Handbuch der Anorganischen Chemie"[25] (German),
    "Comprehensive Inorganic Chemistry,"[26] and "ITP International Review of Science, Inorganic Chemistry"[27]
    General sources of information on inorganic systems including references to syntheses.

All these are encyclopedic in nature and contain numerous references to synthetic procedures, reactions, and physical and chemical properties of the elements and their compounds. In the case of Gmelin each element is treated in several volumes. They are all excellent sources of numerical data and of references to the original literature.

2.6. ANALYTICAL METHODS

Among the more widely used sources of general information concerning principles, practice, techniques, and specific methods are the following.

1. "Standard Methods of Chemical Analysis"[28]
    A survey of major techniques and a systematic presentation of methods arranged by technique and by type of sample.
2. "Chemical Analysis"[29]
    A multivolume exhaustive discussion of principles and techniques of analytical chemistry with a systematic presentation of major techniques with illustrative applications.
3. "Comprehensive Analytical Chemistry"[30]
    Detailed discussions of major techniques, both chemical and instrumental.
4. "Handbook of Analytical Chemistry"[31]

Fundamental data, tabulations of methods by elements, surveys of methods for analysis of complex materials.

5. "Encyclopedia of Industrial Chemical Analysis"[32]
   Discussions of basic techniques and detailed methods for analysis of complex materials organized according to constituent and the nature of the sample.

2.7. TECHNIQUES

1. "Practical Organic Chemistry"[33]
   This text covers a wide variety of techniques and apparatus such as distillation, solvent extraction, sublimation, temperature control, pressure measuring devices, recrystallization, filtration, and drying of solvents, among others.

2. "Experimental Inorganic Chemistry"[34]
   Detailed information on techniques such as crystallization, filtration, distillation, sublimation, temperature measurement and control, manometers, drying agents, melting and boiling points, viscosity, density, surface tension, electrochemical and magnetic measurements, and vapor pressures; information is also given for the apparatus used for the preparation, purification and storage of gases and volatile liquids.

3. "Technique of Organic Chemistry"[35] and "Technique of Inorganic Chemistry"[36]
   Both of these sources are multivolume series covering all the important techniques in detail.

2.8. REFERENCE DATA PUBLICATIONS

"The Journal of Physical and Chemical Reference Data," which began publication in 1972, is devoted to the publication of critically evaluated data compilations. It is the major publication medium of the National Standard Reference Data System, administered by the National Bureau of Standards, Washington, D.C. The scope of the coverage in this journal includes the following areas:

atomic and molecular properties
chemical kinetic parameters
colloid and surface properties
mechanical properties of materials
nuclear properties
solid state properties
thermodynamic and transport properties

Most of the articles contain recommended values, together with uncertainty limits and critical commentaries on methods of measurements. As an example, surface tension data for a large number of organic liquids have recently been compiled.[37]

Prior to issuance of the "Journal of Physical and Chemical Reference Data," data compilations were published in the "National Standard Reference Data Series" (NSRDS). Selected examples are given in the reference list at the end of this chapter.[38-43]

2.9. SADTLER SPECTRAL COLLECTIONS

Sadtler Laboratories has compiled a numbered series of spectra covering a wide range of compounds and the major spectroscopic techniques that are presently used for identification. The various collections are denoted either as Standard Spectra Collections or as Commercial Spectra Collections. The Standard Spectra Collections refer to compounds of well-defined, definite composition, whereas the Commercial Spectra Collections refer to various types of materials of industrial and commercial interest, many of which are not simple compounds. The extent of the major Standard Spectra Collections as of 1975 is as follows:

Infrared Prism Spectra	49,000 spectra
Infrared Grating Spectra	38,000 spectra
Nuclear Magnetic Resonance Spectra (60 MHz)	22,000 spectra
Ultraviolet Spectra	38,000 spectra

In addition to these major Standard Spectra Collections, there are a number of smaller Standard Spectra Collections that are presently of more specialized interest or of more limited scope.

Raman Spectra	4,000 spectra
^{13}C NMR Spectra	4,000 spectra
Fluorescence Spectra	4,000 spectra
100 MHz NMR Spectra	1,000 spectra

Sadtler has also published a number of Special Collections of Spectra covering such areas as inorganic compounds, pharmaceuticals, and biochemicals.

Access to the Standard Spectra Collections is through a series of four independent indices, and a knowledge of the index structure is necessary for effective use of the various collections. The more specialized collections noted above have relatively simple indexing systems in which entries are listed both alphabetically by chemical name, and numerically by compound type. The indexing system of the Standard Spectra Collections is necessarily somewhat more complex, and its use requires some practice.

The indices to the Standard Spectra Collections list entries in four ways:

1. alphabetically by Chemical Abstracts name
2. by chemical formula in a prescribed order dictated by the subscripts for each element in the molecular formula
3. by chemical classes according to the functional groups in the molecule
4. numerically according to the Sadtler number assigned to the Standard IR Prism Spectrum

The choice of a particular index will be determined by what is known concerning the compound in question, but in general, the first three indices will be used most often. Each index consists of a basic index of one or two volumes with periodic supplements that refer to the more recent additions to the particular collection.

The material below is intended as an introduction to the Alphabetical Index, the Molecular Formula Index, and the Chemical Classes Index. This material should be viewed as a summary of the major features of the indices and the manner in which they are used.

Alphabetical Index

Entries are listed alphabetically by the Chemical Abstracts name, with minor changes in symbolism necessitated by the use of a computer printed format. The changes include the following:

1. Parentheses, brackets, and braces are replaced by slashes.
2. Superscript symbols are printed in line and are denoted by a preceding lozenge (□).
3. The term "prime" is replaced by PR, an asterisk or a left brace.
4. Greek letters are replaced by the corresponding English characters, and lowercase letters are capitalized.

Chemical Classes Index

The Chemical Classes Index is based on the Chemical Abstracts order of precedence, and the various classes have been subdivided to permit more rapid location of a particular compound. Each functional group is assigned a number between 0 and 99, and up to three functional groups may be coded for each compound. Compounds are listed in order of increasing values of the three functional group symbols, with the first symbol being the most significant and the third the least significant. Blanks are taken as zeros. Thus a compound with the symbol   1   12   14   would be listed before a compound with the symbol 2   11   13   which in turn would be listed before a compound with the symbol 2   12   12.   This information is given in the first three columns under the heading "Functionality" in the index. The fourth column under this heading lists the total number of functional groups in the molecule up to 9, and the fifth column gives a number that indicates the general structure of the molecule. Additional information concerning the molecule is provided by a listing of the molecular formula for each entry.

Molecular Formula Index

The Formula Index differs from most in that for convenience the total number of each kind of atom present in the molecule is given in a tabular form with the headings of each column arranged in the usual order:

C   H   Br   Cl   F   I   N   O   P   S   Si

TABLE 2.1
Reference and Data Sources

Subject	References
1. *Physical properties*	
melting point, boiling point, density, viscosity, surface tension, refractive index, dipole moment, crystalline form	6–11, 37, 38, 44–49
2. *Thermodynamic properties*	
a. Standard enthalpy of formation, ΔH_f^0; Standard free energy of formation, ΔG_f^0; absolute entropy, S^0; equilibrium constants; heat of solution, combustion, neutralization . . .	40, 47, 48, 50–55
b. phase diagrams	56–61
c. solubilities (aqueous or nonaqueous systems)	45, 62–64
3. *Kinetic data*	
rates of chemical reactions, energies of activation	39, 65–67
4. *Electrochemical data*	
a. electrical conductance, emf and transference numbers, polarography	42, 44, 45, 46, 48, 68–73
b. electrode potentials	48, 70, 74
5. *Synthetic methods*	
a. organic	15–18, 21, 22, 33, 75, 76
b. inorganic	23–27, 34, 36, 77
c. polymers	19, 20
6. *Spectroscopy*	
I.R. (list of basic frequencies for functional groups and how they vary with structure)	78, 79, 80, 81
UV-visible	82, 83
NMR	84, 85
7. *Analytical chemistry*	
including volumetric and gravimetric analysis, coulometric and potentiometric titrations, polarography, selection of indicators.	28–32, 86–88
8. *Techniques*	
a. organic: distillation, recrystallization, temperature control, determination of physical properties, purification . . .	33–35
b. inorganic: sublimation, low and high temperatures, vacuum transfer, zone refining, use of inert atmospheres, ion exchange, high pressure, solvent extraction, spectra, magnetic measurements . . .	24, 34, 36, 77
9. *Miscellaneous*	
spot tests (inorganic)	89
spot tests (organic)	89, 90
Polymer Handbook	91
chemical literature source (books, monographs . . .)	1–5
melting point tables	92, 93
physical methods in organic chemistry	94
crystal structures	95, 96
bond lengths, bond angles	97
thermophysical properties	98
metal ligand heats	99
basic tables in chemistry	100

Zeros are used for missing elements to emphasize the special nature of the entries, but are omitted in the actual indices. Thus the compound p-toluenesulfonic acid, 2-chloroethyl ester, which has the empirical formula $C_9H_{11}ClO_3S$ would be represented by the numbers:

$$9 \quad 11 \quad 0 \quad 1 \quad 0 \quad 0 \quad 0 \quad 3 \quad 0 \quad 1 \quad 0$$

and would precede the corresponding 2-fluoroethyl ester, that has the empirical formula $C_9H_{11}FO_3S$ and would correspond to the numbers:

$$9 \quad 11 \quad 0 \quad 0 \quad 1 \quad 0 \quad 0 \quad 3 \quad 0 \quad 1 \quad 0$$

Zeros included in these examples are suppressed in the actual indices.

2.10 ANNOTATED TABLE OF REFERENCE AND DATA SOURCES

Reference and data sources in the various areas of chemistry are given in annotated form in Table 2.1. This should be useful as a guide, but the reader should consult the complete bibliography for more information on a particular subject.

REFERENCES TO CHAPTER 2

1. C. R. Burman, "How to Find Out in Chemistry: A Guide to Sources of Information," 2d ed., Pergamon Press, Oxford, 1966.
2. E. J. Crane, A. M. Patterson, and E. B. Marr, "A Guide to the Literature of Chemistry," Wiley, New York, 1957.
3. G. M. Dyson, "A Short Guide to the Chemical Literature," Longmans, Green, London, 1960.
4. R. T. Bottle (ed.), "Use of the Chemical Literature," Butterworths, London, 1962.
5. C. C. Waddington and E. T. Marquis, "Chemical Literature, An Introduction," Indiana University Press, Bloomington, Indiana, 1961.
6. R. C. Weast (ed.), "Handbook of Chemistry and Physics," 56th ed., Chemical Rubber Company, Cleveland, Ohio, 1975. (A new edition is published annually, although changes are introduced only very slowly.)
7. J. A. Dean (ed.), "Lange's Handbook of Chemistry," McGraw-Hill, New York, 11th ed., 1973.
8. E. W. Washburn (ed.-in-chief), "International Critical Tables of Numerical Data, Physics, Chemistry and Technology," prepared under the auspices of the International Research Council and the National Academy of Sciences, by the National Research Council of the U.S.A., McGraw-Hill, New York, 1926, vols. 1–7.
9. G. W. C. Kay and T. H. Laby, "Tables of Physical and Chemical Constants," 13th ed., Wiley, New York, 1966 (first published in 1911).
10. J. Timmermans, "Physico-Chemical Constants of Pure Organic Compounds," Elsevier, New York, vol. 1, 1950, vol. 2, 1965. J. Timmermans, "Physico-Chemical Constants of Binary Systems," Interscience, New York, vol. 1, 1959; vol. 2, 1959; vol. 3, 1960; vol. 4, 1960.

11. H. Borchers, H. Hausen, K.-H. Hellwege, and Kl. Schäfer (eds.), "Landolt-Börnstein, Numerical Data and Functional Relationships in Physics, Chemistry, Astronomy, Geophysics and Technology," 6th ed., 28 vols., 1950–1977. New Series (since 1961) 36 volumes.

12. "Beilstein's Handbuch der Organischen Chemie," Springer, Berlin, 1883– .

13. A. J. Gordon and R. A. Ford, "The Chemist's Companion, A Handbook of Practical Data, Techniques and References," John Wiley and Sons, New York, 1972.

14. "Elsevier's Encyclopaedia of Organic Chemistry," Elsevier, Amsterdam, 1946– .

15. L. F. Fieser and M. Fieser, "Reagents for Organic Synthesis," Wiley, New York, 1967– (multivolumes).

16. S. R. Sandler and W. Karo, "Organic Functional Group Preparations," Academic Press, New York, vol. 1, 1969; vol. 2, 1971; vol. 3, 1972.

17. J. Houben-Weyl, "Die Methoden der Organischen Chemie," G. Thieme, Stuttgart, 1974– (multivolumes).

18. W. Theilheimer, "Synthetic Methods of Organic Chemistry," S. Karger, Basel, 1947– (multivolumes).

19. W. R. Sorenson and T. W. Campbell, "Preparative Methods of Polymer Chemistry," 2d ed., Interscience, New York, 1968.

20. "Macromolecular Syntheses," Wiley, Interscience, New York, vols. 1–5. 1967–1977; collected vols. 1–5, 1977.

21. "Organic Syntheses," Wiley, New York, 1975, vols. 1–55, (an annual publication); collected vols. I-V.

22. I. Heilbron, H. M. Bunbury, A. H. Cook, and D. H. Hey, "Dictionary of Organic Compounds," 4th ed. Oxford University Press, Oxford, 1965. 5 vols. In addition there are several supplements (the 9th supplement was published in 1973) and also a formula index.

23. "Inorganic Syntheses," McGraw-Hill, New York, vols. 1–15, 1939–1974.

24. G. Brauer (ed.), "Handbook of Preparative Inorganic Chemistry," 2d ed., Academic Press, New York, 1963, vols. 1 and 2.

25. "Gmelin's Handbuch der Anorganischen Chemie," Verlag Chemie, Berlin, 1951.

26. J. C. Bailar, Jr., H. J. Emeléus, R. Nyholm, and A. S. Trotman-Dickenson (eds), "Comprehensive Inorganic Chemistry," Pergamon Press, Oxford, 1973. vols. 1–5.

27. H. J. Emeléus (ed.), "ITP International Review of Science, Inorganic Chemistry, Series One," Butterworths, London, 1975, vols. 1–19.

28. F. J. Welcher (ed.), "Standard Methods of Chemical Analysis," 6th ed., Van Nostrand, New York, 1956. 3 vols. with parts.

29. P. J. Elving and I. M. Kolthoff (eds.), "Chemical Analysis," Interscience, New York, 1941– (multivolumes).

30. C. L. Wilson and D. W. Wilson, "Comprehensive Analytical Chemistry," Elsevier, Amsterdam, 1959– , 5 vols. in several parts.

31. L. Meites (ed.), "Handbook of Analytical Chemistry," McGraw-Hill, New York, 1963.

32. F. D. Snell and C. L. Hilton (eds.), "Encyclopedia of Industrial Chemical Analysis," Interscience, New York, 1966– .

33. A. I. Vogel, "A Text-Book of Practical Organic Chemistry, including Qualitative Organic Analysis," 3d ed., Wiley, New York, 1956.

34. R. E. Dodd and P. L. Robinson, "Experimental Inorganic Chemistry. A Guide to Laboratory Practice," Elsevier, Amsterdam, 1954.

35. A. Weissberger, "Technique of Organic Chemistry," 3d ed. Interscience, New York, 1959– , vol. I, parts 1–4; vol. II; vol. III, parts 1 and 2; vol. IV–VII; vol. VIII, parts 1 and 2; vols. IX and X.

36. H. B. Jonassen and A. Weissberger (eds.), "Technique of Inorganic Chemistry," Interscience, New York, 1963– .

37. J. J. Jasper, "The Surface Tension of Pure Liquid Compounds," *J. Phys. Chem. Ref. Data* **1** No.(4), 841 (1972).

38. R. D. Nelson, Jr., D. R. Lide, Jr., and A. A. Maryott, "Selected Values of Electric Dipole Moments for Molecules in the Gas Phase," NSRDS-NBS-10, 1967.

39. S. W. Benson and H. E. O'Neal, "Kinetic Data on Gas Phase Unimolecular Reactions," NSRDS-NBS-21, 1970.

40. D. R. Stull, H. Prophet, *et al.*, "JANAF Thermochemical Tables," 2d ed. NSRDS-NBS-37, 1971.

41. B. de B. Darwent, "Bond Dissociation Energies in Simple Molecules," NSRDS-NBS-31, 1970.

42. W. J. Hamer and H. J. DeWane, "Electrolytic Conductance and the Conductance of the Halogen Acids in Water," NSRDS-NBS-33, 1970.

43. V. B. Parker, "Thermal Properties of Aqueous Uni-Univalent Electrolytes," NSRDS-NBS-2, 1965.

44. G. J. Janz and R. P. T. Tomkins, "Nonaqueous Electrolytes Handbook," Academic Press, New York, vol. 1, 1972.

45. G. J. Janz and R. P. T. Tomkins, "Nonaqueous Electrolytes Handbook," Academic Press, New York, vol. 2, 1973.

46. A. K. Covington and T. Dickinson (eds.), "Physical Chemistry of Organic Solvent Systems," Plenum Press, New York, 1973.

47. B. J. Zwolinski, *et al.*, "API (American Petroleum Institute Research Project) 44 Tables: Selected Values of Properties of Hydrocarbons and Related Compounds," vols. 1–6, 1972.

48. G. H. Aylward and T. J. V. Findlay, "Chemical Data Book," 2d ed., Wiley, New York, 1966 (properties of inorganic and organic compounds: thermodynamic data, ionic radii and ionization energies, dipole moments, bond energies and bond lengths, ionic conductances. stability constants of complex ions).

49. A. L. McClellan, "Tables of Experimental Dipole Moments," Freeman, San Francisco, 1963.

50. F. D. Rossini, D. D. Wagman, W. H. Evans, S. Levine, and I. Jaffe, "Selected Values of Chemical Thermodynamic Properties," Circular 500 of the National Bureau of Standards, U.S. Department of Commerce, N.B.S., 1952.

51. D. D. Wagman, W. H. Evans, V. B. Parker, I. Halow, S. M. Bailey, and R. H. Schumm, "Selected Values of Chemical Thermodynamic Properties," N.B.S. Technical Note 270–3, 1968.

52. D. D. Wagman, W. H. Evans, V. B. Parker, I. Halow, S. M. Bailey, and R. H. Schumm, "Selected Values of Chemical Thermodynamic Properties," N.B.S. Technical Note 270–4, 1969.

53. V. B. Parker, D. D. Wagman, and W. H. Evans, "Selected Values of Chemical Thermodynamic Properties," N.B.S. Technical Note 279–5, 1971.

54. V. B. Parker, D. D. Wagman, and W. H. Evans, "Selected Values of Chemical Thermodynamic Properties," N.B.S. Technical Note 270–6, 1971.

55. R. H. Schumm, D. D. Wagman, S. Bailey, W. H. Evans, and V. B. Parker, "Selected Values of Chemical Thermodynamic Properties," N.B.S. Technical Note 270–7, 1973.

56. E. M. Levin, H. F. McMurdie, and F. P. Hall, "Phase Diagrams for Ceramists," American Ceramic Society, Columbus, Ohio, Part I, 1956; Part II, 1959.

57. W. D. Robertson, "Binary Phase Diagrams of Halide Salts," U.S. Atomic Energy Commission Contract AT (30–1)–2723, Yale University, Yale Report no. 2723, vols. 1 and 2, June 1966. U.S. Department of Commerce Clearing House, Federal Science Office, Technical Information, National Bureau of Standards, Springfield, Virginia.

58. R. E. Thoma, Oak Ridge National Laboratory, ORNL–2548; Contract No. W–7405–eng–26.

59. N. K. Voskresenskaya (ed.), "Handbook of Solid-Liquid Equilibria in Systems of Anhydrous Inorganic Salts," Israel Program for Scientific Translations, Jerusalem, 1970. Vols. 1 and 2.

60. "Physico-Chemical Constants of Fused Salts," by the Committee of Fused Salt Chemistry, the Electrochemical Society of Japan, 1964.

61. G. J. Janz, "Molten Salts Handbook," Academic Press, New York, 1967.

62. H. Stephen and T. Stephen (eds.), "Solubilities of Inorganic and Organic Compounds," Pergamon Press, Oxford, 1963, vol. I, parts 1 and 2 (Binary Systems); vol. II, parts 1, 2, and 3 (Ternary and Multicomponent Systems).

63. A. Seidell, "Solubilities of Inorganic and Organic Compounds," 2d ed. Van Nostrand, New York, 1919. Vols. 1 and 2.

64. W. F. Linke, "Solubilities of Inorganic and Metal-Organic Compounds," vols. 1 and 2 (a revision and continuation of Seidell). Van Nostrand, New York, 1958.

65. "Tables of Chemical Kinetics, Homogeneous Reactions," National Bureau of Standards, Circular 510, 1951.

66. "Tables of Chemical Kinetics, Homogeneous Reactions," Supplement 1 to N.B.S. Circular 510, 1956; Supplement 2 to N.B.S. Circular 510, 1960.

67. "Tables of Chemical Kinetics, Homogeneous Reactions," (Supplementary Tables), N.B.S. Monograph 34, vol. 1, 1961 and N.B.S. Monograph 34, vol. 2, 1964.

68. R. A. Robinson and R. H. Stokes, "Electrolyte Solutions," 2d ed. revised, Butterworths, London, 1965.

69. B. E. Conway, "Electrochemical Data," Elsevier, Amsterdam, 1952.

70. R. Parsons, "Handbook of Electrochemical Constants," Butterworths, London, 1959.

71. I. M. Kolthoff and J. J. Lingane, "Polarography," Interscience, New York, 1952.

72. L. Meites, "Polarographic Techniques," Interscience, New York, 1955.

73. G. W. C. Milner, "Principles and Applications of Polarography," Longmans, London, 1957.

74. W. M. Latimer, "Oxidation Potentials," 2d ed., Prentice-Hall, Englewood Cliffs, N.J., 1952.

75. "Organic Reactions," Wiley, New York, 1942—76, vols. 1-23.

76. S. Coffey (ed.), "Rodd's Chemistry of Carbon Compounds: A Modern Comprehensive Treatise," 2d ed., Elsevier, Amsterdam, 1964, vols. 1-4.

77. W. L. Jolly, "The Synthesis and Characterization of Inorganic Compounds," Prentice-Hall, New York, 1970.

78. L. J. Bellamy, "The Infra-Red Spectra of Complex Molecules," 2d ed., 1958; L. J. Bellamy, "The Infra-Red Spectra of Complex Molecules," Halstead Press, London, 1975.

79. "Sadtler Standard Spectra: Prism Standard Infrared Spectra," Sadtler Research Laboratories, Philadelphia, Pa, vols. 1—24.

80. H. A. Szymanski and R. E. Erickson, "Infrared Band Handbook," 2d ed., IFI/Plenum, 1970, vols. 1 and 2 (contains infrared absorption bands and references).

81. C. J. Pouchert, "Aldrich Library of Infrared Spectra," 2d ed., Aldrich Chemical Co., Milwaukee, Wisconsin, 1975.

82. K. Hirayama, "Handbook of Ultraviolet and Visible Absorption Spectra of Organic Compounds," Plenum Press Data Division, New York, 1967.

83. "Sadtler Standard Spectra: Ultraviolet Spectra," Sadtler Research Laboratories, Philadelphia, Pa, vols. 1—86.

84. "Sadtler Standard Spectra: Nuclear Magnetic Resonance Spectra," vols. 1—34. Sadtler Research Laboratories, Philadelphia, Pa; "Sadtler Commercial Spectra: Nuclear Magnetic Resonance Spectra," vols. 1—16, Sadtler Research Laboratories, Philadelphia, Pa.

85. N. S. Bhacca, et al. (compilers), "NMR Spectra Catalog," Varian Associates, Palo Alto, California, 1962—1963, 2 vols.

86. A. Vogel, "A Text-Book of Quantitative Inorganic Analysis," 3d ed., Longmans, London, 1961.

87. F. D. Snell and C. T. Snell, "Colorimetric Methods of Analysis," Van Nostrand, New York, 1948—1961. (six volumes).

88. I. M. Kolthoff and P. J. Elving (eds.), "Treatise on Analytical Chemistry," Interscience, New York, 1959— .

89. F. Feigl, "Qualitative Analysis by Spot Tests, Inorganic and Organic Applications," 3d English ed., Elsevier, Amsterdam, 1946.

90. F. Feigl, "Spot Tests in Organic Analysis," 7th English ed., Elsevier Amsterdam, 1966.

91. J. Brandrup and E. H. Immergut (eds.), "Polymer Handbook," 2d ed., Interscience, New York, 1975.

92. W. Utermark and W. Schicke, "Melting Point Tables of Organic Compounds," 2d supplemented ed., Interscience, New York, 1963.

93. "Handbook of Tables for Identification of Organic Compounds," 3d ed., Chemical Rubber Co., Cleveland, 1967.

94. F. L. J. Sixma and H. Wynberg, "A Manual of Physical Methods in Organic Chemistry," Wiley, New York, 1964.

95. J. Donnay and G. Donnay (eds.), "Crystal Data," American Crystallographic Association Monograph, nos. 5, 1963, and 6, 1967; 3d ed., J. Donnay and H. Ondik (eds.), U.S. National Bureau of Standards and Joint Committee on Powder Diffraction Standards, vols. 1, 1972, and 2, 1973.

96. R. W. G. Wyckoff (ed.), "Crystal Structures," 2d ed. Interscience-Wiley, New York, vols. 1–5 (1966).

97. L. E. Sutton (ed.), "Tables of Interatomic Distances and Configuration in Molecules and Ions," The Chemical Society (London), Special Publication 11, 1958.

98. S. Touloukian (ed.), "Thermophysical Properties of Matter," The TPRC Data Series, vols. 1–7, 1970; vols. 8–13, 1972.

99. J. J. Christensen, D. J. Eatough, and R. M. Izatt, "Handbook of Metal Ligand Heats and Related Thermodynamic Quantities," 2nd ed. revised, Marcel Dekker, New York, 1975.

100. R. Keller, "Basic Tables in Chemistry," McGraw-Hill, New York, 1967.

Recording and Presentation of Experimental Results

3.1. THE LABORATORY NOTEBOOK

All data and observations recorded in the laboratory should be kept in a *bound* notebook. The particular type to be used may be specified by the requirements of individual institutions, but the "research" style notebook which permits carbon copies to be made on detachable sheets has much to recommend it. In a teaching laboratory, the copies may be removed and submitted each period to serve as a record of progress. In a research environment, the duplicate sheets may be kept in a safe place to guard against accidental destruction of the records. In any event, data should *not* be recorded elsewhere for later recopying in the notebook, and in particular loose scraps of paper are not permissible for records, because they are often lost.

The need for properly kept notebooks and the emphasis on adequate note-keeping have several origins. Normally, reports will be based on the data in the notebook and, the more complete the data, the more likely it is that a good report will be prepared and that discrepancies and unexpected results will be traceable to their sources. In addition, it may be necessary to refer from one experiment to another. In a broader sense, a notebook is essential in any research or industrial laboratory, where it may be necessary to review data months or years after they were taken; hence full details are necessary. A table of contents is of great help in locating data at a later date. The notebook represents the primary record of your work, and may have to be referred to by employers or colleagues as well as yourself. It may also represent legal evidence in case of patent claims, for example.

The following are *minimum* requirements for a notebook. Many industrial laboratories require much more.

1. Each page should be dated, numbered and signed. The experiment being done should be indicated clearly.
2. Observations should be complete; that is, in addition to numerical data, *all* observations should be recorded, including color changes, temperatures, and weights. Any unusual occurrences should be noted, since at some later date they may be significant in explaining why the experiment gave certain results. Full details of a procedure need not be elaborated, but each step should be noted as it is done. The statement "I did such and such experiment" is not sufficient. The equipment used should be stated (e.g., instruments which were used, grades and purities of chemicals for critical applications). Sketches of the apparatus used often are helpful. In the case of a partnership experiment, if your partner records the data, state this in your notebook (including partner's name). However, you should still have your own general observations. When data are to be plotted, it is very helpful to do so in a rough manner directly in the notebook as the measurements are taken. In this way serious inadequacies will be obvious at once and additional measurements can be made as necessary.
3. Mistakes should be crossed out with a single line, but should remain legible.
4. The notes must be neat enough, and orderly enough, for someone else to follow them.

On the opposite page is a reproduction of a typical page from a good notebook. It is somewhat cluttered and hard to read, but all observations have been recorded and it can be followed with reasonable effort. Note that the subject of the work is clearly indicated, and that the method used is referenced.

3.2. REPORTS

Reports are an essential product of all experimental work. Excellent research has little value if its findings cannot be communicated to others in an effective way. Thus, skill in reporting data is at least as important as skill in any experimental technique in the laboratory. Reports will vary in their requirements, depending upon the subject, and the intended readers. The outlines below may be subject to modification by the particular requirements of individual institutions.

The preparation of a formal, or full report provides experience in logical, coherent, scientific writing, and requires an extensive evaluation and interpretation of the results. It should be modeled on published articles, but the exact content of the report will depend upon the nature and purpose of the particular experiment. Some general guidelines for suitable format are given below. Keep in mind that the purpose of a report is to inform the reader. It should be clear and concise, but also complete enough that a reader will know exactly what was done and how to repeat it, if necessary. The data are discussed so as to bring out their reliability, their

6/16/77   Charles Brown

Standardization of HCl prepared 6/14/77

Used Na_2CO_3 as std. and followed procedure in manual.

The Na_2CO_3 was found outside the drying oven but was still warm. Put some into a weighing bottle and let it cool in desiccator before weighing. Samples put into 250 ml. flasks.

Balance #3

Sample #	1	2	3
Initial wt. (g)	18.3465	18.1011	17.9145
Final wt. (g)	18.1011	17.9145	17.7063
Wt of sample (g)	0.1454	0.1866	0.2082
	0.2454		

Samples dissolved in ≈ 50 ml. H_2O (grad. cylinder) and titrated as in manual using Methyl orange - bromcresol green indicator.

Sample 1

initial buret reading	0.48 ml.	accidently used 4
color became greyish	46.68 ml.	drops of indicator

Sample heated ≈ 2 min with boiling + color went back to blue.
Color turned slowly greyish. _NO_ sharp change.   46.78 ml.
added ½ drop more — now greyish orange   46.82 ml.
another ½ drops — clearly dirty orange   46.87 ml
                    THIS ENDPOINT IS _NO GOOD_

INSTRUCTOR SAID SOLUTION SHOULD NOT HAVE BEEN WARM

SAMPLE 2   misread buret!
            initial buret reading   1.05 ml
            color became greyish   ~~34.6 ml~~ 35.61 ml
Sample heated as before — color turned back to blue.
LET SAMPLE COOL TO ROOM TEMP, THIS TIME
    Color became greyish suddenly   35.80 ml
    added ½ drop — definite orange color !   35.87 ml
CORRECT ENDPOINT AT 35.80 ml
            | TOTAL VOLUME   34.75 ml. |

25

relation to theory, and their significance with regard to the problem studied. A student experiment should be treated as a formal scientific investigation. While the report should be complete, it need not be long. Do not "pad" the report with extraneous material. It is helpful to prepare an outline of the report before beginning to write it. The final copy should be typed or written carefully in ink.

Technical writing conventionally uses an impersonal passive construction such as "The sample was distilled" rather than "I distilled the sample". Use the past tense; the report describes what was done. Style is less vital than clarity, but the rules of correct grammar must be followed.

A technical report (for example, a published paper) typically consists of a title, an abstract, an introduction, an experimental section, results, discussion, conclusions, and references. There often are variations; e.g., a separate section may be present on theory if it must be treated extensively; there may be appendices; results and discussion may be combined; the conclusion section may be absent; and references may appear as footnotes. Although the exact format is less important than the content, the format must be logical and clear. The usual sections of a formal report are described below. Further details on writing reports are available from the sources cited at the end of this section.[1−3]

Title.   This should be a brief, clear description of the subject of the report, e.g., "Kinetics of the Diazotization of *p*-Toluidine."

Abstract.   This is a concise statement of the major results obtained; it is best prepared after the rest of the report has been completed. It should consist of *only* one or two complete sentences. The following is an acceptable example:

The kinetics of the diazotization of *p*-toluidine were studied in aqueous perchloric acid at $25°C$. The reaction was found to be first order in *p*-toluidine and in nitrous acid, with a rate constant of $7.3 \pm 0.2 \times 10^{-7}$ ℓ mol^{-1} s^{-1}.

Introduction.   This section acquaints the reader with previous work on the problem, explains relevant theories and equations, and indicates the purpose of the experiment. Other work on which this study was based, and proposals and conclusions made by prior investigators mentioned here, should be acknowledged by references. For experiments taken from the laboratory manual, this section normally would be very brief. The introduction given in the text should not be reproduced although it may be used as a guideline.

Experimental.   This gives the experimental procedure used, in enough detail that other workers could reproduce the work. All pertinent details should be given, e.g., solvents used in recrystallization, lengths of heating times, temperatures, observations of color changes. Sources and grades or purity of reagents, and models of instruments used for critical measurements often must be included to permit the quality of the work to be evaluated. Some judgement is necessary regarding what should be included. Standard procedures such as titrations and filtration need not be described in detail, but sometimes specific points, such as the grade of filter paper used, must be specified.

If a procedure given in the laboratory manual or other standard source is followed, this section should simply contain a complete reference to the source. However, any variations should be indicated. (For example, "The procedure described by Smith and Jones (2) was followed, except that no temperature control was used.")

Results. This section contains the experimental results. Raw data, such as buret readings or balance readings should not be included, but final volumes and weights employed should be. When possible, data should be presented in tables and plots. These should be labeled clearly, units should be included, and all numerical results should be accompanied by an indication of their limits of error.

Experiments involving synthetic work should include a report of the actual and percent yield (error estimate not required). The percent yield is the actual yield expressed as a percentage of the theoretical yield, based on the reactant present in limiting quantity and assuming complete reaction. All products must be characterized by melting or boiling points, infrared spectra or other appropriate means. IR and NMR charts should be attached to the report and properly labeled; the most significant peaks should be assigned either on a separate sheet or directly on the chart. Any indications of impurities suggested by these characterizations should be discussed in the discussion section. Other observations are included as appropriate to the explanation of the results.

If further results are calculated from the experimental data they may appear here, or they may be part of the later discussion section. You must decide in each case which is more appropriate. A sample calculation must be included in the report, but to maintain continuity it is often preferable to place it in an appendix. Give some explanation to guide the reader through the arithmetic. In addition, wherever the calculations appear, *propagation of errors* should be considered. A formal, statistical error analysis is necessary in some cases, although not for most reports. Chapter 4 should be consulted for details.

The appropriate number of significant figures should not be exceeded in the numerical results. It is common practice to report values such that all figures in it are accurate except the last one. Do not include more than the justified number of significant figures either in an experimental value or in a calculated value. Rules concerning the significant figures to be carried in a calculation are given in Chap. 4.

Discussion. The validity and meaning of your results are examined in this section, and as much information as possible is extracted from them. As far as possible, results should be explained with reference to pertinent theories, they should be compared to literature or calculated values, and discrepancies and/or errors should be discussed (hence calculations and propagation of error can be put here, or they may appear elsewhere and their significance discussed here). The error analysis can be used to suggest how the experiment could be made more accurate. In comparing numerical data, cognizance should be taken of the uncertainty limits. For example, values of 430 and 520 are in reasonable agreement if the error limits of each are ±10%. However, values of 430 and 440 are not in good agreement if the limits are ±1%. Faults in the experiment and suggestions for improvement should also be discussed in this section. All literature values used should be properly cited. Often observations must be explained or questions answered.

Conclusions.   This is a brief summary of the conclusions that can be drawn from the results and discussion. Unless there are a considerable number of points to be summarized, this section probably is not necessary. Conclusions of the sort "Such and such technique is a good method for measuring so and so" are childish and unacceptable.

References.   References should be given to all literature sources of information used in the report. A form used by standard journals should be followed, for example for books, name (with initials) of author(s), title of book, publisher, year, and chapter or page number, and for journals, name of author(s), title of journal (properly abbreviated), volume (underlined), page, year (in parentheses). (In a few cases, e.g., some handbooks, no author will be given.) Each reference is indicated by a number which is given in the text when that source is referred to, e.g.:

1. F. A. Cotton and G. Wilkinson, "Advanced Inorganic Chemistry," 3rd edition; Wiley & Sons, New York, 1972.
2. A. B. Smith and C. D. Jones, *J. Chem. Educ., 51,* 1111 (1974).

References to 3.2

1. "Handbook for Authors of Papers in the Journals of the American Chemical Society," American Chemical Society Publications, Washington, D.C., 1967.
2. L. F. Fieser and M. Fieser, "Style Guide for Chemists," Reinhold, New York, 1960.
3. F. H. Rhodes, "Technical Report Writing," McGraw-Hill, New York, 1961.

3.3. THE PRESENTATION OF DATA

When significant amounts of quantitative data are presented in a report or publication, it is most effective to use tables and/or graphs.[1] Tables permit the actual numbers to be seen most clearly, while graphs are superior for showing trends and changes in the data. Graphs also permit interpolation (the estimation of a value between two data points, such as the value of the vapor pressure at 325°K in Fig. 3.1) or extrapolation (the estimation of a value beyond the experimentally

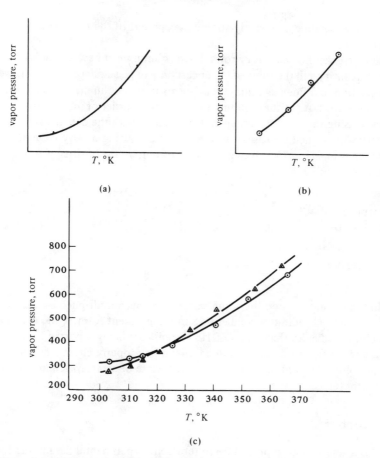

Figure 3.1. Methods of presenting graphical data; vapor pressure as a function of temperature. (a) the incorrect use of dots, (b) the use of open circles, (c) the use of open circles and triangles to distinguish data from different sources (e.g., △ represents investigator A and ○ represents investigator B in Table 3.1).

measured quantities, such as the value of the vapor pressure at 375°K in Fig. 3.1) to be made with ease. A third approach is to represent the data with an empirical equation which describes the graph. This last method is most useful for exact interpolation or extrapolation. Some of the more important features of these methods are described below.

3.3.1. Tables

Table 3.1 illustrates the presentation of data in *tabular form*. Note the following.

1. Tables should be numbered for ease of reference and to avoid confusion.

2. A concise heading, accurately stating the content of the table, should be provided.

3. Data are given in columns or rows. Each column must have a heading stating the quantity and the value of the uncertainty (e.g., standard deviation, average deviation, confidence limits) that appears in that column, and the units employed. Data in the columns are arranged in order of increasing or decreasing values of the independent variable, with the decimal points vertically aligned for ease of reading. Data consisting of very large or very small numbers can be given as a power of ten, in one of three ways, as shown in the example below.

Concentration (mol/ℓ) $(\pm 0.1 \times 10^{-3})$		Concentration (mol/ℓ), $\times 10^3$ (± 0.1)		Concentration $(10^{-3}$ mol/ℓ) $(\pm 0.1 \times 10^{-3})$
1.1×10^{-3}		1.1		1.1
1.2×10^{-3}	$\equiv$	1.2	$\equiv$	1.2
1.3×10^{-3}		1.3		1.3

In the second method the actual values have been multiplied by the value given in the heading to yield entries in a convenient form, while the third column indicates that values of 10^{-3} mol/ℓ are being reported.

4. Data taken from a literature source should be properly cited. Any unreliable data should be indicated as such, with the use of an asterisk and a suitable footnote.

3.3.2. Graphs

It is a very common procedure to represent experimental data in the form of a graph. Sometimes a graph is used to replace a table of data and to draw attention to significant features of the data that may not be readily apparent. For example, a

TABLE 3.1

Vapor Pressure of CCl_4 as a Function of Temperature

Temp. (°C ±0.05)	Investigator A pressure (Torr ±1 torr)	Investigator B[a] pressure (Torr ±1 torr)
73.40	735	728
65.30	542	538
59.65	460	457
59.90	368	365
40.40	280	284
30.65	215	220

[a] The temperature bath was not regulated electronically; temperatures were read as the bath cooled.

graph may often reveal trends in the data, such as the presence of maxima, minima, or points of inflection. Graphs are also particularly useful in presenting a ready comparison between various sets of data either from one laboratory or from laboratories of different investigators. A graph also acts as an automatic averaging method, i.e., by drawing the "best" smooth curve among the data, they are being averaged. A further use of the graphical approach is to check the fit of the data to a particular functional form. It is usually worthwhile spending some time to consider whether an equation can be rearranged into a linear form. Not only is a linear plot the easiest to use for extrapolation or interpolation, but useful information may often be obtained from the slope and/or intercept of the plot. As an example, the data from Table 3.1 may be used to obtain a value for the ΔH_{vap} of CCl_4 by plotting $\ln P$ vs. $1/T$ and measuring the slope.

It is good practice to plot the data roughly as the experiment is being performed, as well as to record the actual numerical results being observed. This approach will not only indicate whether additional measurements should be taken, but it will also indicate which points have a very large deviation from the curve defined by the other data, and thereby indicate the need for repeating those particular measurements.

Graphical data can be differentiated (by drawing tangents to a curve and determining the slope), integrated (by determining the area under the curve), interpolated, or extrapolated, but one must realize that these procedures cannot always be performed with great accuracy. For instance, the inaccuracy in the slope or intercept obtained from a graphical analysis is significantly larger than that of the experimental data points from which the tangent is constructed. Sometimes such techniques may be usefully replaced by computer programs, e.g., when there are a large number of calculations to be done. Computer analysis can also be of assistance in obtaining slopes, areas under curves, and the uncertainties in values from nonlinear regression analysis, which is very tedious to do manually.

Several considerations are important for good graphical representation.

1. Graphs should be numbered and each graph should have a caption which briefly and clearly describes its content, such as "The Vapor Pressure of Carbon Tetrachloride as a Function of Temperature".

2. If the graph is used for anything more than to illustrate trends or to compare data (i.e., if the tabulated data are not given, or if slopes, intercepts, or interpolated values are taken from the graph) then it is imperative that the graph in no way limits the precision of the experiment. The scales chosen on the axes and the quality of the paper must be such that no reading errors are introduced. However, the scales should not be so expanded that the experimental deviations or scatter become unmanageable.

3. The dependent variable is normally the ordinate (y axis) and the independent variable is normally the abscissa (x axis).

4. Both axes of the graph must be labelled clearly and the units must be included (e.g., Vapor Pressure, torr, and Temperature, °K). Scale values

should be given at reasonable and regular intervals for easy reference (not necessarily at every major division). As mentioned previously for tabular presentation of data, very large or small numbers can be handled by using the standard scientific notation e.g., 6.023×10^{23}.

5. The final graph (i.e., curve and data points) should be drawn with a fine ball-point pen or drafting pen. It is useful to draw the curve faintly, at first with a sharp pencil, and then to trace the "best" curve with ink.

6. The points must be presented in such a way that they can be clearly distinguished from the smooth curve as well as from each other. There are various ways of representing data points on the graph such as open circles ○, closed circles ●, open triangles △ or ▽, closed triangles ▲ or ▼, open squares □, closed squares ■, or other geometric figures. The use of these different types of symbols serves as an aid in distinguishing the data obtained from different techniques or from different investigators. A single dot is not satisfactory, because it is hard to see. The curve should not obscure the data points. Some authors prefer to interrupt the curve where it intersects the data points. Some of the methods used are illustrated in Fig. 3.1. A rectangle is a convenient means of indicating the uncertainties in an experimental point, because the size of the rectangle can be drawn to show the range of values, the probable error, or some other measure of the uncertainty. Error bars are an extreme case where essentially all the error is in one variable.

7. It is important to draw the "best" *smooth* curve (i.e., the one with the least scatter) through the data points. A transparent ruler is useful for a straight line graph, while French curves are helpful to obtain smooth curvature. Some care must be taken to overcome the natural tendency to overemphasize the end points; because of experimental difficulties these are often the *worst* points.

8. Often, experimental data will be compared with a theoretical equation. In this case, the values predicted from the equation may be represented by a line while the experimental data points will show how well the theoretical model predicts the experimental results. The use of the broken line method mentioned above is often convenient in such a situation.

9. Many equations contain a logarithmic term and graphs for such equations may be plotted on log or semilog paper, thus eliminating the need to compute the logarithm of each data point. Semilog paper is used when one of the coordinates is a logarithmic function (e.g., $\log P$ vs. $1/T$ in a vapor pressure experiment), while log-log paper is used when both coordinates are logarithmic functions (e.g., $\log P$ vs. $\log V$ as in the Freundlich equation for gas adsorption), or when the range of values is very large.

10. In some applications the area under the curve is desired. These areas can be measured in a number of ways:

carefully cutting out the graph and weighing the paper
using a planimeter
counting the number of squares under the curve.

11. It is preferable not to include too much information on one graph because it will be very difficult to interpret, and sometimes two or more graphs may be necessary. Sometimes there will be problems of overlapping points and it is important to clarify these on the graph. An example is shown in Fig. 3.1.

When making comparisons of data with tabulated values Percent Departure (P.D.) graphs are sometimes used, where:

$$\text{P.D.} = \left[\frac{\text{Experimental Value} - \text{Reference Value}}{\text{Reference Value}} \right] \times 100$$

An example of this type of plot is illustrated in Fig. 3.2, where percent departures for the data in Table 3.1 [Investigators A and B] are given at different temperatures.

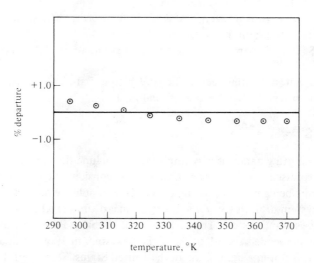

Figure 3.2. Percent departure graph showing the percent departures of investigators B and A for the data in Table 3.1.
—— investigator A [Reference value] ; o investigator B.

3.3.3. Empirical Equations[2,3,4]

For compactness it is often convenient to represent the results of experimental data in the form of an equation. For example, the vapor pressure (*P*) of a liquid varies with the temperature (*T*) according to the relationship:

$$\log P = \frac{a}{T} + b$$

where $a = \Delta H_v/2.303R$ (ΔH_v is the enthalpy of vaporization) and b is a constant.

The form of this equation is $y = mx + b$, with $x = 1/T$. Thus a plot of $\log P$ vs. $1/T$ will be a straight line, and the slope will yield ΔH_v. Sometimes we know the functional form, while in some cases we must discover it by plotting the data and comparing the shape of the curve with known functions.

If the form of the equation is not known, a purely empirical equation, e.g., a power series $y = a + bx + cx^2$ may be fitted to the data. The values of the coefficients in polynomial equations may be obtained by performing a least squares analysis. Some of the methods used may be found in Refs. 2–4. It is important to attach some error limits to these coefficients when they are reported.

References to 3.3

1. "Handbook for Authors of Papers in the Journals of the American Chemical Society," American Chemical Society Publications, Washington, D.C., 1967. Available from ACS, Washington, D.C.
2. P. D. Lark, B. R. Craven, and R. C. L. Bosworth, "The Handling of Chemical Data," Pergamon Press, Oxford, 1968. p. 25.
3. D. S. Davis, "Nomography and Empirical Equations," 2d ed., Reinhold, New York, 1962.
4. P. G. Guest, "Numerical Methods of Fitting Data," (Regression Analysis included), Cambridge University Press, London, 1961.

3.4. UNITS

When reporting data it is very important to include the units in which the quantity is expressed. At the present time there is considerable variation in the systems of units being used in the literature. Some people are advocating the so-called SI system of units (Système International d'Unites) while others prefer to use the cgs-esu-emu system. The International Union of Pure and Applied Chemistry recommended the adoption of the SI system in 1969,[1] but at the present time it is not in common use in the United States. Some useful comments concerning the various sets of units have been made by Adamson.[2]

The basic physical quantities used in establishing a set of units are given in Table 3.2.

The meter (m) is defined as 1,650,763.73 wavelengths in vacuum corresponding to the transition $2p_{10}-5d_5$ of krypton 86; the kilogram (kg) is defined as the mass of the prototype kilogram stored at Sevres, France; the second (s) is defined as the duration of 9,192,631,770 periods of the radiation corresponding to the transition between the two hyperfine levels of the ground state of cesium 133; the kelvin (K) for temperature is defined as 1/273.16 of the thermodynamic temperature of the triple point of water; and the ampere (A) for electric current, is defined as the current which, if flowing in two infinitely long parallel wires in vacuum separated by 1 m, would produce a force of 2×10^{-7} N/m of length between the wires.

TABLE 3.2

Basic Physical Quantities in SI and cgs Units

Basic physical quantity	Symbol	Name and symbol of SI unit	Name and symbol of cgs unit
length	l	meter (m)	centimeter (cm)
mass	m	kilogram (kg)	gram (g)
time	t	second (s)	second (s)
electric current	I	ampere (A)	ampere (A)
thermodynamic temperature	T	Kelvin (K)	centigrade (C)
amount of substance	n	mole (mol)	mole (mol)

All the other physical quantities are to be regarded as being derived from these basic quantities and have dimensions that can be derived from them.

Some common examples using both sets of units are given in Table 3.3.

Widely Used Chemical Units

Because the SI System has not been fully implemented at this time, some mention should be made of common units widely used in chemistry. Thermodynamic and energy data are still commonly reported in calories or kilocalories as well as in joules: spectroscopic data often use cm^{-1} for energy and frequency and electron volts (eV) are employed for ionization potentials; pressure is commonly given in torr (formerly mm of Hg). In this text the cgs units will generally be used, although the most common SI units will be given in parentheses. Some conversion factors to SI units from other units are summarized in Table 3.4.

The SI commissions have also advocated the use of fractions and multiples for prefixing the various quantities instead of using special names such as micron (10^{-6} m) or angstrom (10^{-8} cm). These are given in Table 3.5.

General Physical Constants

The National Bureau of Standards has published a current list of Fundamental Physical Constants[3] and a selection of those most commonly used is given in Table 3.6. These new values have been endorsed by CODATA for general international use. The Defined Values and Conversion Factors are given in Table 3.7.

Concentration Units

Different types of concentration units are commonly encountered in experimental chemistry and those most commonly used are summarized in Table 3.8, together with some comments. Because the quantities of material generally encountered in analytical work are normally milliliters rather than liters, it is often convenient to work in terms of millimoles. Notice that the number of mol/ℓ is the

TABLE 3.3

Typical Physical Quantities in cgs and SI Units

Physical quantity	cgs			SI		
	Name	Symbol	Definition	Name	Symbol	Definition
force	dyne	dyn	$g\, cm\, s^{-2}$	newton	N	$m\, kg\, s^{-2}$
pressure	atmosphere	atm	$dyn\, cm^{-2}$	pascal (newton per square meter)	Pa	$m^{-1}\, kg^{-1}\, s^{-2}\, (= N\, m^{-2})$
energy	erg	erg	$dyn\, cm^{-1}$	joule	J	$m^2\, kg\, s^{-2}\, (= Nm)$
power	watt	W	$erg\, s^{-1}$	watt	W	$m^2\, kg\, s^{-3}\, (= J\, s^{-1})$
electric charge	electro-static unit	esu	$g^{1/2}\, cm^{3/2}\, s^{-1}$	coulomb	C	A. s
volume	liter	ϱ	$1.000028 \times 10^3\, cm^3$	cubic meter		m^3
concentration	mole per liter		$mol\, cm^{-3}\, 10^3$	mole per cubic meter		$mol\, m^{-3}$
molar entropy, molar heat capacity	cal per degree mole		$cal\, deg^{-1}\, mol^{-1}$	joule per mole Kelvin		$J\, K^{-1}\, mol^{-1}$
dynamic viscosity	centi-poise	cp	$g\, cm^{-1}\, s^{-1}$	newton-second per square meter		$N\, s\, m^{-2}$

TABLE 3.4

Conversion Factors to SI Units

1 dyne	=	10^{-5} N	1 Debye	=	3.336×10^{-3} cm
1 atm.	=	101.325 kN m^{-2}	1 erg	=	10^{-7} J
1 torr	=	133.322 N m^{-2}	1 cal	=	4.1840 J

TABLE 3.5

Common Prefixes

Multiple	Prefix	Symbol
10	deka	da
10^2	hecto	h
10^3	kilo	k
10^6	mega	M
10^9	giga	G
10^{12}	tera	T
10^{-1}	deci	d
10^{-2}	centi	c
10^{-3}	milli	m
10^{-6}	micro	μ
10^{-9}	nano	n
10^{-12}	pico	p

same as the number of mmol/ml, and that the number of g/mol is the same as the number of mg/mmol.

The normality (N) concentration scale is not recommended in this text, although it still appears (mainly in older works).

References to 3.4

1. "Manual of Symbols and Terminology for Physicochemical Quantities and Units," *Pure Appl. Chem.* **21**, 1 (1970).
2. A. W. Adamson, "A Textbook of Physical Chemistry," Academic Press, New York, 1973.
3. E. R. Cohen and B. N. Taylor (compilers under the auspices of the CODATA Task Group on Fundamental Constants), "Fundamental Physical Constants," officially adopted by CODATA and published in: *J. Phys. Chem. Ref. Data*, **2**, 663 (1973); CODATA Bulletin No. 11 (Dec. 1973); Dimensions/NBS Jan (1974); NBS Special Publication 398 (August 1974).

3.5. CALCULATIONS

The calculations that are required to convert experimental data into the desired quantities vary in complexity, but only rarely in undergraduate experiments

TABLE 3.6

Fundamental Physical Constants

Constant	Symbol	Value	Uncertainty*	Units: Système Internat. (SI)	Units: Centimeter-gram-second (cgs)
Speed of light in vacuum	c	2.997 924 58	1.2	$\times 10^8$ m/s	$\times 10^{10}$ cm/s
Elementary charge	e	1.602 189 2	46	10^{-19} C	10^{-20} cm$^{1/2}$ g$^{3/2}$ s^{-1}†
		4.803 250	21	...	10^{-10} cm$^{1/2}$ g$^{3/2}$ s^{-1}‡
Avogadro constant	N_A	6.022 045	31	10^{23} mol^{-1}	10^{23} mol^{-1}
Atomic mass unit	u	1.660 565 5	86	10^{-27} kg	10^{-24} g
Electron rest mass	m_e	9.109 534	47	10^{-31} kg	10^{-28} g
Proton rest mass	m_p	1.672 648 5	86	10^{-27} kg	10^{-24} g
Neutron rest mass	m_n	1.674 954 3	86	10^{-27} kg	10^{-24} g
Faraday constant	F	9.648 456	27	10^4 C/mol	10^3 cm$^{1/2}$ g$^{1/2}$ mol^{-1}†
Planck constant	h	6.626 176	36	10^{-34} J · s	10^{-27} erg · s
$h/2\pi$	$\hbar$	1.054 588 7	57	10^{-34} J · s	10^{-27} erg · s
Charge to mass ratio for the electron	e/m_e	1.758 804 7	49	10^{11} C/kg	10^7 cm$^{1/2}$/g$^{1/2}$†
Rydberg constant	R_∞	1.097 373 17	83	10^7 m^{-1}	$10^5 c$ cm^{-1}
Bohr magneton	μ_β	9.274 078	36	10^{-24} J/T	10^{-21} erg/G†
Gas constant	R	8.314 41	26	10^0 J · K^{-1} mol^{-1}	10^7 erg · K^{-1} mol^{-1}
Molar volume, ideal gas, s.t.p.	V_m	22.413 83	70	10^{-3} m^3 mol^{-1}	10^3 cm^3 mol^{-1}
Boltzmann constant	k	1.380 662	44	10^{-23} J/K	10^{-16} erg/K
Gravitational constant	G	6.6720	41	10^{-11} N · m^2/kg^2	10^{-8} dyn · cm^2/g^2

* Based on 1 std. dev.; applies to last digits in preceding column.
† Electromag. system.
‡ Electrostatic system.

3.5. Calculations

TABLE 3.7

Defined Values and Conversion Factors

Atomic mass unit (u)	1/12 the mass of an atom of the ^{12}C nuclide
Standard acceleration of free fall	9.806 65 m/s^2, 980.665 cm/s^2
Standard atmosphere	101,325 N $\cdot$ m^2, 1,013,250 dyn/cm^2
Thermochemical calorie	4.184 J, 4.184 x 10^7 ergs
Liter	0.001 cubic meter
Mole (mol)	amount of substance comprising as many elementary units as there are atoms in 0.012 kg of ^{12}C
Inch	0.0254 m, 2.54 cm

will the amount of calculation be large. In a few experiments in Vol. 2 (e.g., Exp. 18) and in much research work, digital computers can be useful. Computers may range from small, desk-sized programmable calculators to large, high-speed, high-capacity machines, and all must be programmed to perform the required calculations.

Familiarity with programming a computer is valuable in view of the extensive use that these machines now have in science. Many students will obtain this experience in other courses, but those who do not may program the calculations for one or more of the experiments here. Books[1-5] are available for guidance in programming, and the language requirements and limitations of the machine used must be considered.

The texts by Dickson,[1] by Golde,[2] and by McCracken[3] provide thorough instruction in Fortran programming as applied to scientific computation.

When the program is available on magnetic tape, punch cards, or in the computer memory, data may be entered and results obtained with little effort. If the program must first be written, considerably more effort and skill is required. Many programs for standard chemical calculations are available in the common computer languages (e.g., Fortran, Basic[1]). A number of relatively simple prepared Fortran programs of interest to chemists and a survey of Fortran programming are given by Isenhour and Jurs,[4] and a number of Fortran programs of varying complexity oriented toward physical chemistry are given by de Maine and Seawright.[6] Detar[7] has presented for chemists a number of rather advanced Fortran programs of use primarily in research problems.

Pre-written programs which reduce the tedious calculations are helpful with a small number of the experiments that we describe, but, if used, they should be accompanied by enough manual calculations to ensure that the student understands the theory involved.

In any use of a computer, it is important to remember that unless the program so instructs, the computer shows no judgment about the quality of the data given to it. Questionable data are given as much weight as reliable data; therefore the experimenter should evaluate his experimental results carefully before

TABLE 3.8

Units of Concentration

Unit	Symbol	Definition	Comments
molality	m	$\dfrac{\text{moles of solute}}{\text{kg of solvent}}$	note that number of moles = $\dfrac{\text{wt.}}{\text{M. wt.}}$ m is independent of temperature
molarity	M	$\dfrac{\text{moles of solute}}{\text{liters of solution}}$	M varies with T as the volume changes. For dilute aqueous solutions $m \approx M$. Note that the number of moles of solute in a given volume of solution is $n = M \times V$ (in liters)
mole fraction	X	$\dfrac{\text{moles of component}}{\text{total no. of moles}}$	used mainly in theoretical expressions
normality	N	$\dfrac{\text{gram equivalents}^a}{\text{liters of solution}}$	N depends upon the particular reaction (e.g. $KMnO_4$ will have a normality which is different in acidic and basic solutions and depends upon how far the manganese is reduced).
weight percent	wt %	$\dfrac{\text{weight of solute} \times 100}{\text{total weight of solution}}$	total weight of solution consists of solvent plus all solutes; very little use in volumetric work.
parts per million	ppm	parts by weight per million e.g. grams of solute per million grams of solution, or $\dfrac{\text{weight of solute} \times 10^6}{\text{total weight of solution}}$	wt % is parts per hundred; ppm is analogous, as is parts per billion

a The equivalent weight of a material is the weight that would react with, or be produced by 7.999 g of oxygen or 1.00 g of hydrogen. In redox reactions it can be defined as the molecular weight divided by the change in oxidation state, e.g.

$$2MnO_4^- + 5H_2C_2O_4 + 6H^+ \rightarrow 2Mn^{+2} + 10CO_2 + 8H_2O; \text{ eq. wt of } KMnO_4 = \frac{\text{mol. wt}}{5}$$

$$MnO_4^- + e^- \rightarrow MnO_4^{2-}; \text{ Eq. wt of } KMnO_4 = \text{mol. wt}$$

$$MnO_4^- + 3e^- + 2H_2O \rightarrow MnO_2 (s) + 40H; \text{ eq. wt of } KMnO_4 = \frac{\text{mol. wt}}{3}$$

using them. If necessary, programs that give preferential weighting to some of the data can be devised. The results given by the computer should be considered carefully in relation to expected or realistic results, to guard against entering data incorrectly (e.g. using the wrong units).

Many sophisticated instruments are now being constructed with built-in microcomputers that may control the operation of the instrument and/or perform calculations on the data as it is obtained. Often these computers are pre-wired for a fixed set of operations. In other cases, the instrument may be interfaced with a more elaborate, separate computer which can be programmed to perform a variety of functions. Although the problems of interfacing and programming such systems are specialized and are not gone into in this book, the student should be aware of this important field.

Excellent discussions on interfacing small digital computers to laboratory experiments are given in the texts by Perone and Jones[8] and by Wilkins et al.[9] and in the series edited by Mattson, Mark, and MacDonald.[10]

References to 3.5

1. T. R. Dickson, "The Computer and Chemistry," W. M. Freeman, San Francisco, 1968.

2. H. Golde, "FORTRAN II and IV for Engineers and Scientists," MacMillan, New York, 1966.

3. D. D. McCracken, "A Guide to FORTRAN IV Programming," Wiley, New York, 1965.

4. T. L. Isenhour and P. C. Jurs, "Introduction to Computer Programming for Chemists," Allyn and Bacon, Boston, 1972.

5. C. Wilkins, C. Klopfenstein, and P. C. Jurs, "Introduction to Computer Programming for Chemists, Basic Version," Allyn and Bacon, Boston, 1975.

6. P. A. D. de Maine and R. D. Seawright, "Digital Computer Programs for Physical Chemistry," Macmillan, New York, 1963, 1965, vols. I and II.

7. D. F. Detar (ed.), "Computer Programs for Chemistry," W. A. Benjamin, New York, 1968–1969, vols. I–III.

8. S. P. Perone and D. O. Jones, "Digital Computers in Scientific Instruments," McGraw-Hill, New York, 1973.

9. C. L. Wilkins, S. P. Perone, C. F. Klopfenstein, K. C. Williams, and D. E. Jones, "Digital Electronics and Laboratory Computer Experiments," Plenum Press, New York, 1975.

10. J. S. Mattson, H. B. Mark, Jr., and M. C. MacDonald, Jr. (eds.), "Computer Fundamentals for Chemists," Marcel Dekker, New York, 1972–1973, vols. I–III.

Treatment of Experimental Uncertainties

4.1 INTRODUCTION AND DEFINITIONS

In many experiments the chief goal is to obtain one or more numerical results. Regardless of whether these numbers are needed for the analysis of a sample, to examine some theoretical prediction, or for some other application, it is obvious that the value of the information that can be obtained from them depends to a large degree on how well they have been determined. Any measurement, no matter how carefully performed, contains some degree of uncertainty because of experimental errors. It is therefore imperative that when a result is reported, some indication of the reliability of that result also be reported. In addition to providing an estimation of the overall error in an experimental quantity, a detailed error analysis can tell the investigator where the major sources of error are and hence where the effort should be concentrated to improve the experiment. It is pointless, for example, to concentrate on improving the temperature control if the major source of error is in a volume measurement. Similarly, studying the sources and magnitudes of errors in experiments aids students in deciding how carefully measurements have to be made. While it is better to err on the side of being too careful, there are clearly times when, for example, a triple beam balance will serve as well as an analytical balance. This chapter is concerned with how to evaluate the limits of error of experimental quantities. Before proceeding, however, it is necessary to define several terms.

A. *Discrepancy*. This is defined as the difference between two measured values of a quantity, such as that obtained by a student and a literature value, or the differences among results obtained by different students. This term is often confused with the term *error* when making comparisons with literature values. Values found in handbooks or other literature sources are not really "true" values, and until proven otherwise, there is no reason to assume that an experimental value is any less "true" than that in a handbook. Handbook values are the results of experimental measurements themselves — often done with older, poorer instruments — and as such they are also uncertain. Furthermore, the literature values may have been obtained under different conditions, or the composition of the samples used may have been different.

B. *Error.* There are two correct meanings of this word. The first denotes the difference between the "true" (not literature) value and the measured one. It is extremely rare, however, that the "true" value is known, and this often leads people to use the term *error* when they really mean *discrepancy*. The second definition is the estimated uncertainty in an experimental result, and it is expressed in terms of quantities such as average deviation and standard deviation. There are several types of errors with which we are concerned.

1. *Systematic Errors.* These errors are sometimes called *constant errors* because they often cause all the measurements to be in error by the same amount. In any event they cause all the measurements to be either too high or too low. Examples include incorrect calibrations, impurities in the materials, and interfering side reactions. Often these errors can be identified, and by proper experimental design and care reduced to satisfactorily low values. In many cases, however, systematic errors are not obvious, and consequently difficult or even impossible to treat.

2. *Random Errors.* These errors are inherent in any measurement and are just as likely to cause the result to be too high as to be too low. Regardless of how carefully the measurements are made, there will be some variability between successive measurements of a quantity. Errors of this type are called *random errors*, or sometimes *experimental* or *accidental errors*. They can be caused by many things such as errors in judgment, fluctuating conditions (e.g., temperature, or line voltages) and "noise" in an instrument.

3. *Illegitimate Errors.* In addition to the above errors, there are some that can be avoided, and that the reader of a report must be permitted to assume are absent. These include blunders, such as spillage or mistakes in reading the instruments, errors in computation, and so-called chaotic errors, which are caused by disturbances (e.g. line fluctuations) which become *unreasonably* large compared with the natural random errors.

C. *Accuracy and Precision.* Although these two quantities actually measure two different types of errors, they are often confused with each other. The *precision* is a measure of the reproducibility of a result, or how well several individual determinations agree with each other. Thus, high precision implies small random errors. On the other hand, a result is said to be *accurate* if it agrees well with the "true" value, and this implies the absence of systematic errors. Because the "true" value is almost never known, the precision is often taken as an indication of the accuracy, although this is a somewhat dangerous practice. It is quite possible to have a result with high precision and low accuracy or vice versa, as well as a result which has both high precision and accuracy or, unfortunately, neither.

When they can be detected, systematic errors can be minimized by a calibration procedure followed by application of the appropriate correction factor(s) to the data. Therefore they will not be considered further here, although their possible presence in all work should be kept in mind. Random errors are treated by statistical methods as outlined below. Space limitations prevent a detailed discussion of the theory, and for more details a book on the subject[1,2] should be consulted.

4.2. RANDOM ERROR IN A MEASURED QUANTITY

In most experiments the results are found to be distributed about a mean value in a way that corresponds closely to a Gaussian (sometimes called "normal") frequency distribution. Such a distribution is represented by a bell-shaped curve, in which large deviations from the mean occur much less frequently than small ones. A *deviation* is defined as the difference between the ith measured value, x_i, and the most probable value, $\bar{x}$, that is

$$\delta x_i = x_i - \bar{x} \tag{4.1}$$

The most probable value of the quantity is taken to be the one for which the sum of the squares of these deviations is a minimum. According to Eq. (4.1) the sum of the squares of the deviations is

$$\sum_{i=1}^{n} (\delta x_i)^2 = \sum_{i=1}^{n} (x_i - \bar{x})^2 = \sum_{i=1}^{n} (x_i^2 - 2\bar{x}x_i + \bar{x}^2)$$

$$= \sum_{i=1}^{n} x_i^2 - 2\bar{x} \sum_{i=1}^{n} x_i + n\bar{x}^2 \tag{4.2}$$

where n is the total number of measurements. To determine the quantity $\bar{x}$ which makes this sum a minimum, Eq. (4.2) is differentiated with respect to $\bar{x}$ and the derivative is set equal to zero, that is

$$\frac{\partial}{\partial \bar{x}} \sum_{i=1}^{n} (\delta x_i)^2 = -2 \sum_{i=1}^{n} x_i + 2n\bar{x} = 0 \tag{4.3}$$

Solving this for $\bar{x}$ yields the result:

$$\bar{x} = \frac{\sum_{i=1}^{n} x_i}{n} \tag{4.4}$$

Notice that the quantity on the right hand side of Eq. (4.4) is simply the arithmetic average or mean value of the quantity. Hence, the justification for using the mean as the most probable value is that the sum of the squares of the deviations is then a minimum. The student should demonstrate for himself that the sum of the deviations about the mean is equal to zero.

The width of the Gaussian curve gives an indication of the precision of the measurement, with a very precise experiment yielding a very narrow curve. Although there are a number of methods used to describe the width of the curve, one of the quantities most frequently employed is the *standard deviation* which, for a finite number of measurements, is given by the expression:

$$s = \sqrt{\frac{\Sigma(\delta x_i)^2}{n-1}} = \sqrt{\frac{\Sigma x_i^2 - n\bar{x}^2}{n-1}} \tag{4.5}$$

The term $n-1$ in the denominator is known as the degrees of freedom, and is equal to the number of functionally independent deviations. When $n = 1$, of course there can be no deviations. When $n = 2$, two deviations may be computed, but they are equal in magnitude with opposite signs. Similarly, for n measurements n deviations may be computed, but only $n-1$ of them are independent, because their sum is always zero. (The actual relationship is that the derivative, with respect to the mean, of the sum of the squares of the deviations equals zero, as shown in Eq. (4.3).) Many authors use n rather than $n-1$ in the denominator. This simplification arises from the actual definition of the standard deviation from classical probability theory and is correct only if the true value of the quantity being measured is known. If, as is usually the case, the deviations about the *calculated* mean must be used to determine s, then $n-1$ is the correct divisor. As the number of determinations increases it is expected that the calculated mean will approach the true value, and accordingly the difference between the two divisors will become negligible.

The significance of the standard deviation becomes more apparent when the actual form of a Gaussian distribution is examined. The Gaussian distribution function is[1]

$$P = \frac{1}{s\sqrt{2\pi}} \exp -\frac{(\bar{x} - x_i)^2}{2s^2} = \frac{1}{s\sqrt{2\pi}} \exp \frac{-\delta^2}{2s^2} \tag{4.6}$$

where P is the probability that a measurement, x_i, will deviate from the mean by $\delta = \bar{x} - x_i$. The probability that a result will have a deviation lying in the range of $-a$ to $+b$ may be found by integrating $Pd\delta$ over this range. The values of such integrals may be found in various statistical tables as well as in the Handbook of Chemistry and Physics. In the case of $a = b = s$, integration yields a result of 0.682, that is 68% of the results will fall within ±1 standard deviation of the mean. Conversely about one measurement in three will fall outside this range. These ideas are summarized in Fig. 4.1.

Although the standard deviation can be related directly to the Gaussian curve, several other parameters are often used to describe the precision of a measurement. Some of those commonly used are discussed briefly below.

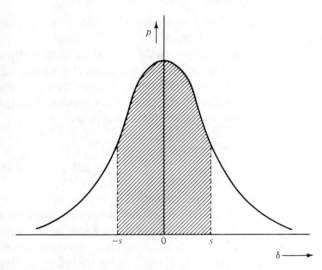

Figure 4.1.   The Gaussian distribution function. The probability that a deviation will fall within ±1 standard deviation of the mean is the area under the curve in this region, which is about 68%.

Variance.   The variance is the square of the standard deviation, and is often discussed in statistical treatments of data. Its chief advantage is that it is additive, that is, the variances in two quantities can be combined to yield the total variance in a third quantity. As we shall see, the standard deviation itself does not usually have this property. Nevertheless, the error in a quantity is normally reported as the standard deviation rather than as the variance.

Average Deviation.   This is the favorite choice of many people because of its ease of calculation. It is merely the sum of the absolute value of the deviations divided by the total number of measurements, i.e.,

$$a = \Sigma |\delta x_i|/n \qquad (4.7)$$

Although simple to calculate, it does not have the direct theoretical significance that the standard deviation has, and is quite dependent on the number of measurements.

Range.   The range of the measurements is merely the difference between the largest and smallest value, and it has virtually no theoretical significance.

Limit of Error.   When a quantity is expressed as, for example, 37.64 ± 0.13, the quantity ±0.13 is often referred to as the limit of error. However, the term "limit of error" is rather ambiguous. Generally it is assumed to be a deviation that has a low probability of being exceeded. Unfortunately there is little agreement as to what is meant by "low," although 1% is often assumed. The term limit of error will be used here to refer to *any* expressed uncertainty. As will be indicated later, the ambiguity is best removed by stating explicitly what the quantity represents, for example whether it is a standard deviation or 95% confidence limits.

Several of the above quantities can be related to each other through proportionality constants. This is done in most books on error and will not be given here. It should be noted, however, that these proportionalities hold only for the true Gaussian distributions.

4.3. PROPAGATION OF ERRORS

To this point the discussion has been concerned with determining the error associated with a single measured quantity. Generally, however, the quantity of interest is one that must be calculated from two or more experimentally determined quantities. Although it is possible to make several determinations of such quantities and to obtain the limits of error by the methods used above, this is not generally desirable. What is preferable is to examine how the errors in the individually measured quantities are propagated to yield the total error in the final result. This procedure has the additional advantage of indicating which quantities are responsible for most of the error and hence should be the focus of attention for improving the experiment. Furthermore, the simple statistical methods outlined above are not really appropriate for the small number of determinations usually made in an experiment. If less than four measurements have been made, then it is best to compare the observed standard deviation (or average deviation) with that estimated by the methods given below, and to use the larger of the two.

Consider the function $F(x, y, z)$, where x, y, and z are experimentally measured quantities with errors e_x, e_y, and e_z, respectively. We wish to determine the error in F, e_F. It can be shown[1] that if x, y, and z are independent of each other, then the *most probable* propagated error in F is given by the equation

$$e_F = \sqrt{ \left(\frac{\partial F}{\partial x} \right)^2 e_x^2 + \left(\frac{\partial F}{\partial y} \right)^2 e_y^2 + \left(\frac{\partial F}{\partial z} \right)^2 e_z^2 } \qquad (4.8)$$

This is the equation that should be used for most of the calculations encountered in the laboratory. It does not depend on a Gaussian distribution, although calculation of e_x, etc., often does.

It should be noted that the errors used in Eq. (4.8) should be compatible, that is, if e_x is a standard deviation, all should be standard deviations, and the result will then be the standard deviation in F; it is not correct, for example, to mix average deviations with standard deviations. Frequently, however, the errors used are estimates based on such things as the uncertainty in reading a buret. In this case the quantity calculated by Eq. (4.8) might best be described as the estimated propagated error. However, it is often treated as if it were the estimated standard deviation. Although this is not strictly correct, it is not a serious problem and will generally lead to a slight overstatement of the error limits. This is because there is a reasonably high probability (32%) that a deviation will be greater than the standard deviation, and most investigators assign reading errors and others that are even more stringent. Nevertheless, if the standard deviation in F is desired, then efforts should be made to estimate the standard deviations in x, y, and z. Finally, notice that if the errors are expressed as standard deviations, then squaring both sides of Eq. (4.8) demonstrates the additivity of the variance mentioned in the previous section.

Although the use of Eq. (4.8) is uncomplicated, it is useful to demonstrate its application to several common functions. In doing so it is also useful to define a term known as the fractional or relative error. The errors used thus far have been *absolute errors* and have the units of the quantity being measured. The *relative* (or *fractional*) *error* is the absolute error divided by the average value of the quantity and, it is dimensionless. Thus, if the length of a box is determined to be 25 ± 1 cm, the absolute absolute error is 1 cm, while the relative (fractional) error is 0.04 or more commonly 4%.

1. *Simple sum and difference.* $F = x \pm y$. Because in this case the partial derivatives equal unity, the absolute error in the quantity is equal to the square root of the sum of the squares of the absolute errors in x and y. That is,

$$e_F = \sqrt{e_x^2 + e_y^2} \tag{4.9}$$

2. *Products and Quotients.* $F = xy/z$. In this case $\partial F/\partial x = y/z$, $\partial F/\partial y = x/z$, and $\partial F/\partial z = -xy/z^2$ and application of Eq. (4.8) yields

$$e_F = \sqrt{\left(\frac{y}{z}\right)^2 e_x^2 + \left(\frac{x}{z}\right)^2 e_y^2 + \left(\frac{-xy}{z^2}\right)^2 e_z^2} \tag{4.10}$$

Although this expression is correct, it is cumbersome to use. It can be simplified by dividing both sides by $F = xy/z$, i.e.,

$$\frac{e_F}{F} = \sqrt{\left(\frac{e_x}{x}\right)^2 + \left(\frac{e_y}{y}\right)^2 + \left(\frac{e_z}{z}\right)^2} \tag{4.11}$$

Recalling the definition of relative error, we can now see from Eq. (4.11) that the relative error in F is equal to the square root of the sum of the squares of the relative errors in x, y, and z.

3. *Powers.* $F = x^m y^n$. Application of Eq. (4.8) leads to the conclusion that for $F = x^m y^n$ the relative error in F is given by the square root of the sum of the terms consisting of the squares of the relative errors of the quantities multiplied by the square of their powers, that is

$$\frac{e_F}{F} = \sqrt{m^2 \left(\frac{e_x}{x}\right)^2 + n^2 \left(\frac{e_y}{y}\right)^2} \tag{4.12}$$

4. *Logarithm.* $F = \ln x$. Application of Eq. (4.8) leads to the result that the absolute error of the natural (base e) logarithm of x is equal to the relative error of x, that is,

$$e_F = \frac{e_x}{x} \tag{4.13}$$

Notice that if the error in $F = \log x$ is required the result is multiplied by 0.434, because $2.303 \log x = \ln x$.

The above list is not meant to be all inclusive, but to serve only as a sampling. Generally it is easier to apply Eq. (4.8) directly to the function than it is to locate the desired result, especially if it is remembered that sums and differences involve absolute errors, while products and quotients are most easily expressed in terms of relative errors.

It is logical to ask why Eq. (4.8) is used rather than the simpler expression obtained directly from differential calculus, namely

$$e_F = \left| \frac{\partial F}{\partial x} e_x \right| + \left| \frac{\partial F}{\partial y} e_y \right| + \left| \frac{\partial F}{\partial z} e_z \right| \tag{4.14}$$

The error calculated using Eq. (4.14) is referred to as the *maximum* error. It is the error that would be obtained if the errors in the individual quantities happened to occur such that there was no cancellation. We have assumed, however, that the errors are independent, that is the error e_x has no relation to the error e_y. Thus, we expect that there will be some cancellation of errors. For example, the error in x may cause F to be too large, while that in y may cause F to be too small. The overall error is expected statistically to be somewhat less than the sum of the separate contributions, which means that for independent errors Eq. (4.14) overstates the error. Equation (4.8) compensates for this exaggeration by noting that the cross terms obtained when Eq. (4.14) is squared will vanish if the errors are independent. For most of the applications to be encountered the errors will be independent, hence Eq. (4.8) is the correct one to use. In spite of the above, many laboratory manuals, especially the more elementary ones, use Eq. (4.14) to estimate the propagated error, presumably because it is slightly easier to calculate. If this approximation is used, it should be clearly stated that the *maximum* propagated error has been calculated.

As an example of the evaluation of a propagated error, consider the uncertainty in the calculated molarity M of a solution prepared by dissolving a certain weight W of NaCl in a measured volume V of solution. The molarity is

$$M = \frac{W/m}{V}$$

where m is the molecular weight.

49

The data are as follows:

$W = 0.5678$ g

$V = 100$ ml, measured with a volumetric flask

$m = 58.443$ g/mol, from tabulated data.

Two types of errors will be considered — the *maximum error*, which would be obtained in the worst case when all uncertainties act in the same direction, and the statistical or most probable error, which was discussed above.

The balance reading error is ±0.0001 g, but a balance must always be read twice in obtaining a weight. Hence the weighing error is itself a propagated error in the difference of two quantities; i.e., $W = W_2 - W_1$, where W_1, and W_2 are the two balance readings.

The *maximum* error in the weight from Eq. (4.14) (because $\partial w/\partial w_i = 1$), is

$$e_{w_{max}} = \pm |e_{w_1}| + |e_{w_2}| = \pm 0.0002 \text{ g,}$$

while the maximum *relative* error is

$$e'_{w_{max}} = \pm \frac{0.0002 \text{g}}{0.5678 \text{g}} \times 100\% = \pm 0.035\%$$

The statistical or most probable error in W, however, is given from Eq. (4.9),

$$e_{w_{stat}} = \pm\sqrt{(e_{w_1})^2 + (e_{w_2})^2} = \pm 0.00014 \text{ g}$$

The statistical relative error is

$$e'_{w_{stat}} = \pm \frac{0.00014 \text{g}}{0.5678 \text{g}} \times 100\% = \pm 0.025\%.$$

A larger source of error comes from the purity of the NaCl. This would introduce a systematic, not a random error, and could be corrected for if the purity were known. We shall not consider this correction at this point.

The tolerance of a 100-ml volumetric flask is ±0.08 ml. If it is assumed that the reading error is less than this value, then the volume error e_v is ±0.08 ml, or $e'_v = \pm(0.08/100) \times 100 = \pm 0.08\%$ relative error.

There is also an uncertainty in the tabulated value of the molecular weight. Unfortunately, error limts often are not stated for such data, although they could be found by reference to official atomic weight values. As a working guide, one could assume an uncertainty of ±1 in the last digit; i.e., $e_m = \pm 0.001$ g/mole. This yields a relative error $e'_m \pm 0.002\%$. As we shall see, this is negligible. Because only multiplication and division are involved the *relative* error in M can be found easily. The *maximum* relative error is e_F/F, and from Eq. (4.14) applied to this case,

$$e'_{M_{max}} = \pm(|e'_{w_{max}}| + |e'_v| + |e'_m|)$$

$$= \pm(0.035\% + 0.08\% + 0.002\%) = \pm 0.117\% \approx \pm 0.12\%$$

The *statistical* relative error, from equation (4.11), is

$$e_{M_{stat}} = \pm \sqrt{(e'_{w_{stat}})^2 + (e'_v)^2 + (e'_m)^2}$$
$$= \pm \sqrt{(0.025\%)^2 + (0.08\%)^2 + (0.002\%)^2}$$
$$= \pm 0.084\% \approx \pm 0.08\%$$

Notice that in calculating the maximum error, the maximum error in the weight must be used, while the statistical error in weight is required for the overall statistical error.

One can see from the above that the error in the molecular weight value is negligible with respect to the other quantities and that the largest error arises from the volume measurement. The results could be improved by calibrating the flask. Hence an error analysis, in addition to indicating the reliability of results, often will show how the experiment can be improved.

4.4. STRAIGHT-LINE PLOTS

Frequently two measurable quantities such as x and y are related to each other by a linear equation involving two unknown constants, m and b, such that $y = mx + b$. (Even in cases in which the measured quantities are not directly linearly related, they can very often be put into a linear form by use of reciprocals, square roots, or logs.) A plot of y vs. x will yield a straight line with a slope m and intercept b. The question of what are the "best" values of m and b, then arises. We will now discuss two methods of determining m and b, as well as the errors associated with them.

4.4.1. Graphical Method

Normally one plots y vs. x as a set of points. Because each point has error associated with it, the plot should actually be made using a rectangle with the dimensions of $2e_x$ and $2e_y$ centered about the appropriate point (x, y). (Some investigators prefer to use circles or ellipses rather than rectangles.) The "best" line is then drawn in the usual fashion (i.e., with the scatter about the line minimized). The limits of error are determined by drawing the lines with the maximum and minimum slope passing through the rectangles. The difference between the two limiting slopes and intercepts is taken as an estimate of twice the limit of error in m and b, respectively. The process is illustrated in Fig. 4.2.

4.4.2. Method of Least Squares

The object in this method is to find the straight line for which the sum of the squares of the distances between the data points and the line is a minimum. The usual derivation assumes that all the error is in y; that is, the error in x is negligible. This is not an unreasonable assumption, because usually one quantity can be

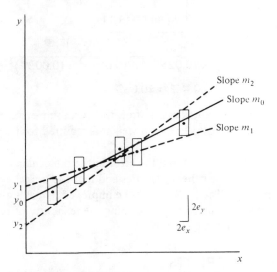

Figure 4.2. The graphical method of obtaining the "best" slope and intercept of a straight line. The "best" intercept is $y_0 \pm (y_1 - y_2)/2$ and the "best" slope $m_0 \pm (m_2 m_1)/2$.

determined much more accurately than another. The derivation, which may be found in several books,[1,2] will only be outlined here. The student is urged to go through the steps as an aid in understanding the method of least squares. The procedure is analogous to that used in an earlier section to show that the average value of a measured quantity was the one for which the sum of the squares of the deviations was a minimum. The deviations (often called residuals) are given by

$$\delta y_i = y_i - (mx_i + b) \tag{4.15}$$

where m and b respectively are the "best" values of the slope and intercept, and y_i is the actual observed value of y when $x = x_i$. Equation (4.15) is then squared and the result is summed over all of the n points. The resulting expression is differentiated with respect to m and b, and the derivatives are set equal to zero. The two equations which result are then solved for m and b. The final expressions obtained are

$$m = \frac{n\Sigma(x_i y_i) - \Sigma x_i \Sigma y_i}{n\Sigma x_i^2 - (\Sigma x_i)^2} \tag{4.16}$$

$$b = \frac{\Sigma x_i^2 \Sigma y_i - \Sigma x_i \Sigma(x_i y_i)}{n\Sigma x_i^2 - (\Sigma x_i)^2} \tag{4.17}$$

Note that Σx_i^2 and $(\Sigma x_i)^2$ are not the same. In the first case each value of x is squared before taking the sum; while in the second case the values of x are first added together and then the sum is squared.

To compute the standard deviation in m and b, the standard deviation in y must first be computed from the expression

$$s_y = \sqrt{\frac{\Sigma(\delta y_i)^2}{n-2}} \tag{4.18}$$

Notice that the number of degrees of freedom appearing in the denominator is $n - 2$ and not $n - 1$ as in Eq. (4.5). This is because in the least squares treatment another relationship between the δy_i values is required, namely two derivatives are set equal to zero instead of only one. The numerator may be evaluated directly by calculating all the deviations according to Eq. (4.15), but this procedure is usually too cumbersome. The numerator may be determined indirectly from the expression[1]

$$\Sigma(\delta y_i)^2 = \Sigma y_i^2 - \frac{(\Sigma y_i)^2 \Sigma x_i^2 - 2\Sigma x_i y_i \Sigma x_i \Sigma y_i + n(\Sigma x_i y_i)^2}{n\Sigma x_i^2 - (\Sigma x_i)^2} \tag{4.19}$$

or in terms of m and b

$$\Sigma(\delta y_i)^2 = \Sigma y_i^2 - b\Sigma y_i - m\Sigma x_i y_i \tag{4.19a}$$

Equation (4.19) was obtained by substituting Eqs. (4.16) and (4.17) into the expression for the sums of the squares of the deviations and collecting terms. Comparison of Eq. (4.19a) with the square of Eq. (4.15) illustrates that a number of terms cancelled.

The standard deviation in m and b are given by

$$s_m = s_y \sqrt{\frac{n}{n\Sigma x_i^2 - (\Sigma x_i)^2}} \tag{4.20}$$

$$s_b = s_y \sqrt{\frac{\Sigma x_i^2}{n\Sigma x_i^2 - (\Sigma x_i)^2}} \tag{4.21}$$

These last two equations were obtained by applying Eq. (4.8) directly to the expressions for m and b, that is Eqs. (4.16) and (4.17). As an example, consider the case of s_b. According to Eq. (4.8) the standard deviation in b is

$$s_b = \sqrt{\left(\frac{\partial b}{\partial y_1}\right)^2 s_y^2 + \left(\frac{\partial b}{\partial y_2}\right)^2 s_y^2 + \cdots + \left(\frac{\partial b}{\partial y_n}\right)^2 s_y^2} = s_y \sqrt{\sum_{j=1}^{n} \left(\frac{\partial b}{\partial y_j}\right)^2} \tag{4.22}$$

The variable y_j occurs in two of the terms in the expression for b, namely once in the sum Σy_i and once in the sum $\Sigma x_i y_i$. Therefore from Eq. (4.17)

$$\frac{\partial b}{\partial y_j} = \frac{\Sigma x_i^2 - x_j \Sigma x_i}{n\Sigma x_i^2 - (\Sigma x_i)^2} \tag{4.23}$$

According to Eq. (4.22) this expression must be squared and then summed from $j = 1$ to $j = n$, that is

$$\sum_{j=1}^{n} \left(\frac{\partial b}{\partial y_j}\right)^2 = \sum_{j=1}^{n} \frac{(\Sigma x_i^2)^2 - 2x_j \Sigma x_i \Sigma x_i^2 + x_j^2 (\Sigma x_i)^2}{[n\Sigma x_i^2 - (\Sigma x_i)^2]^2} \qquad (4.24)$$

$$= \frac{n(\Sigma x_i^2)^2 - 2\Sigma x_j \Sigma x_i \Sigma x_i^2 + \Sigma x_j^2 (\Sigma x_i)^2}{[n\Sigma x_i^2 - (\Sigma x_i)^2]^2}$$

The indices i and j are "dummy" indices, that is they only serve as identification tags. Hence $\Sigma x_i = \Sigma x_j$ and $\Sigma x_i^2, = \Sigma x_j^2$, so that Eq. (4.24) reduces to

$$\sum_{j=1}^{n} \left(\frac{\partial b}{\partial y_j}\right)^2 = \frac{n(\Sigma x_i^2)^2 - 2(\Sigma x_i)^2 \Sigma x_i^2 + \Sigma x_i^2 (\Sigma x_i)^2}{[n\Sigma x_i^2 - (\Sigma x_i)^2]^2}$$

$$= \frac{\Sigma x_i^2 [n\Sigma x_i^2 - (\Sigma x_i)^2]}{[n\Sigma x_i^2 - (\Sigma x_i)^2]^2} = \frac{\Sigma x_i^2}{n\Sigma x_i^2 - (\Sigma x_i)^2} \qquad (4.25)$$

Substitution of this into Eq. (4.22) yields Eq. (4.21), as desired.

Although these calculations look quite formidable, many of the sums are required in several places and need be calculated only once. Unfortunately, because small differences between relatively large numbers are often required, all the figures must be retained. Obviously computers and modern calculators have greatly simplified this process in recent years.

It should be noted that the method of least squares is not limited to straight lines, and a more complete discussion can be found in textbooks.[2] The procedure for other functional forms is the same as that outlined above. In general, if there are k quantities for which the best value is desired, then there will be k derivatives to set equal to zero and hence $n - k$ degrees of freedom. Generally, however, it is easier first to try to put the equation into a linear form than it is to derive the necessary nonlinear equations.

There are several methods that can be used to check how well the data fit the assumed curve (e.g., linear, quadratic, etc.). One of the most commonly used parameters is the correlation coefficient, r, which for a linear equation is given by:

$$r = \frac{n\Sigma x_i y_i - \Sigma x_i \Sigma y_i}{\sqrt{[n\Sigma x_i^2 - (\Sigma x_i)^2][n\Sigma y_i^2 - (\Sigma y_i)^2]}} \qquad (4.26)$$

The value of r can range between 0, which indicates no correlation, and 1, which indicates a perfect correlation. Experimental scatter will generally yield a nonzero value of r even if there is no correlation, so there are tables which indicate the probability that a certain number of *uncorrelated* data points will yield a value of r which exceeds some given value. (Such a table is reproduced in Ref. 2.) Although correlation coefficients are fairly commonly reported, they should be used with caution. The linear correlation coefficient above, for example, increases as the slope increases.[3] It is therefore best actually to plot the data, as indicated below.

The increased availability of computers and modern calculators has made the method of least squares increasingly popular in the analysis of data. Because the method is purely numerical, it is possible to obtain the parameters of a straight line (or other function) without actually plotting the data. This approach is dangerous for at least two reasons. First, if any curvature is present − either because of a breakdown in the theory or some systematic experimental error − the method of least squares given here will miss it, and the values obtained will therefore be erroneous. On the other hand, this curvature often becomes important only when one of the variables becomes quite large or small. A preliminary plot will show which data points remain reasonably linear, and *these* points can then be treated by the method of least squares. Second, sometimes the straight line obtained from a least squares analysis does not resemble the "best" line drawn by eye. This is often caused by an overemphasis of the errors on one end of the scale, in which case a weighted least squares treatment should be used.[4,5,6] The usual procedure is to weight each value of y_i by the inverse of the variance in its measurement. The overemphasis at one end of the scale is particularly a problem when y is the logarithm of a quantity (for example, pH, or log p in a Classius-Clapeyron plot, or log c in a first order kinetics plot). According to Eq. (4.8) the variance in log y is merely $s_{\log y_i}^2 = s_{y_i}^2/(2.303\, y_i)^2$, so the weighting factor is proportional to y_i^2. Regardless of what procedure is used, however, the data should be plotted to verify that the scatter is indeed random.

4.5. CONFIDENCE LIMITS AND UNCERTAINTY OF A MEAN

As was mentioned in the introduction, to obtain the maximum amount of information from quantitative experimental data, it is necessary to indicate the degree of uncertainty present. It is not sufficient, however, merely to report a result as, for example, 37.64 ± 0.13, because it is then not clear exactly what the ±0.13 represents. To avoid confusion it should be stated clearly what is meant by the assigned error limits, whether, for example, they are the average or standard deviation of a number of determinations, the error propagated through a calculation, or even just a guess. Many investigators prefer to give the 95% confidence limits described below.

Previously it was shown that for a Gaussian distribution 68% of the results will lie within ±1 standard deviation of the mean. In a similar way it may be shown that 95% of the results will lie within ±1.96 standard deviations of the mean. Now suppose that after obtaining the mean, $\bar{x}$, and standard deviation, s, for n measurements of a quantity, the set of experiments is repeated to obtain n more measurements. In general these new measurements will have a slightly different mean and standard deviation than the first. If the process is repeated several more times, a whole set of means will be obtained, which will themselves form a Gaussian distribution about some grand mean. It may be shown by application of Eq. (4.8) that the standard deviation of this distribution of means is equal to $s/\sqrt{n}$. Therefore it is expected that 95% of the means will fall within ±1.96 $s/\sqrt{n}$ of the true mean.

The 95% confidence limits are therefore $\bar{x} \pm 1.96 \, s/\sqrt{n}$. The quantity $1.96 \, s/\sqrt{n}$ is often referred to as the 95% confidence interval. If the sample is small, some additional uncertainty in the value determined is expected. This problem has been treated mathematically; accordingly, for each value of n there is an appropriate factor, conventionally designated as t, such that the 95% confidence limits become

$$\bar{x} \pm ts/\sqrt{n} \tag{4.27}$$

Values of t for various degrees of freedom $(n-1)$, as well as for confidence limits other than 95%, are tabulated in texts on statistics and in handbooks. Note, however, that many tables list the value of t according to the probability that a value will *exceed* rather than fall within the range given by Eq. (4.27).

The 95% confidence limits provide an objective, unambiguous method of reporting the uncertainty in a result. Thus, the example given earlier could be reported as 37.64 ± 0.13 (95% confidence limits, $n = 5$).

4.6. SIGNIFICANT FIGURES

The question now arises as to how many figures should be used to report a result. It is poor practice and quite meaningless to use more figures than the data justify. Unfortunately, there is no universal agreement about how many figures are significant. If there is uncertainty in the second decimal place, it makes no sense to report a result to four decimal places. There is, however, less agreement about the status of the third decimal place in this case. Unless otherwise stated, the last digit is always considered to be uncertain and, depending on the investigator, the next to last digit may also be slightly uncertain (± 3 or less). Again, the best way to avoid ambiguity is to report the limits of error with the quantity. These assigned limits of error, however, are only estimates, and must never be reported to more than two significant figures. Some of the major rules for the use of significant figures are given below. Note, however, that *the best method of determining the number of significant figures is from the calculated uncertainty*, and the suggestions given in rules 4–6 are only rough guidelines.

1. All digits reported are defined as significant, except possibly for the digit zero. The digit zero is always significant unless it is merely marking the place of the decimal point, in which case it may or may not be significant. In the case of a number less than one, the zeros preceding the first nonzero digit are never significant, but any zeros to the right of this digit would be significant. For numbers of 10 or more the significance of the zero is ambiguous. *This ambiguity is completely removed by using exponential notation.* Thus if 1000 is reported as 1.00×10^3, it is clear that there are three significant figures (i.e., only two zeros are significant.)

    Examples. The following numbers all contain four significant figures:

    1050 (better written as 1.050×10^3)
    1.050×10^4
    0.01050

2. When performing arithmetical calculations, retain one or two digits beyond the last significant figure to avoid "round off" errors.
3. When rounding numbers, if the first (i.e., the leftmost) digit to be dropped is less than 5, leave the last retained digit unchanged. If the first dropped digit is greater than 5 or 5 followed by any nonzero digits, increase the last retained digit by 1. If the first digit to be dropped is *exactly* 5, round the result to the nearest even number.

    Examples. When rounded to four figures, the following numbers become 37.64.

37.6382	37.6450
37.63501	37.6350
37.6449	

But 37.645001 would be rounded to 37.65.

4. When performing addition or subtraction, the final result should contain the same number of decimal places as the component with the fewest decimal places.

    Examples.

$$
\begin{array}{r}
25.6 \\
3.84 \\
6.739 \\
\hline
36.179
\end{array}
$$

Since 25.6 has only one decimal place, the answer should be reported as 36.2. Notice that if the numbers had been rounded before addition, the result would have been 36.1 instead of the correct value 36.2.

$$
\begin{array}{r}
17.256 \\
-17.21 \\
\hline
0.046
\end{array}
$$

Because 17.21 has only two decimal places, the answer should be reported as 0.05.

5. When performing multiplication or division the result should have a *relative* uncertainty that is about the same as the least precise component. This can be achieved usually, but not always, by retaining the same number of significant figures in the result as there are present in the component with the fewest significant figures.

    Examples.

$$6.52 \times 14 \times 3.50 = 319.4800$$

If the answer is rounded to two digits, as indicated by the number 14, the answer becomes 3.2×10^2. The relative error in the final result is then about 1/32, which is on the same order as 1/14.

$$10.50 \times 0.97 = 10.1850$$

The uncertainty in the answer should be on the order of 1/97 or 1%. The answer should therefore be reported as 10.2, although this is more figures than contained in 0.97.

6. When converting numbers to logarithms, use as many *decimal places* in the mantissa as there are significant digits in the number. When determining antilogarithms, keep as many significant figures as there are decimal places in the mantissa.

    Examples.

$$\log 11.52 = 1.0615$$

The number has four significant figures, and hence there are four decimal places in the mantissa.

$$\text{antilog } 0.005 = 1.01$$

There are three decimal places in the mantissa, hence the number is reported to three significant figures.

4.7. REJECTION OF DATA

Occasionally a result is obtained that deviates greatly from the other values. For the small number of determinations normally encountered, such a result will have a large effect on the value of the mean. The question therefore arises as to whether or not the problem value can be discarded. The first step in such a decision is a careful examination of the laboratory notebook to see if something occurred to make that determination different from the others. Perhaps, for example, a sample was heated for less time than the others, or the source of one of the reagents was different. This is where a well kept notebook becomes valuable. Note, however, that before a result is discarded on the basis of such a comment, an effort should be made to see that the deviation expected from the occurrence is in the correct direction. If, for example, it is desired to determine the amount of water in a sample by weighing the sample before and after placing it in a drying oven, then an insufficient time spent in the oven can only yield a low value for the amount of water present.

If no reason to discard the result can be found in the laboratory notebook, then it should not be discarded unless so indicated by some statistical test. If there are five or more determinations, one commonly used rule is to discard a value if it deviates from the mean of the others by more than four average deviations. A better procedure is to use the so-called Q test described below. Determine the range of all the results, that is the difference between the highest and lowest values. Divide the range into the difference between the questionable result and its nearest neighbor to obtain the quotient, Q. The result may be rejected with 90% confidence if Q exceeds the value below.

Total Number of Observations	Q
3	0.94
4	0.76
5	0.64
6	0.56

One final word of caution. Even if the Q test or some other criterion can be found for rejecting a result, it must be done cautiously. The suspect value may in fact be the best one. Suppose, for example, that in the determination of the amount of water in a sample mentioned previously, a low result was obtained, and the notebook showed this sample had been heated for less time than the others. Although the direction of the error is correct, a small difference in heating time is unlikely to result in a large difference in the final result. Furthermore, if the sample was slowly decomposed by heat into volatile components, then prolonged heating would make the sample appear to contain more water than was actually present. In this case all values would be suspect, but the sample heated least might well give the best result.

4.8. EXAMPLE

The following example will serve to illustrate many of the ideas discussed in this chapter. Consider an experiment in which a student desires to determine the molecular weight of the gas N_2O_4. The method to be used consists of filling a vessel of known volume to slightly more than 1 atm of pressure, allowing the excess to leak into the atmosphere at a known temperature, and finally determining the mass, m, of N_2O_4 left in the container. According to the ideal gas law, the molecular weight, M, will be

$$M = mRT/PV \qquad (4.28)$$

The first experimental problem is to obtain the volume of the container by determining the mass of water the vessel can hold. The following calibration data were obtained at $25.0 \pm 0.1°C$:

wt of container + water	463.14 ± 0.08 g
− wt of evacuated container	145.07 ± 0.02 g
wt of water in container	318.07 ± 0.08 g

The 0.02 represents the student's estimate of the ability to reproduce a reading on the balance used, based on the reading error and condition of the balance. Although the scale actually could be estimated to the nearest 0.01 g, it was felt that the balance was not capable of reproducing a reading this precisely. Accordingly the uncertainty was increased from 0.01 to 0.02 g. It should be noted that this is the uncertainty in reproducing a *reading* on the balance. To obtain the

weight of an object, the balance must be read twice (zeroing the balance involves reading it), so that the uncertainty in the weight of an object is a larger value. For example, the most probable error in the weight according to Eq. (4.8) or Eq. (4.9) is $\sqrt{(0.02)^2 + (0.02)^2} = 0.03$ g. What is desired in this experiment, however, is the weight of water the vessel can hold, which is given by the difference between the balance readings when the vessel is full and empty. Therefore it is not important that the actual weight of the container be known accurately, and the smaller figure of 0.02 may be used if the balance is not adjusted in any way between the required readings. The larger error associated with the filled container also considers the possibility of incompletely filling the container as follows: the radius of the neck was estimated to be 0.5 cm, and the student felt that the container was filled to within 1 mm of the top, so that the possible error in filling is about $\pi \times (0.5)^2 \times 0.1 = 0.079$ cc, or 0.079 g of water. (Notice that this is actually a systematic error, because the vessel cannot be overfilled, but it is convenient to maintain the ± sign.) The overall error in this reading will be a combination of this filling error and the one mentioned previously, and so may be calculated from Eq. (4.8) as $\sqrt{(0.079)^2 + (0.02)^2} = 0.081$ g. Finally the most probable error in the weight of the water is obtained by application of Eq. (4.8) or Eq. (4.9) as $\sqrt{(0.08)^2 + (0.02)^2} = 0.08$ g. This value represents a relative error of $0.08/318 = 0.025\%$. Notice that essentially all of the error is due to the uncertainty in filling, that could be greatly reduced by using a vessel with a neck made of capillary tubing. The density of water at $25.0°C$ is 0.99707 g/ml and the temperature range is small enough so that the variation in this density is negligible. Therefore the volume of the container is

$$V = \frac{318.07 \text{ g} \pm 0.025\%}{0.99707 \text{ g/ml}} = 319.004 \text{ ml} \pm 0.025\%$$

This value could be reported as 319.00 ± 0.08 (estimated propagated error) ml.

The mass of N_2O_4 in the vessel was determined. While the system was approaching equilibrium, the temperature was $24.8 \pm 0.5°C$, and readings of the barometer varied from 757 to 761 mm, or 759 ± 2 mm. The weights obtained were

wt of container + N_2O_4	146.10 ± 0.02
wt of evacuated container	145.06 ± 0.02
wt of N_2O_4 in container	$1.04 \pm \sqrt{(0.02)^2 + (0.02)^2}$

These data can be summarized in Table 4.1.

According to Eq. (4.28) the molecular weight is, therefore:

$$M = \frac{0.08206 \text{ } \ell \cdot \text{atm(Kmol)}^{-1} \times 298.0 \text{ K} \times 1.04 \text{ g}}{1.00 \text{ atm} \times 0.31900 \text{ } \ell} = 79.7242 \text{ g/mol}$$

4.8. Example

TABLE 4.1

Summary of Data for Molecular Weight of N_2O_4

quantity	value	error	relative error
m	1.04 g	0.028 g	0.028/1.04 = 2.7%
T	298.0 K	0.5 K	0.5/298 = 0.17%
V	319.00 ml	0.08 ml	0.08/319 = 0.025%
P	1.00 atm	0.003 atm	2/759 = 0.26%

The least precise quantity contains 3 significant figures and although it is suspected that this is the number to use in the final result, the actual number is best determined from the calculated error. Application of Eq. (4.8) (or Eq. (4.11)) to (4.28) shows that the relative error in the molecular weight is

$$e_M/M = \sqrt{(e_m/m)^2 + (e_T/T)^2 + (e_P/P)^2 + (e_V/V)^2}$$

$$= \sqrt{(2.7)^2 + (0.17)^2 + (0.26)^2 + (0.025)^2}$$

$$= \sqrt{7.29 + 0.029 + 0.068 + 0.00062} = \sqrt{7.40} = 2.7_2\%$$

This leads to an absolute error in the molecular weight of 79.7 x 0.027 = 2.1 g/mol. Thus we see that almost the entire error is in the weight determination. If an analytical balance had been used, the weighing error would have been made negligible, so that if nothing else were changed the propagated error would be only $\sqrt{0.029 + 0.068 + 0.0006} = 0.31\%$. In the actual case the result could be reported as 79.7 ± 2.1 (estimated propagated error).

The sum of the atomic weights for N_2O_4 is 92.011, which is clearly outside the range of experimental error. The experiment was repeated (this time using the analytical balance) three more times with the results, 79.02, 78.95, and 79.23. One might ask if the value previously obtained could be discarded, because it is considerably higher than the others. Application of the Q test is indicated. The range is 79.72 − 78.95 = 0.77, while the difference between 79.72 and its nearest neighbor is 79.72 − 79.23 = 0.49. Therefore $Q = 0.49/0.77 = 0.64$, which is less than the 0.76 required when four values are present, and so all values should be retained. The average molecular weight is thus:

$$\bar{M} = \frac{79.02 + 78.95 + 79.23 + 79.72}{4} = 79.23$$

According to Eq. (4.5) the standard deviation is

$$s = \sqrt{\frac{(0.21)^2 + (0.28)^2 + (0.00)^2 + (0.49)^2}{4 - 1}} = 0.34$$

If it is desired to report 95% confidence limits, the handbook gives t for four measurements (3 degrees of freedom) as 3.18, so that the 95% confidence interval is $3.18 \times 0.34/\sqrt{4} = 0.54$. The result could therefore be reported as 79.23 ± 0.34 (standard deviation, four runs) or 79.2 ± 0.5 (95% confidence limits, $n = 4$).

The discrepancy between the observed molecular weight and the one calculated from the sum of the atomic weights suggests the presence of some systematic error. The calibration of the volume is one possibility, as is the assumption of the ideal gas law. However it is very unlikely that either of these possibilities could produce such a large discrepancy. The student therefore consulted the literature more carefully and found that N_2O_4 dissociates into NO_2 according to the equilibrium:

$$N_2O_4 \rightleftharpoons 2NO_2$$

The observed molecular weight can be used to obtain the degree of dissociation, α, and this in turn can be used to determine the equilibrium constant, K. The derivation of the equations is simple and may be found in several places[7] but does not concern us here. The final equations are

$$\alpha = \frac{92.011 - M}{M} \tag{4.29}$$

$$K = \frac{4\alpha^2 P}{1 - \alpha^2} \tag{4.30}$$

When the equations above are solved for the case of $M = 79.23$, the results are $\alpha = 0.161315$ and $K = 0.10687$. As before, the number of figures which actually should be reported will depend on the limits of error. For convenience only the standard deviation will be considered in the rest of the example.

In the determination of α the uncertainty in the value 92.011 for the molecular weight may be neglected so that only that in M need be considered. As reported before, the standard deviation in M is 0.34 or $0.34/79.2 = 0.43\%$. According to Eq. (4.8) the error in α will be simply the absolute value of $(\partial\alpha/\partial M)e_M$. Because $\partial\alpha/\partial M = 92.0/M^2 = 1.47 \times 10^{-2}$, the error in α is $1.47 \times 10^{-2} \times 0.34 = 5.0 \times 10^{-3}$. Thus α could be reported as 0.161 ± 0.005 (propagated standard deviation, four measurements). Notice that the relative standard deviation in α is 3.1%. This undesirable large increase in uncertainty often occurs when two numbers are subtracted from each other.

According to Eq. (4.8) the error in K is

$$e_K = \sqrt{(\partial K/\partial\alpha)^2 e_\alpha^2 + (\partial K/\partial P)^2 e_p^2}$$

However, the relative error in P is less than one-tenth of that in α and may therefore be neglected, so that the error in K is simply the absolute value of $(\partial K/\partial\alpha)e_\alpha$. In this case $\partial K/\partial\alpha = 8\alpha P/(1 - \alpha^2) + 8\alpha^3 P/(1 - \alpha^2)^2 = 1.32 + 0.03 = 1.35$, so that the propagated error in K is $1.35 \times 0.005 = 0.007$. The value of K may therefore be reported as 0.107 ± 0.007 (propagated standard deviation, four values). A fair amount of work could have been saved by noticing that because α^2 is quite a bit less than 1, the error in the denominator should be much less than that caused by α in the numerator. If the error in $1 - \alpha^2$ had been neglected, it could have been concluded that the relative error in K should be twice the relative error in α, because α is squared in the numerator. Because the relative error in α is 3.1%, the relative error in K should be 6.2%, which yields an absolute error of 0.062 x 0.107 = 0.007, as calculated. These labor saving approximations should always be sought when performing error calculations.

The standard free energy of a reaction is related to the equilibrium constant by the equation

$$\Delta G° = -RT \ln(K) \qquad (4.31)$$

Thus, in the case of the present equilibrium:

$$\Delta G° = -1.987 \; cal(mol . K)^{-1} \times 298.0 \; K \times \ln(0.107)$$

$$= -1.987 \times 298.0 \times (-2.2349) \; cal/mol = 1323.3 \; cal/mol$$

Again the number of figures reported will be determined by the error in $\Delta G°$. (Notice that to avoid any rounding error, one more figure than suggested by rule 6 was used for $\ln(K)$.) The relative error in $\Delta G°$ is the square root of the sum of the squares of the relative errors in T and $\ln(K)$. However the relative error in T is negligible, so that the relative error in $\Delta G°$ is equal to that in $\ln(K)$. According to Eq. (4.13) the absolute error in $\ln(K)$ equals the relative error in K, which is 0.062. Thus, the relative error in $\ln(K)$, and hence $\Delta G°$, is 0.062/2.23 = 2.7%. Therefore $\Delta G°$ may be reported as 1.323 ± 0.037 Kcal/mol (propagated standard deviation, four values), or if preferred 1.32 ± 0.04 (propagated standard deviation, four values).

REFERENCES TO CHAPTER 4

1. Y. Beers, "Introduction to the Theory of Error," Addison-Wesley, Reading, Massachusetts, 1958.

2. H. D. Young, "Statistical Treatment of Experimental Data," McGraw-Hill, New York, 1962.

3. W. H. Davis, Jr. and W. A. Pryor, *J. Chem. Educ.*, **53**, 285 (1976).

4. See for example, D. E. Sands, *J. Chem. Educ.*, **51**, 473 (1974) and references therein.

5. K. P. Anderson and R. L. Snow, *J. Chem. Educ.*, **44**, 756 (1967).

6. E. D. Smith and O. M. Mathews, *J. Chem. Educ.*, **44**, 757 (1967).

7. See for example, F. Daniels, J. W. Williams, P. Bender, R. A. Alberty, C. D. Cornwell, and J. E. Harriman, "Experimental Physical Chemistry," 7th ed., McGraw-Hill, New York, 1970. p. 102.

Laboratory Chemicals

5.1. REAGENTS

A wide variety of chemicals are available commercially for laboratory use, and although such supplies often are regarded in a very matter-of-fact way, it is important to be aware of the differences among the various grades and purities. Common chemicals often can be obtained in a variety of grades, less common ones perhaps only in one grade. Materials of low purity normally will be less expensive, and will be satisfactory for many purposes. It is sometimes more economical to purify a chemical as needed than to purchase a higher grade, and for some purposes even very good commercial materials need further purification. The identity, as well as the amounts, of impurities may be important, and selection of chemicals must consider the use to which they will be put. The most frequently encountered grades of materials are the following.

1. *Technical or Practical.* This represents material that is suitable for most preparative work, but that contains appreciable levels of impurities. Typically, it may be 90–95% pure. Technical grade chemicals are more economical than higher purity materials if the impurities do not interfere, and in many cases they are the only grades available. For special applications, purification will be required. In most instances, the nature and amounts of impurities are not given for technical grade chemicals, or they are given only in a very general way. Often they will be isomers, other reaction by-products, starting materials, decomposition products, or water.

2. *USP.* This grade represents material conforming to standards set in the United States Pharmacopoeia[1] (in Britain, BP). These standards are intended primarily for drug use, and not necessarily to indicate chemical purity, which may be low.

3. *Chemically Pure (C.P.)*.   These usually are reasonably pure materials, suitable for most purposes, but not of the highest quality.
4. *Analyzed Reagent (A.R.)*.   This represents a chemical of high purity, normally with a batch analysis given on the container. The most important impurities will be listed, usually in terms of maximum limits (the actual level may be less). Often, such chemicals are specified as ACS Reagent, meaning that they meet standards of purity set by the American Chemical Society Committee on Analytical Reagents.[2] A primary standard is a chemical of this grade,[3] suitable as a standard in work of high accuracy. (A primary standard has other requirements; e.g., it should be stable and inert to the atmosphere, it should be non-hygroscopic so that weighing is not difficult, and the molecular weight should be high to reduce weighing errors.)
5. *Special Grades*.   Among various special grades are spectroscopic grade, histological grade and pesticide grade. These materials are specially purified for particular uses, e.g., spectroscopic grade solvents are free of impurities having interfering absorption bands in their spectra. They might contain other impurities, however. Ultrapure material is very highly purified for use in trace analysis, or other applications where high purity is needed and may have impurity levels orders of magnitude less than A. R. grade. Such expensive materials should be used only when necessary to accomplish the required tasks.

The stated purity of commercial chemicals must be regarded with caution. Batch analyses may not represent contamination introduced during bottling or decomposition during storage, and the analysis for some impurities may not have been performed. Composition may change, e.g., water of hydration may be lost or gained, depending on temperature and humidity. (Water is a very common impurity in virtually all chemicals which are not specially dried.) Impurities are easily introduced into an opened bottle by improper care of the cap, use of spatulas, and return of chemicals to the wrong bottle. Consequently,

1. Unused chemicals should *never* be returned to the reagent bottle. Take only what is needed, and discard small excesses. This procedure should be followed with stock solutions as well as with solid and liquid reagents.
2. Spatulas, or other tools should never be used to remove analyzed reagents which will be employed for analytical or other critical purposes from the reagent bottle. Pour small amounts of the materials into a suitable container (e.g., the top of the bottle) and discard any excess. Similarly, a pipet must never be used to take a sample directly from a bottle of reagent or stock solution. Pour a small amount of the solution into a beaker, pipet a sample from this amount, and discard the excess.

References to 5.1

1. "Pharmacopoeia of the United States of America," 18th ed., Mack Publishing, Easton, Pa., 1970.

2. "Reagent Chemicals, American Chemical Society Specifications," 4th ed., American Chemical Society, Washington, D.C., 1968.

3. "Reference Materials," 3d ed., NBS Miscellaneous Publication 260, Government Printing Office, Washington, D.C., 1968.

5.2. WATER

The most widely used chemical in the laboratory is water. Ordinary tap water contains many impurities, the kinds and amounts being highly variable from place to place, and consequently tap water is not suitable for use in most work. Of course, it is used for washing apparatus, although a rinse with purer water generally is desirable.

The quality of water often is measured by its electrical conductivity, because many impurities produce ions in solution. Simple devices for indicating water purity are based upon the measurement of ionic conductivity. However, it is possible to have water of low conductivity but with significant nonionic impurities. Two general means of purification are encountered commonly.

5.2.1. Distillation

A simple distillation will improve the purity of water considerably, because the bulk of the impurities normally are nonvolatile salts. Distilled water is generally pure enough for most purposes, but it may still contain some impurities:

1. Some nonvolatile materials may be carried over as spray in the boiler.
2. Volatile impurities may still be present – e.g., NH_3.
3. Other impurities (e.g., metal ions) may be leached from the pipes or storage tanks.
4. The distilled water may reabsorb gaseous impurities from the air. Normally CO_2 will be present (this will give an acidic solution of about pH 6) as well as O_2 and N_2.
5. Impurities such as silica can be extracted from glassware.

Gases can be removed either by bubbling another gas (e.g., N_2) through the water for a few minutes, or by boiling for a short time. It is sometimes necessary to remove O_2 in this way when working with easily oxidized materials, and CO_2 must be removed for work in carbonate-free systems. Such water will reabsorb gases from the air unless protected from the atmosphere either with a tightly closed container or with a suitable absorption tube.

For critical work, a greater degree of purification may be necessary. Examples are for trace analyses, electrical conductivity studies, and reactions in which trace metal ions can have catalytic action. A more elaborate distillation can be used, for example, a second distillation on a small scale (doubly distilled water). High quality distilled water may be prepared in the following way. The water is first treated with alkaline $KMnO_4$ to oxidize organic materials, and is then distilled. Only the middle third of the distillate is collected. This is then redistilled after slight acidification with H_2SO_4. The apparatus must be scrupulously clean, and the water must be protected against atmospheric contamination.

5.2.2. Deionization

Various substances can exchange ions in their structures for other ions in solution. Typically, these ion exchangers are insoluble organic resins containing functional groups such as $-COO^-$, $-SO_3^-$, or $-NR_3^+$ which have ions of the opposite charge associated with them by electrostatic forces. Such an ion may be replaced readily by another ion from solution. Ion exchange or demineralizer columns composed of resins containing both positive and negative functional groups associated with OH^- and H^+ ions will exchange these for other ions if impure water is allowed to flow through the column.

$$e.g., \ H^+ \ resin + Na^+(aq) \rightarrow Na^+ \ resin + H^+(aq)$$
$$OH^- \ resin + Cl^-(aq) \rightarrow Cl^- \ resin + OH^-(aq)$$
$$H^+(aq) + OH^-(aq) \rightarrow H_2O$$

(Note that it is not necessary that the ions exchanged have the same numerical charge. Two monovalent ions can replace one bivalent ion, or vice versa.) Such deionized water often is better than ordinary distilled water in many respects, although nonionic impurities will not be removed. For many purposes such nonionic substances are not objectionable.

Various deionizing units are available. In some, water flows down the column from an upper reservoir. In most, water flows into the column at the bottom and pure water is collected at the top. In this design, it is important not to mistake the drainage outlet on the bottom for the pure water outlet.

Deionizer columns must be replaced when exhausted. This condition is indicated by a change in color with some columns, while others have a meter which indicates the quality of the output water in terms of conductivity.

5.3. COMPRESSED GASES

Many chemicals which exist normally as gases are frequently provided in pressurized cylinders in a variety of sizes. The gases may be highly compressed (e.g., O_2, N_2, He, H_2), or they may be liquified because of the pressure (e.g., NH_3, CO_2) or they may be dissolved in an inert liquid (e.g., C_2H_2/acetone).

Full cylinders of permanent gases such as O_2 and N_2 are at pressures of $2000-2500 \ lb/in^2$ at room temperature. Cylinders of liquified gases are at lower pressures, but the pressures are still high enough to require caution in handling. The outlet valve on a compressed gas cylinder is designed primarily for on-off use. Such valves are not satisfactory for practical control of the gas flow, and an auxiliary valve must be used. The cylinders themselves are heavy and unstable, giving further cause for care when using them.

Cylinders are color coded for different gases, but different companies use different codes. Therefore, the labels must be checked before using a cylinder. Cylinders of different gases also have different outlet connections to reduce the chances of attaching the wrong cylinder to a control valve. Do not try to use a valve that does not seem to fit the cylinder.

5.3.1. Safety Precautions

High pressure gas cylinders can be extremely dangerous. If the valve breaks they can behave as rockets with sufficient thrust to cause severe damage. Many gases in cylinders are flammable, explosive, poisonous, or corrosive. The following rules for handling gas cylinders must be followed:

1. Cylinders always must be supported with a cylinder clamp or chain to prevent falling. A falling cylinder can crush a foot, or the valve can be broken off with more serious consequences.
2. *Never* open the main cylinder valve unless a control valve has been attached. The main cylinder valve usually is opened with a small hand wheel on top of the cylinder. With more hazardous gases, such as Cl_2, the hand wheel is omitted and a wrench must be used.
3. To move a cylinder, place the protective cap over the valve and use a hand truck. Never move a cylinder with a control valve attached.
4. Do not warm cylinders above room temperature.
5. Always keep the main cylinder valve closed when the cylinder is not in use.

5.3.2. Valves and Controls

The most commonly used control valves or regulators are diaphragm type reducing valves. These can be used to obtain gas at a fairly constant, low outlet pressure from a high pressure source. Many controls have two gauges, one to monitor the internal cylinder pressure, and the second to indicate the outlet pressure. A typical reducing valve is illustrated in Fig. 5.1a and b.

A needle valve without gauges is often used with corrosive gases; most reducing valves also have a needle valve as the final control. A needle valve allows gas to escape at a slow, controllable rate, but does not control the outlet pressure. Hence, if the outlet tube is closed, pressure in it will increase to the value on the high pressure side, but if the outlet tube from a reducing valve is closed, the ultimate pressure will reach only that of the regulator setting. The construction of a needle valve is illustrated in Fig. 5.2.

To use a reducing valve:

1. Be sure the reducing valve is closed — handle turned *counterclockwise* — i.e., it is unscrewed.

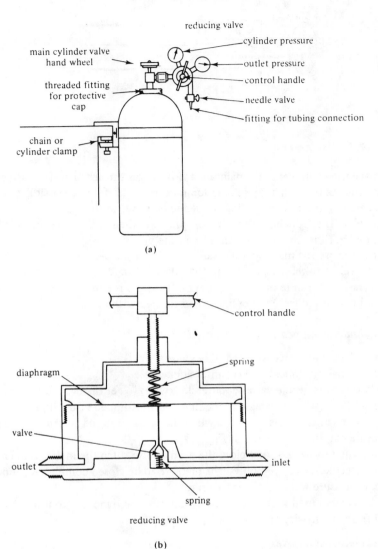

Figure 5.1. (a) A typical gas cylinder with regulator, (b) a reducing valve or regulator.

2. Open the main cylinder valve by turning it counterclockwise. (Note that this direction is opposite to that of the reducing valve.) The gauge will now give the cylinder pressure. (A hissing sound indicates a leak, and the main valve should be closed.)

3. The reducing valve is opened by turning it clockwise. If the outlet needle valve is closed (clockwise), the outlet pressure can be read on the second gauge. Be careful when opening the needle valve, because a sudden burst of pressure can

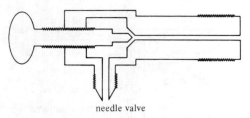

needle valve

Figure 5.2. A needle valve.

damage the equipment. To maintain a flow of gas through a system, as when bubbling gas through a liquid reaction mixture, it is quite permissible to open the needle valve before opening the reducing valve, so that gas flow starts gradually. The reducing valve controls pressure, the needle valve controls rate of flow, but the flow also depends upon the pressure.

4. After use, close the main cylinder valve (clockwise). Open the reducing and needle valves to release pressure in the valves. Then, close the reducing valve to release all pressure on the diaphragm. If the gas is corrosive, the regulator must be removed from the cylinder and flushed with dry air.

To use a cylinder equipped with only a needle-valve:

1. Be sure the needle valve is closed (clockwise).
2. Open the main cylinder valve (counterclockwise).
3. Slowly open the needle valve (counterclockwise) to get the desired flow rate.
4. Always have a pressure release in the line. (A tube dipped into a liquid (e.g., mercury) will allow gas to escape when the pressure in the system exceeds the pressure exerted by the liquid (Fig. 5.3.)
5. To stop the flow of gas, first close the needle valve, then the main cylinder valve. Needle valves need only light pressure to be closed completely, and excess pressure will ruin the valve seat.
6. If the valve was used with a corrosive gas, it must be removed from the cylinder and flushed with dry air after use.

With all regulators and valves:

1. *Never* use oil or grease on any valve or regulator or on the threads.
2. A suitable trap (e.g., an empty flask) should always be placed between the gas cylinder and any liquid through which the gas may be bubbled. If the gas dissolves rapidly, it is quite possible for liquid to be sucked back and damage the valve in the absence of a trap.
3. Beware of inadvertent pressure increases in closed systems. Note that an open system may become closed by the clogging of bubblers, outlet tubes, or other parts of the apparatus. The pressure release system in Fig. 5.3 is essential for needle valves in all but the simplest applications, and it is desirable for reducing valves.

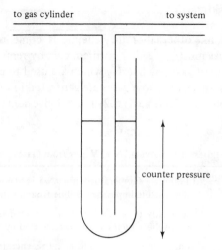

Figure 5.3. A simple pressure-release device to limit the pressure that can build up in a system.

Use of Lecture Bottles:

Lecture bottles are the smallest size cylinder, about 2 in by 15 in, and deserve special comment. They are the most convenient source of small amounts of gas. Some of those containing non-hazardous gases are hand-wheel controlled, and the cylinder valve itself is sensitive enough to be used to control the gas flow, although a regulator can be used if desired. Others must be equipped with an auxiliary valve, and usually a small needle valve is adequate. On these valves, the main control is the small hexagonal nut directly at the base of the needle valve. It is the smallest hexagonal nut on the cylinder, and if a fitted wrench is not attached, be very careful that only this, and not any of the other (usually two) hexagonal fittings are turned. The needle valve rotates with the hexagonal fitting.

The sequence of operation is as before: close the needle valve, open the main valve and slowly open the needle valve. To stop the flow of gas, close the needle valve and then the main cylinder valve.

To summarize opening and closing directions (Directions are given looking at the valve from above):

main cylinder valves: open — counterclockwise; close — clockwise (open two or three full turns, or all the way)

needle valves: open — counterclockwise; close — clockwise (open slowly to the extent needed to give desired flow rate)

reducing valves: open — clockwise (reverse of above); close— counterclockwise

5.3.3. Purification of Gases

Compressed gases, like other laboratory chemicals, come in a variety of purity grades. Sometimes the gas must be further purified, e.g., by removal of traces of O_2, H_2O, or CO_2 from inert gases such as N_2 which are used to protect sensitive materials. In some cases, special-purpose gases are generated in the laboratory, and also require drying before use. Tables 5.1, 5.2, and 5.3 give some simple methods

TABLE 5.1

Methods of Removing Water Vapor from Gases

Liquid N_2 trap	At liquid N_2 temperatures most water vapor and other condensible impurities will be frozen from low-boiling gases
Molecular sieves	Used mostly to remove moisture from carrier gases for inert atmosphere boxes; easily regenerated by heating
Activated alumina or silica gel	Good; chemically inert and easily regenerated. Indicating kinds show when capacity is exhausted
Chemical drying agents, such as those below	The gas should be passed through the drying agent; a solid absorber must be granular or coated on a granular support
P_2O_5	Very effective, but powder form difficult to use with gases
$Mg(ClO_4)_2$	Very effective; may react explosively with oxidizable vapors
KOH, NaOH	Good; surface coating of hydroxide limits capacity. Removes acid vapors
H_2SO_4	Good; gas must be bubbled through liquid. Removes basic vapors
$CaSO_4$ (Drierite)	Fairly good; inert, cheap, regenerable and indicating type available
$CaCl_2$	Not as effective as the rest, but cheap

TABLE 5.2

Methods of Removing O_2 from Gases

Method	Comments
Hot copper turnings	Copper turnings heated to 350°C–400°C react to form CuO
Oxygen-absorbing solutions such as those below	O_2 will react with several substances in solution; water vapor will be introduced into the gas, and other substances also will react
alkaline pyrogallol	Pyrogallic acid in 50% aqueous KOH
chromous sulfate or chloride	Produced by reducing the chromic salt with Zn in an acid solution
Solid granular absorbers (available commercially)	O_2-reactive compounds on a granular support – e.g., "Oxosorb"

TABLE 5.3

Methods of Removing CO_2 from Gases

Method	Comments
KOH, NaOH or $Ba(OH)_2$ solutions	Introduce water vapor
KOH or NaOH pellets	Remove both CO_2 and H_2O; surface is quickly coated with reaction products
Solid granular absorbers	KOH or NaOH coated on a granular support – e.g., "Ascarite"

for removing H_2O, O_2, and CO_2 from some other gases. The methods are by no means complete or universally applicable. The chemical nature of the gas must be considered; e.g., KOH could not be used to dry CO_2; H_2SO_4 could not be used for NH_3. The usual experimental arrangement is to pass the gas stream through a tube packed loosely with a solid absorber, or to bubble the gas through a liquid absorber in such a way that the bubbles are well dispersed (Fig. 5.4a). Carry-over of dust and spray must be prevented, usually with a loose plug of glass wool. If a liquid absorbent is used, an empty trap such as that shown in Fig. 5.4b should be placed before and after the wash bottle in the event that liquid is sucked out of, or blown from, the wash bottle.

5.3.4. Compressed Air

Many laboratories have a compressed air line. While this sometimes has special uses, one of the common, simple applications is to blow glassware dry. Unfortunately, most air supplies are highly contaminated by oil from the

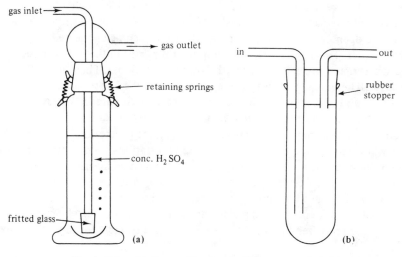

Figure 5.4. (a) Gas-washing bottle, (b) trap to isolate (a).

compressor, and moisture, and sometimes other contaminants may be present. Hence, unless the air is filtered in some way, it is not desirable to use it for drying glassware when cleanliness is important.

5.4. SPECIAL HANDLING TECHNIQUES

5.4.1. Controlled Atmosphere Boxes

Many compounds are air-sensitive — that is, they react with a component of the atmosphere (oxygen or water in particular) — and must be prepared and handled in air-free (oxygen-free, moisture-free) systems. Volatile substances are most readily handled in a vacuum system (Chap. 14). However, it is common practice to perform manipulations of air-sensitive, moisture-sensitive or highly toxic materials in an inert atmosphere box (also called a dry box or glove box). This is simply an airtight box made of metal, plastic, or fiberglass, with a transparent window and ports equipped with gloves that extend into the box to permit manipulations to be carried out. The box typically is filled with nitrogen or another inert gas. When the system has been flushed, a positive pressure of nitrogen prevents inflow of air. In some systems, the gas is recirculated through a purification system, consisting of tubes of molecular sieves and silica gel or other purifying agents, but for many purposes this is not done. When moisture sensitivity is of concern, trays of a good desiccant such as P_2O_5 are often placed in the box. An auxiliary side-port which can be flushed separately is usually provided that enables equipment to be transferred to or from the box without contaminating the internal atmosphere. Electricity, vacuum, and other utilities may be provided in the box.

An inexpensive replacement for a glove box for materials which are moderately sensitive to air or H_2O is a glove bag, a polyethylene bag equipped with gloves, a large opening for apparatus, and a tube which can be connected to a nitrogen cylinder. To use, the required equipment is placed in the bag, the opening is closed loosely, and the bag flushed with N_2 for about 5 min. Then the N_2 is squeezed out. The opening is closed more securely, and the bag inflated to a comfortable working pressure (it should be only partially inflated). A heavy, flat object placed in the bottom of the bag is useful to hold it in place.

Sometimes very simple methods will give adequate protection in manipulating materials which are not too reactive. For example, filtration of a moisture sensitive compound can be carried out by passing nitrogen through an inverted funnel suspended over the filter funnel. A shield of plastic taped to the rim of a large inverted funnel through which nitrogen gas is allowed to flow provides an area of low water and O_2 concentration in which various manipulations can be carried out. Obviously, these "nitrogen-blanket" techniques are not suitable for highly sensitive materials.

5.4.2. Transfer Techniques

In addition to glove boxes, specialized techniques have been developed that permit the manipulation of air sensitive materials easily and safely under an argon or nitrogen blanket with complete protection from the atmosphere. Special glassware for manipulation of air sensitive materials is available commercially.[1] The transfer of air sensitive solutions is the most commonly encountered problem, and a simple technique suitable for laboratory scale operations on the 0.01−1 mol level is described below. More sophisticated techniques for the manipulation of air-sensitive liquids and solids are given by Shriver[2] and in Brown's text.[3]

Hypodermic syringes provide a convenient and flexible means to transfer up to 100 ml of solution with virtually no exposure to the atmosphere. The syringes are used in conjunction with septa, specially designed, self-sealing rubber stoppers sometimes known as "sleeve stoppers" or "serum caps," which are available from most laboratory supply houses. Septa are available in a range of sizes ranging from those small enough to seal an NMR tube to those large enough to seal 24/40 and 29/42 standard taper joints, but the most widely used sizes are those designed to fit 8, 9, and 10 mm o.d. glass tubing. In use, the septum is inserted in the glass tubing, and the skirt is folded back and wired in place. The glass tubing receiving the septum may be built permanently into a reaction flask, or it may be attached to a suitable standard taper joint.

Glass surfaces adsorb sufficient moisture under ordinary laboratory conditions to interfere with some reactions. The adsorbed moisture can best be removed by heating the glassware in an oven to 125°C for 30 min. Assemble the glassware while hot, while flushing it with a stream of dry nitrogen. Coat all standard taper joints with a light film of suitable grease to prevent seizure of the joints. If a suitable oven is not available, the glassware may be assembled cold and dried by heating with a Bunsen burner or a hot air blower while it is flushed with a stream of dry nitrogen.

During use the reaction vessel and any associated glassware are maintained under a blanket of dry argon or nitrogen at a slight positive pressure to prevent entry of the atmosphere. A cylinder of dry, high purity gas fitted with a pressure regulator supplies the gas, and a mercury or mineral oil bubbler serves as a vent to prevent pressure buildup. Before connecting the gas line to the reaction vessel, be sure that the pressure regulator is set to the full off position and the needle valve is closed. Connect the gas line, slowly open the needle valve and adjust the regulator to the lowest pressure that will give a gentle flow of gas through the assembled apparatus. The bubbler must be connected to a hood or to a suitable train if toxic or dangerous gases are evolved.

Liquids in amounts up to 100 ml can be transferred readily using a previously dried syringe fitted with a flexible needle which is approximately 1 ft longer than necessary to reach the bottom of the reagent container. Such a long flexible needle is advantageous because it permits the syringe to be held in an upright position without the need to tip the reagent bottle during the transfer, and thereby prevents attack by the solvent on the septum. The needle point should be tapered to minimize damage to the septum and it is advantageous to insert the needle through an existing puncture. A thin film of grease on the septum will facilitate insertion of the needle and prolong the life of the septum.

The syringe transfer is initiated by pressurizing the reagent bottle. When the transfer involves only a few milliliters of liquid the bottle may be pressurized by first injecting a volume of dry gas slightly larger than the volume of liquid to be removed. A more convenient method, especially when the transfer involves a larger volume of liquid, is to connect the reagent bottle to the vented reaction system using a hypodermic needle wired to a length of plastic or rubber tubing. When not in use, the needle is inserted in a rubber stopper. The syringe is filled slowly until it contains a slight excess of the reagent, and the excess reagent along with any entrained gas is returned by pushing the plunger in as the syringe is held upright with the needle upward, and the plunger down. The transfer is completed by placing the needle in the reaction vessel and slowly emptying the syringe.

References to 5.4

1. Ace Glass, Inc., P.O. Box 688, Vineland, NJ; sells a line of glassware specifically designed for manipulation of air- and moisture-sensitive compounds.

2. D. F. Shriver, "The Manipulation of Air-Sensitive Compounds," McGraw-Hill, New York, 1969.

3. H. C. Brown, "Organic Syntheses via Boranes," Wiley, New York, 1975, chap. 9.

Glass Tubing and Elementary Glassworking

6.1. INTRODUCTION

This section is intended to illustrate some of the simple techniques used in glassworking which are frequently encountered in the laboratory. For a discussion of more sophisticated manipulations, reference should be made to one of the many publications on the subject.[1-3]

It is sometimes necessary to cut, bend, or draw out a piece of glass tubing. The tubing usually encountered in a laboratory may be "soft" (sodalime) glass or borosilicate glass. Sodalime glass softens at a lower temperature ($\sim$400°C) than borosilicate glass and can be worked easily with a Bunsen burner flame. Borosilicate glasses soften at higher temperatures ($\sim$600°C), and the manipulations described here can be accomplished with the hot flame of a Meker burner. For more elaborate work, a gas-oxygen torch is essential. Although soft glass can be worked at lower temperatures than borosilicate glasses, it is more susceptible to thermal stresses than borosilicate glasses and therefore is much more difficult to use, except for very simple bends or connections. The most common type of borosilicate glass is Pyrex 7740 (or simply "Pyrex"). The various kinds of glass can be differentiated by their respective refractive indices. Pyrex 7740 has a refractive index of 1.474, which is the same as a 16:84 (by volume) mixture of methanol and benzene. Thus, immersion of a dry (water will cause the solution to become cloudy) piece of glass into this mixture will render the piece almost invisible if it is Pyrex 7740. Soft glass (index 1.512) and Vycor (96% Silica glass, useful at high temperatures, index 1.458) for example, will remain visible in this solution.

6.2 CUTTING GLASS TUBING

Tubing up to 25 mm in diameter or thin rod can be cut as follows: Make a single, narrow scratch with a sharp triangular file or "glass-knife" ("scorer") at the point at which the tubing is to be cut by drawing the file once across the tubing with pressure. Do not "saw" the glass. Moisten the scratch and hold the tubing with the scratch away from you with one hand on each side of the scratch, close to it and with the thumbs behind it. To be safe, wrap the tubing in a towel. Pull (with a slight bending motion) the tubing along the long axis of the tube until it breaks. If the break is ragged it may be scraped to near-smoothness with a wire gauze and then fire polished (see section 6.3).

For larger tubing, or for cuts near an end, make a scratch as above and touch the very hot end of a thin glass rod to the edge of the scratch. Thermal shock should start a crack which can be extended around the tube by further applications of the reheated rod. Moistening the scratch may help.

6.3. FIRE POLISHING

The ends of *all* glass tubing and stirring rods should be fire-polished, i.e., held in a flame until the sharp edges have partially melted and become round. A very jagged end can be made more uniform before fire polishing by gently chipping with a wire gauze (**use eye protection**). Prolonged heating will cause the tubing to shrink in diameter, so heat it only long enough to smooth the edges.

6.4. DRAWING A CAPILLARY

To draw glass tubing into a capillary tip, heat the center of the tube in a broad flame while rotating it continuously. When it is quite soft, remove it from the flame while still rotating it and slowly pull the ends apart. It will require some practice to be able to rotate both ends of the tube at the same rate when the glass begins to soften. The heated portion will stretch to a long, narrow tube, which can be cut as required after cooling. The rate at which the tube is stretched controls the final capillary diameter. Pulling too rapidly, especially with insufficiently softened glass, will cause the capillary to break. The most frequent difficulties result from improper rotation and from not heating the glass sufficiently. A sturdier, very fine capillary can be drawn from capillary tubing.

6.5. BENDING GLASS TUBING

The glass is heated as above, and when very soft it is removed from the flame and bent to the angle required. The bend should be a smooth curve; not a sharp angle with the inner side buckled (Fig. 6.1). Uniform rotation, with thorough heating of a sufficient length of tubing, using a Meker burner or a wingtip on a Bunsen burner is required. If the bend becomes constricted, it can be corrected

correct   incorrect

Figure 6.1. Representation of the correct and incorrect forms for a bent glass tube.

sometimes by closing one end of the tube with a cork, heating the bend to the softening point and gently blowing into the open end until the bend is smoothed. However, the use of a glass blowing torch is recommended for this operation, because it is difficult to soften the glass sufficiently with the broader, cooler flame of a Bunsen burner.

6.6 PROPER CLEANUP PROCEDURE

Fragments of glass (cooled to room temperature) should be discarded in a trash can, used only for glass, as soon as possible. **Do not allow glass debris to remain behind after you have completed your work**. The next person to use that bench space may not see the pieces until it is too late!

6.7. CONNECTION OF GLASS TUBING

Often, glass tubing must be connected to a larger size tube (including the necks of reaction vessels such as Erlenmeyer and round-bottomed flasks) or to glass tubing of the same size. These connections can often be accomplished simply by inserting the tubing in a tightly fitting sleeve made of cork, rubber or plastic (e.g., Tygon) as shown in Fig. 6.2.

In all cases, the ends of the glass tubing *must* **be fire polished.** This simple precaution not only significantly reduces the occurrence of cuts and scratches, but it also makes insertion of the glass into the sleeve easier. Tygon tubing can be softened sufficiently to allow the diameter to be expanded by several millimeters if the tubing is held under hot running water and then quickly forced over the end of the glass tubing. When the Tygon sleeve cools it will shrink and harden to make a

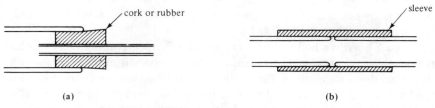

(a)   (b)

Figure 6.2. Methods of joining glass tubing: (a) tubing of different sizes, (b) tubing of similar sizes.

slip-proof and gastight seal. Thin-walled rubber tubing (gas or water tubing) may be made to slip onto glass tubing more easily by lubricating the glass with a small amount of glycerin; soapy water is also useful. Generally, it is not advisable to lubricate glass tubing when it is to be inserted into thick-walled vacuum tubing because the stiffer tubing will tend to slip from the glass if the joint is subjected to movement. In any event, it is wise to fasten securely both ends of the sleeves to the glass tubing with appropriate clamps or simply by tightly wrapping each end with wire.

When inserting glass tubing or a thermometer into a tightly fitting orifice, it may be necessary to apply considerable force to the glass tube or rod. *Never* hold the tubing more than $\frac{1}{4}$ to $\frac{1}{2}$ in from the opening into which it is to be inserted (see Fig. 6.3). Holding the glass tube further from the end is dangerous because the glass may be stressed sufficiently to break and to be driven into the hand holding the tube or stopper. Inserting only $\frac{1}{4}$ to $\frac{1}{2}$ in of glass at a time and rotating it will take a bit longer, but it is the safest way to complete the task. For added safety the tubing should be wrapped with a towel.

Simple connections between the various parts of an apparatus and a reaction vessel are often made using glass tubing and corks (for work at atmospheric pressure) or rubber stoppers (for work at, or slightly above/below atmospheric pressure). If slight positive pressure is to be developed, the rubber stopper should be fastened securely to the flask with wire to prevent it from being ejected from the flask. No more than two-thirds of the cork or rubber stopper should go into the opening (particularly if pressure in the vessel is to be reduced) so that enough of the cork or stopper remains accessible for it to be removed easily.

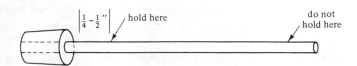

Figure 6.3.   The correct way of inserting glass into an orifice.

6.8. USE OF A CORK BORER

Some rubber stoppers are available with one or two holes already in them, but if these are not available, or if corks are being used, the holes will have to be bored using a cork borer (Fig. 6.4a,b).

A cork borer is a hollow brass tube which is sharpened at one end and which has a handle at the other end (perpendicular to the long axis of the tube) (see Fig. 6.4b). A solid rod which fits in the smallest borer is used to remove the core of cork (or rubber) that may remain in the tube after the hole has been bored. Generally, several sizes of borers are nested in each other, together with the rod, to make a compact, conveniently stored package (see Fig. 6.4a). Select the proper size by

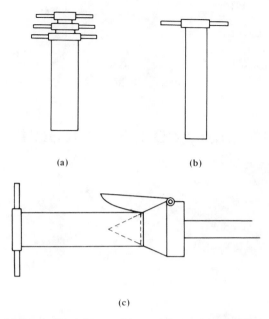

Figure 6.4. (a) A set of cork borers, (b) a cork borer, (c) a cork borer sharpener.

using a borer one size smaller than that which just passes around the exterior of the glass tube.

To be most effective, the cork borer must be sharpened properly. The pointed end of the sharpening cone should be inserted into the borer and the sharpening blade pressed against the body of the borer (see Fig. 6.4c) with sufficient pressure to form turnings when the borer is rotated against the blade. Too much pressure will nick or otherwise damage the borer.

To bore a hole, hold the cork or rubber stopper firmly within a circle formed with the thumb and index finger and press the sharpened end of the borer into the cork or stopper with a twisting motion. When the borer is $\frac{1}{2}$ to $\frac{2}{3}$ through the cork or stopper, remove the borer and start cutting from the other end so that the two holes meet on center. Cutting from both ends ensures clean-cut openings that aid in the insertion of tubing or rods. Corks should be softened by rolling in a device called a cork softener before boring or they may crack.

REFERENCES TO CHAPTER 6

1. For a detailed description of some of the more advanced techniques see "Laboratory Glass Blowing," Bulletin B-72 of the Corning Glass Works, Corning, New York, 1961.

2. E. L. Wheeler, "Scientific Glassblowing," Interscience, New York, 1958.

3. J. D. Heldman, "Techniques of Glass Manipulation in Scientific Research," Prentice-Hall, Englewood Cliffs, N.J., 1946.

Basic Laboratory Equipment and Techniques

A typical chemistry laboratory contains a wide variety of equipment. This section will indicate, in a general way, the kinds of equipment available for many common laboratory operations, some of the considerations to be made in selecting equipment for a particular purpose, and some aspects of the proper usage of each item. More detail on some of the items mentioned below is to be found in other sections of this book.

7.1. LABORATORY GLASSWARE

This term covers typical equipment such as test tubes, beakers, flasks, condensers, adapters, among others. Important items of glassware are illustrated in various parts of the text. The majority of modern laboratory glassware is made of a borosilicate glass (Pyrex®, Kimax®) which softens around 600°C, and has a low coefficient of thermal expansion. Consequently, it is moderately resistant to breakage caused by sudden changes in temperature. Older equipment was made of soda glass (soft glass) which softens at a lower temperature and is much more susceptible to breakage caused by thermal shock. Soft glass is encountered in the laboratory nowadays chiefly as glass rod or tubing, where its lower softening point makes it easier to bend.

Glassware of a high silica content is needed for high temperature use (500–1000°C) or when large temperature differences are encountered. For such applications fused quartz, silica or Vycor® apparatus is used. It is much more expensive than ordinary glass equipment.

Pyrex or Kimax apparatus can be heated and cooled with reasonable confidence that it will not crack. However, any glassware contains some strain. Heating should be spread over a large area (e.g., through use of a wire gauze under a flask heated with a burner) and extremely rapid changes of temperature should be avoided, especially with thick walled equipment. Indeed most thick walled glassware should not be heated with a burner at all.

Glass is very resistant chemically, but it is attacked by hydrogen fluoride or acidic fluoride solutions, and by strongly alkaline materials. Its surface will adsorb many substances strongly, including water. Clean glass surfaces are wet by water, but can be rendered hydrophobic by treatment with a silanizing agent such as trimethylchlorosilane or Desicote®. This treatment is sometimes useful, for example, with very small glassware (micro-apparatus) to avoid leaving significant volumes of solution behind as a film on transferring liquid from one vessel to another. Some suggestions for cleaning glassware are given in Sec. 9.1.

In most laboratory procedures involving glass apparatus, connections between the various components are best made using ground joints. These are of two types:

1. The conical form or standard taper joints, denoted by the symbol ℥ A/B, where ℥ represents Standard Taper, A represents the maximum diameter, in millimeters, of the joint and B represents the length, in millimeters, of the joint (Fig. 7.1(a)). A variety of sizes are available commonly, but 10/30, 19/22, 19/38 and 24/40 are the sizes most often encountered.
2. The hemispherical form or spherical joints, denoted by the symbol ℥ C/D, where ℥ represents Spherical Joints, C represents the maximum diameter of the joint and D represents the diameter of the joint opening (Fig. 7.1(b)).

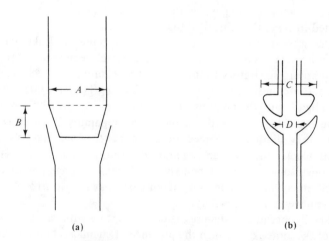

(a)   (b)

Figure 7.1.   (a) Representation of a Standard Taper (℥) Ground Glass Joint, (b) representation of a Spherical Joint (℥).

The spherical joints make more flexible connections than the standard taper joints. This added flexibility requires the use of spring clamps to hold the spherical joints together. Clamps (made from spring steel rod) are also available for use with conical joints, or the joints may be held together with springs or rubber bands.

Ground joints are expensive items, and should be cared for in specific ways. The same care should be given to standard taper stoppers, bottles, and flasks. The effectiveness of ground joints rests upon the fact that they may be joined in an almost perfect fit without lubricant for use at atmospheric pressure. To assure that the ground surfaces are not damaged, they should be kept clean and free from hard particles (e.g., pieces of glass, boiling chips). Contact with corrosive materials such as hydrofluoric acid or concentrated alkali solutions should be minimized. No joints should be left in contact with each other (nor should stoppers be left in joints) when wet with HF or alkaline materials, because they may become stuck together (often denoted as being "frozen"). Joints (and stoppers) should be washed and wiped clean when they do come in contact with such corrosive materials, before they are put together. Solutions containing dissolved solids of any kind should also be washed from joint surfaces, because deposition of the solid between the surfaces of a set of joints can also bind the parts together.

To make a ground glass connection leak-tight (e.g., for use at reduced pressures) the surface of the inner joint should be covered lightly with a film of petroleum jelly, Lubriseal® or silicone grease. Silicone grease and Apiezon® greases are useful for high vacuum applications while petroleum jelly and similar lubricants are not (see Sec. 14.3). Care should be exercised to minimize contamination of the reaction mixture or compound to be distilled with these materials because they are difficult to eliminate once introduced into a system.

Small strips of grease are applied and spread by pressing the joints together and rotating them until the joint is uniformly covered with grease, as evidenced by transparency. A common failing is the use of too much grease. The minimum amount needed to cover the joint is all that should be used.

Note that glass stopcocks also must be greased to prevent leaking; these are discussed in more detail in Sec. 14. Teflon® plugs which require no lubrication are preferable for burets, separatory funnels, and other equipment which is not used under vacuum. (Care must be taken to avoid deforming Teflon® plugs by leaving them screwed tightly into their barrels.) Thin Teflon® sleeves which fit over the inner joints also are available for use with standard taper apparatus. These sleeves are very useful because they eliminate the need for grease, but they are relatively expensive.

Most stopcock lubricants can be removed with organic solvents ($CHCl_3$, toluene) although some effort may be required to do a thorough job. Silicone greases are extremely adherent and may require brief soaking in strongly alkaline solutions to remove them completely.

Occasionally, because of the use of too little grease with vacuum systems, overheating of the joints, or through the presence of chemicals between the glass surfaces, a joint may "freeze." When separating joints, stoppers, or stopcocks, avoid applying excessive force. The following procedures should be followed.

1. If the joint is stuck because of the presence of chemicals, soaking with hot soapy water or another appropriate solvent may loosen it. Liquids that are capillary active, such as isoamyl hexanoate, can be extremely effective. Allow a few drops to soak into the joint overnight. Even if the presence of chemicals is not obvious, this is the best initial approach.
2. The outer joint may be heated with a stream of hot water or steam, or by a heat gun so that it expands slightly with respect to the inner joint. Gentle heating with a small flame may also be tried. **Use a towel to prevent burns or injury in case of breakage. Be *very* sure that the apparatus is washed completely free of flammable materials, and that it is open to the atmosphere before heating.**
3. *Lightly* tapping the outer joint at the lip may be useful in separating frozen joints.
4. If the above fail, or if the apparatus contains hazardous materials which cannot be removed, consult the instructor.

7.2. PLASTIC WARE

For materials that attack glass or are adsorbed by it, plastic apparatus can be useful. Plastic is encountered frequently as beakers, testtubes, centrifuge tubes, Erlenmeyer flasks, graduated cylinders, funnels, and stock bottles, although many other items also are available. Polyethylene (fairly soft), polypropylene (harder and higher melting), and polytetrafluoroethylene or Teflon® (very inert) are among the common materials. The polymeric hydrocarbons are useful especially for fluoride and caustic solutions, but cannot withstand much heat, and are subject to attack by some organic solvents. Teflon® equipment has a much higher temperature limit and chemical resistivity to organic reagents than the others, but it is also much more expensive.

7.3. PORCELAIN WARE

Some apparatus has been made traditionally of porcelain. Examples are evaporating dishes (broad, round bottomed vessels designed for evaporation of liquid), crucibles, and Büchner funnels. The use of these items is discussed elsewhere (see Sec. 10). Glazed porcelain can be regarded as similar to glass in its chemical nature, but unglazed areas on an item are porous and will absorb materials. Porcelain can be used at higher temperatures than glass e.g., ignition with a Meker burner.

7.4. IDENTIFICATION AND LABELING

All containers must be marked so that their contents can be identified. It cannot be stressed too strongly that the common practice of keeping track of the contents of several containers mentally, based on appearance or position, is an open invitation for mistakes. Below are several common means of labeling glassware.

1. Gummed labels or tape. (Transparent cellophane tape can be used to protect labels.)
2. Grease pencils (china-marking pencils), which write directly on dry glass or plastic. They work especially well if the item is slightly warm. Marks are easily removed by wiping with a paper towel moistened with an organic solvent (e.g., acetone). Because they are subject to obliteration by solvents and rubbing, grease pencil markings are useful chiefly for short-term labels.
3. Felt-tipped pens will write on dry glass or plastic. Marks are easily removed from glass with organic solvents.
4. Lead pencils can be used to write directly on the frosted glass areas on most flasks and beakers. These marks can be removed with an eraser, but otherwise are fairly permanent.
5. Permanent fired inks are especially useful for items heated to high temperatures.

Labels may describe the contents of a vessel, or may be simply a code number, because a full identification written on the vessel may be unwieldy. If code numbers are used, full explanations of their meaning must be given in the laboratory notebook, or they are as bad as no label at all. Vessels to be left in a common area (e.g., drying oven or refrigerator) must be labeled with your name as well as the identity of the contents. It is a frequent practice in many laboratories to discard immediately the contents of any unlabeled vessel found in a common area.

7.5. WASH BOTTLES (OR SQUEEZE BOTTLES)

A flexible plastic bottle with a tube reaching to the bottom is a convenient device for delivering a directed stream of liquid which can be used to wash reagents from the walls of flasks, or for slowly adding liquid when filling volumetric apparatus (see Sec. 9.2), and for many other purposes. A wash bottle of distilled water is extremely useful in much laboratory work, and bottles with other solvents are useful in particular cases. Wash bottles should be labelled to indicate contents.

7.6. STIRRING EQUIPMENT

The contents of flasks, beakers, and other containers can be mixed effectively by swirling the vessel manually. Other means of stirring are often necessary, however. The major techniques are as follows:

1. *Stirring rod.* This is simply a length of solid glass rod of convenient size with the ends fire-polished (see Sec. 6.3). It is useful for stirring, breaking up lumps of solid (be careful not to put it through the bottom of the flask) and for directing the flow of liquid when pouring solutions from a beaker or flask into another vessel as discussed in Sec. 10.1.
2. *Mechanical stirring paddles and stirring motors.* Continuous stirring for any length of time is impractical by hand, and mechanical stirring is then useful.

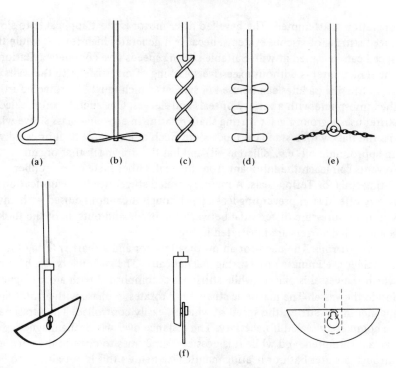

Figure 7.2. Some common designs of stirring paddles: **(a)** a simple bent glass rod, **(b)** propeller-glass or metal, **(c)** glass spiral, **(d)** multiple paddles for efficient stirring of deep vessels, **(e)** loops of wire (Nichrome® or other inert material) attached to a glass shaft, **(f)** a demountable Teflon® paddle on a glass shaft; particularly well suited for use in round-bottom flasks. Various paddle sizes are available for different size flasks.

The stirring paddle may have a variety of shapes, of which a bent rod is the simplest. Some are shown in Fig. 7.2. They may be made of glass, metal, or in some types of glass with Teflon® blades. The stirring paddle is driven with a small electric motor (or occasionally with an air-driven motor). The paddle is attached to the motor with a chuck, although sometimes a rubber tubing connection between the motor shaft and the paddle can be used. This arrangement tends to "whip" dangerously unless the stirring speed is slow, or the stirring shaft is supported with a bearing, however. The full motor speed is not often practical with mechanical stirring paddles. Some motors have geared shafts rotating more slowly than the direct motor shaft, others have speeds controlled with an adjustable bevelled drive, while still others control the speed of the motor with a built-in rheostat. Most stirring motors (but not synchronous motors) also can be controlled with a variable transformer (Variac®, see below). It must be remembered that stirring motors operated at less than their rated voltage with a rheostat or transformer control develop a comparatively low torque, and may stall easily. A stalled motor may overheat and burn out, so that some care is necessary before unattended operation is attempted. The bevelled drive motor is most applicable to slow speed stirring of viscous systems, because it generates high torque, while the speed can be varied at will. Suitable stirring speeds give complete agitation of the flask contents without excessive splashing of materials onto the walls.

Stirring paddles can be used in systems which must be protected from the atmosphere with a variety of seals and sleeves. One such type of sealed stirrer uses a ground glass stirring shaft rotating in a ground glass sleeve which fits the flask with a standard taper joint. Such a shaft must be lubricated with an appropriate oil (e.g., silicone oil) and has the drawback that organic solvents can leach the lubricant from the seal. Other systems are rubber O-ring seals, or Teflon seals. A mercury sealed stirrer needs no lubrication and is very effective in preventing leakage, although it cannot be used with any significant pressure differential between the inside and outside of the flask. Some sealed stirrers are illustrated in Fig. 7.3.

3. *Magnetic stirring.* The need for an overhead motor and a shaft entering the container is eliminated by stirring magnetically. The stirrer acts as a base on which the vessel is placed, while stirring is accomplished with a bar magnet inside the vessel. The magnetic stirring bar rotates in phase with a rotating magnet in the stirrer, the speed of which is easily controlled. The vessel can be completely sealed if necessary. The distance over which the magnetic force can act is limited, and while it is possible sometimes to operate the stirring through a water bath or heating mantle (see below) this is not always so. Magnetic stirring often is not practical with viscous materials. Stirring bars typically are cylindrical magnets encased in glass, polyethylene, or Teflon®. The polymer cases usually are molded with ridges for more effective spinning, and offer some cushioning of the bar against the walls of the vessel. Home made bars can be made of iron (e.g., the shaft of a nail) sealed in glass. Unless they are small compared to the flask, cylindrical bars will not rotate properly in round-bottomed flasks, and for these ellipsoidal bars are available. Other special shapes also are made. Bars should rotate rapidly enough to produce a vortex in the liquid, but if rotation is too rapid the bars will bounce and be quite ineffective; they even may break the container.

7.7. HEATING EQUIPMENT

Some aspects of heating and temperature control are discussed in Sec. 13.2. Some of the equipment available for different heating purposes in the laboratory is described briefly here.

1. *Burners.* A simple Bunsen burner (Fig. 7.4a) consists of a lower inlet for gas, which passes upward through the barrel and is mixed with air that enters at the bottom of the barrel. The size and the character of the flame are determined by the rate of flow of gas which is controlled with the gas valve,

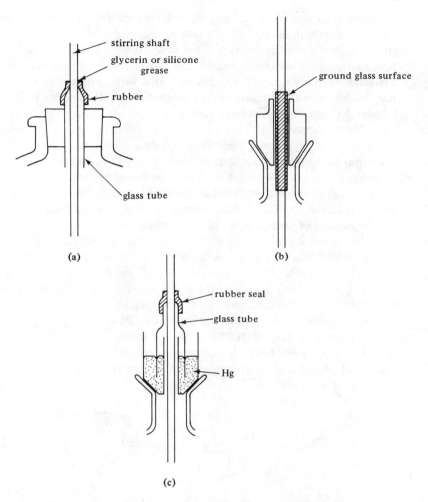

Figure 7.3. Some sealed stirrers: (a) Kyrides type, (b) ground glass sealed type, (c) mercury sealed type.

and with the amount of air admitted. The air flow is adjusted by rotating the burner barrel or a sleeve around its lower end. The flame consists of an inner bluish cone (the reducing part of the flame) surrounded by an outer yellow region (the oxidizing part). The hottest region is just above the tip of the blue cone. Insufficient air produces a yellow, or even a smoky flame. A typical Bunsen burner is about 5 in tall; a micro-burner less than half this size is useful for gentle heating of small objects. A slightly more elaborate design, a Tirrell burner (Fig. 7.4b), has a lower control wheel to adjust the gas flow more precisely than can be achieved with a simple Bunsen burner. Burners of these types are used frequently to heat flasks or beakers containing reaction mixtures. A wire gauze should be used to spread the heat over the glass vessel. Direct heating with the flame is permissible only if it is brushed back and forth constantly over the bottom of the vessel. When a burner is used for heating, thought always must be given to the nature of the materials present. **Flammable organic solvents must *not* be heated with a burner flame. Care must be taken that flammable solvents are not used in the vicinity of lighted burners, because the vapors can be ignited easily.**

Meker burners (Fig. 7.4c) are useful for intense heating; e.g., ignition of a precipitate in quantitative analysis and bending of glass tubing (see Sec. 6). These burners are larger than Bunsen burners; they have both air and gas controls, and a large flat grid above which the flame projects. The hottest zone is quite close to the grid. With full gas flow, very intense heating of objects such as crucibles is possible. Meker burners generally are not suitable for routine heating of liquids except in large amounts.

Glass blowing torches are used for working glass when more elaborate manipulations than are possible with a Meker burner are necessary. These torches burn gas with oxygen. Controls for both gas and oxygen flow are provided, so that both the size of the flame (amount of gas) and temperature (amount of oxygen) can be controlled. These torches are ignited by turning the gas on slightly, igniting it, and adjusting the size of the flame and the oxygen flow. To extinguish the torch, the oxygen is turned off *before* the gas.

2. *Hot plate*. Electrically heated hot plates are very useful when burners are undesirable, but it must be remembered that **many flammable vapors can be**

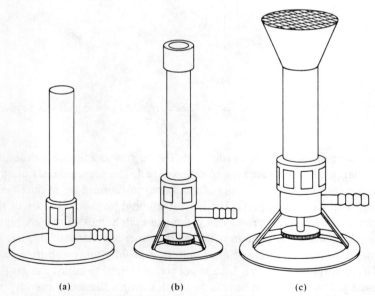

Figure 7.4. Laboratory burners: **(a)** Bunsen, **(b)** Tirrell, **(c)** Meker.

ignited by a hot plate. These devices are practical only to heat flat-bottomed vessels (see (3) below). A simple hot plate does not permit magnetic stirring, but combination hot plate-magnetic stirrers are available.

3. *Water and oil baths.* A beaker or other vessel filled with heated water or oil (paraffin oil, silicone oil, stearin) is useful when direct heating of a vessel is not practical, e.g., for round-bottomed flasks which cannot be heated directly with a hot plate. The bath may be heated with a hot plate, a small immersion heater, or, in certain circumstances, with a Bunsen burner. These baths distribute the heat evenly, and even simple systems such as a beaker of water or oil on a hot plate can provide approximate temperature control. More elaborate baths are discussed in Sec. 13.2.

4. *Sandbaths.* Fine, dry sand in a metal container is an effective heat transfer medium. Sandbaths are useful chiefly when temperatures greater than those easily attained with an oil bath are needed. They also offer greater safety than a liquid bath should the reaction vessel break.

5. *Steambaths* (Fig. 7.5). Organic liquids often are heated on a steambath, because this eliminates open flames or heated surfaces on which vapors can ignite. It also offers an automatic limit to the temperature that can be attained, and clearly is not useful if temperatures above $100°C$ are required. A steam bath is a metal container with a series of concentric, removable rings in the top. To heat a flat-bottomed container, rings are removed until the central hole is just a little smaller than the bottom of the container to be heated. For safety, the container should be held with a clamp, but it should rest with its weight on the steam bath. A round-bottomed flask should be allowed to extend into the bath so that a large area can be exposed to the steam. The size of the steam bath opening should be slightly smaller than the diameter of the flask. Clamping is desirable to prevent the flask from tipping. Steam is passed at a moderate rate from the laboratory supply lines into the upper spout on the steambath. The lower spout is led to a drain.

6. *Heating mantles.* Round-bottomed flasks are heated most conveniently by electrical heating mantles. These are bowl-shaped glass-fiber insulated networks of heating elements, sometimes mounted in a metal case. Heating mantles are easily damaged and require some care in use, especially to prevent overheating. The mantle used must be of the proper size to fit the flask to be heated, and empty flasks never should be heated. Both of these requirements are to ensure that heat is conducted away from the mantle fast enough to prevent its overheating. A heating mantle should not be operated directly from the power line voltage, because this may cause it to overheat. A variable transformer or other control device should be used. The setting of the transformer is adjusted by trial and error to give the degree of heating required, starting from the low settings.

Most heating mantles have a thermocouple in the form of a coil of wire attached to them (see Sec. 13.1). When necessary, the internal mantle temperature can be monitored using this thermocouple and a potentiometer.

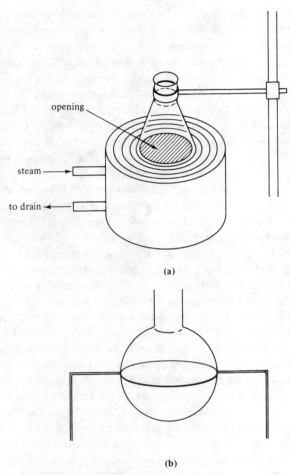

Figure 7.5. Steam bath: **(a)** proper use with a flat-bottomed vessel, **(b)** proper use with a round-bottomed flask.

Great care should be taken not to spill chemicals on heating mantles because unpleasant fumes may be evolved on heating, and the mantle may be damaged. Flammable liquids and their vapors can be ignited by a hot mantle. Chemicals of any kind should not be added to flask while it is in a mantle. Water can cause electrical short-circuits, and for this reason a mantle should always be supported on a ring stand above the surface of the laboratory bench so that spills of water will not affect it. There is a real hazard of electrical shock from misuse of heating mantles.

7. *Heating tapes*. Irregularly shaped glassware can be heated with flexible, fiberglass-insulated heating tapes which can be wrapped around the apparatus. The precautions necessary when using heating mantles should be followed.

8. *Heat guns.* A stream of heated air is useful for warming "frozen" ground glass joints and for other purposes. A hot air or heat gun, which resembles a hand-held hair drier, is a convenient source of such air. **Some heat guns can provide air heated to several hundred degrees, and they must be used with appropriate care.**

9. *Infrared heating.* An infrared or heat lamp is useful for drying materials and for evaporating liquids. Normally, the lamp is suspended a few inches above the system to be heated. The amount of heating can be controlled roughly by adjusting the distance between lamp and sample. Heat lamps are convenient for evaporating liquids, because the heat can be directed to the top surface of the liquid and no bumping will occur. **However, the use of such lamps should be restricted to nonflammable materials. The use of a heat lamp with flammable organic solvents can easily result in ignition of the organic material if it comes in contact with the hot lamp.**

10. *Drying ovens.* Electrically heated ovens are commonly employed for drying glassware and chemicals. Most such ovens are adjustable from room temperature to about 200°C. The actual temperature normally is read with a thermometer. For removal of water from stable materials a setting of 110°C is suitable, but chemicals should be dried well below their melting points, and certainly below their decomposition temperature. **Materials that may evolve flammable vapors should not be placed in a drying oven** (see Sec. 10.5).

11. *Muffle furnaces.* For heating to high temperatures (up to 1200°C), special electrically heated furnaces called muffle furnaces are used. They can be set to the temperature desired (often by trial and error methods), and have a temperature read-out device. A chief use of muffle furnaces is for the ignition of precipitates in quantitative analysis (see Chap. 18).

12. *Tube furnaces.* When it is necessary to heat a cylindrical reaction tube, a furnace having a cylindrical opening is often used. These furnaces can be mounted vertically or horizontally. Heating of tube furnaces may be controlled by regulating the input voltage with a variable transformer. More elaborate controllers are available (see Sec. 13.2).

7.8. CONTROL DEVICES FOR ELECTRICAL EQUIPMENT

As was indicated in the sections on heating and stirring, it often is necessary to control the input voltage to motors, heating mantles, and other devices. Control is usually accomplished with a variable transformer (Variac,® Powerstat®). The output voltage is controlled with a dial that in various models is calibrated in terms of per cent of input line voltage, or directly in volts. Although usually used to reduce voltage, most variable transformers can give a slightly higher output than input (e.g., 130 V with operation from a 115-V line). Small controllers usually have an output current capacity of 5 A but various larger sizes are available.

Slightly less expensive control devices are based on an on-off cycle of variable duration. The power applied depends on the fraction of time the devices are on.

These controls are satisfactory for use with heating equipment, but not with motors.

The least expensive means of reducing and controlling the power applied to a motor or heater is to pass the current through a rheostat (variable resistor). The voltage reduction depends on the current drawn as well as the resistance. Rheostats convert part of the power to heat.

Other, more sophisticated electronic power controls may be encountered. Some of these are intended to operate electric motors at low speeds without the reduction of torque that limits the applicability of variable transformers and rheostats.

7.9. FLEXIBLE TUBING

Water, gas, and vacuum connections to apparatus usually are made with flexible rubber or plastic tubing. The most common plastic tubing is polyvinylchloride (Tygon®). Tubing is available in various sizes and wall thicknesses. Thick-walled tubing is necessary for vacuum applications (see Sec. 14.3). When connecting such tubing to other equipment, a tight fit is important. Rubber and polyvinylchloride tubing will stretch significantly, and will go over glass or metal tubing somewhat larger than its own inner diameter. A slightly loose fit can sometimes be made acceptable by wiring or clamping the tubing in place. Side-arms on many pieces of apparatus intended for connection with rubber tubing are ridged, and it is important that the end of the tube extends well beyond the ridge. Lubrication with a little water is helpful when attaching the tubing. After a long time, rubber and polyvinylchloride tubing is difficult to remove from glass or metal connections. In such cases it is best to cut the tubing to remove it.

Both rubber and polyvinylchloride can be attacked by chemicals. Some solvents will cause rubber to soften or swell, while sulfur-containing compounds may be leached from some rubbers. These compounds can "poison" catalysts and have other harmful effects. Polyvinylchloride is also attacked by some solvents which leach plasticizers from it, causing it to become opaque and inflexible. Although these faults are rarely important in routine work, their existence should be recognized.

Although connections may appear to be tight, the fact that these materials can stretch means that an increase in internal pressure may cause a connection to come loose. Consequently, all connections on unattended apparatus employing rubber or polyvinylchloride tubing to carry compressed gases and especially to carry water should be wired or clamped. Rubber in particular deteriorates with age, and tubing should be checked for signs of cracks before use.

7.10. TUBING CLAMPS

The flow of gas or liquid through rubber or plastic hose can be controlled with clamps, but they should *not* be used to attempt to control flow if the pressure inside the tubing is significant, or if it can become large. There are two types of clamps:

1. *Pinch clamps*. These are spring type clamps used for on-off control. The best closure can be obtained by bending the tubing before clamping, as in Fig.7.6b.
2. Screw clamps can be used to restrict flow as well as to stop it. One is illustrated in Fig. 7.6c.

Hose clamps, circular screw-tightened devices, are used to fasten rubber or plastic tubing securely to glass or metal apparatus.

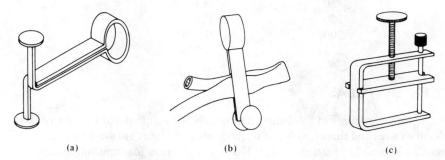

(a)   (b)   (c)

Figure 7.6. Tubing clamps: (a) a pinch clamp, (b) a pinch clamp in use, (c) a screw clamp.

7.11. GRINDING AND PULVERIZING

Solids may be crushed and ground to a powder with a mortar and pestle. Many mortars are made of porcelain, but this is comparatively soft and can cause contamination of samples. For very abrasive substances, or when contamination is undesirable, harder substances such as agate or Mullite® are employed. Lumps of material should be crushed by pressing with the pestle, not by striking them. (Use care; pieces may be ejected from the mortar.) Grinding is achieved by moving the pestle around the inside of the mortar so that the material is rubbed and crushed between the pestle and the walls of the mortar. Too much material at one time cannot be ground effectively.

Ball mills may be used for grinding large amounts of material. These consist of a cylindrical vessel, often porcelain, containing porcelain or steel grinding balls. The solid is placed in the cylinder with the balls, the lid is attached, and the cylinder is rotated on motor-driven rollers. Another form of automatic grinding and mixing device is the Wig-L-Bug® dental amalgamator often used for preparing samples for infrared spectroscopy (see Sec. 15.6).

Particle size sometimes is specified for powders. This may be given in terms of actual particle dimensions (e.g., 100 μm) or in mesh size. Mesh size refers to the screen through which the powder will pass, and is given in terms of number of standard wires per inch of screen. A 100-mesh powder will pass through a screen having 100 wires per inch. Frequently a range is given; for example, 50–100 mesh powder will pass through a screen having 50 wires per inch, but not through one having 100 wires per inch. There is a limited range of particle sizes present. The higher the mesh size, the finer the powder. The following is a partial list of sieve openings for various ASTM mesh designations.

Mesh	Approximate sieve opening (mm)
5	4.0
10	2.0
18	1.0
50	0.3
100	0.15
200	0.074
325	0.044

7.12. SPATULAS

Small amounts of solid chemicals usually are manipulated with a spatula. Various sizes and shapes with flat or scoop-shaped blades are available. Many spatulas are made of stainless steel. The primary reason for mentioning these very obvious tools here is to point out their frequent misuse. Spatulas are prime causes of contamination. Although they are very handy to use for stirring solutions, they can be attacked by strong acids and thereby contaminate the solution. It is good practice never to use a spatula as a stirring rod. If the chemicals in reagent bottles are "caked" a spatula is convenient for removing them, but this practice never should be adopted when using primary standards or analytical reagents for special purposes. The greatest care should be taken that only a clean, dry spatula is used to avoid contamination of the reagent.

7.13. INDICATOR PAPERS

Papers impregnated with test reagents are available for a variety of purposes. The best known example is litmus paper, which turns red when exposed to an acidic solution, and blue when exposed to a basic solution. However, much more information can be obtained from multirange pH papers. General purpose pH papers will indicate the pH value of a solution over the pH range of 1–14. The papers take on a color characteristic of the pH, as indicated by a color chart provided with them. They are accurate to roughly ±1 pH unit, but short range pH papers can give values accurate to a few tenths of a pH unit. The attainable accuracy is limited if colored solutions must be tested.

In addition to pH, many other test papers are available. For example, oxidizing agents can be detected by starch-iodide paper, which turns blue in their presence. Papers sensitive to the presence of specific elements of groups of elements also can be obtained.

Most test papers should be used by touching them with a glass rod dipped into the solution; immersion of the paper directly may result in extraction of the reagent, producing a false color and/or contamination of the solution. Some test strips are designed for dipping, however. Gases or vapors sometimes must be tested. In these cases, most test papers should be moistened before being exposed. When testing vapors from a testtube, flask, or other container, care should be taken that the paper is not responding to contact with droplets of liquid around the mouth of the vessel, or to spray, rather than to the vapors themselves.

7.14. DRYING TUBES (CALCIUM CHLORIDE TUBES)

Reactions that are sensitive to moisture are usually protected with a drying tube. (Very sensitive materials may need more elaborate protection, such as high vacuum systems (see Sec. 14) or inert atmosphere glove boxes or glove bags (see Sec. 5.4.1). Drying tubes may be made of glass or plastic and are attached to the system with a rubber tube or stopper. One end of the tube ordinarily is open, although a stopper containing a glass tube can be inserted if a gas is to be passed through the tube. Plastic drying tubes have two removable ends with a nipple for rubber hose on each. To fill a drying tube, a loose plug of glass wool should be pushed to the end of the tube, which is then filled with a granular drying agent (see Sec. 10.5), followed by a final plug of glass wool. Care must be taken not to pack the glass wool plugs or the drying agent so tightly that easy passage of gas is prevented. Freshly packed drying tubes must be used to ensure that the desiccant is effective.

Tubes of a U-shape, with stopcocks on the top of each arm, are occasionally found in gas drying trains.

7.15. DROPPERS AND DISPOSABLE PIPETS

Medicine droppers are very useful for removing small amounts of liquid from a precipitate or to separate small volumes of immiscible liquids, and for adding small amounts of liquid reagents. Standard medicine droppers are often inconveniently short, so that longer droppers, called disposable pipets, are more useful. These are available commercially, but they can be made easily from 6-mm glass tubing by heating it as described previously in the flame of a Meker burner (or a Bunsen burner equipped with a wing tip if soft glass tubing is used). When soft, the glass is removed from the flame and pulled. When cool, the thin section is broken as desired. Two pipets can be obtained if the process is performed correctly (see Sec. 6.4).

It is generally assumed that a drop of liquid from a medicine dropper or a pipet is roughly 0.05 ml in volume, but the volume of a drop can vary greatly with the tip. Drop size can be calibrated by counting the number of drops required to make up a convenient volume (e.g., 2 ml) in a 10-ml graduated cylinder.

Rubber bulbs are used with these droppers. Care must be taken that liquid is not drawn into the bulb where it can be contaminated. Natural latex bulbs are somewhat better than other types in this regard.

7.16. THE CATHETOMETER

Often it is necessary to measure accurately the change in level of a liquid in a tube — e.g., with some manometers, capillary rise measurements, dilatometry, among others. This is usually accomplished with a cathetometer. A typical cathetometer consists of a horizontal telescope attached to a movable mounting on a vertical metal rod. Usually a sensitive gear drive is employed for fine control. The telescope cross-hairs are aligned with the meniscus or other feature to be measured, and the position of the telescope is read from a scale engraved on the support rod. The telescope can be moved to follow the motion of the meniscus, and the difference between the original position and the new one gives the distance moved. In any use of the cathetometer, it is essential to have the rod perfectly vertical and the telescope horizontal; leveling screws in the base are provided. A firm and vibration free support should be employed.

Balances and Weighing Procedures

Weighing is one of the most common quantitative procedures used in the laboratory, and a variety of balances is available. The type used depends on the accuracy required. Generally, high accuracy is obtained at the cost of greater time required to perform the weighing. It is poor practice to use an analytical balance when a rough balance will suffice, for example, in most synthetic work. Virtually all of the common balances which are encountered depend on the principle of the lever and fulcrum, although some special purpose balances are based on mechanical deformation (spring and torsion balances), or on the measurement of the current required to restore a movable coil to its null position (electrobalances), or other principles. The basic design of a mechanical balance is illustrated by the two-pan equal-arm balance, shown schematically in Fig. 8.1a. The unknown mass M_1 is determined by the value of the known mass M_2 which must be added to the other pan to establish balance. This is an example of a first order lever; the moments about the fulcrum must be equal to establish balance and, if the arms are of equal length, this means the masses M_1 and M_2 must be equal. In practice, fine adjustment is made with a movable "rider" on the beam, or by a movable chain attached to one end of the beam. In some triple beam balances the effective arm length is changed by sliding weights along the beam. The most common type of balance is the analytical balance, giving results to 10^{-4} g ($\pm$ 0.1 mg) or better.

Most analytical balances used today are of the single pan type. Their principle of operation is basically the same as the above, but with modifications as illustrated in Fig. 8.1b. They do not have equal arms, and when the pan is empty M_2 and M_3 are such that the balance is in its null position. In practice M_2 consists of a set of removable weights at the front of the beam. The weight of the unknown M_1 is found from the portion of M_2 which must be removed to restore balance. In

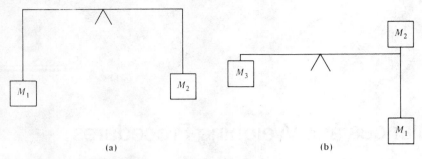

Figure 8.1. (a) Principle of the two-pan, equal-arm beam balance, (b) principle of the single-pan analytical balance.

practice slightly less than the required amount of M_2 is removed, and the extent of unbalance is indicated on an optical scale calibrated in weight units. The primary advantage of single pan balances is speed because the weights are selected quickly with a dial. Another advantage is that the sensitivity is independent of the weight, in contrast to the two-pan balance, because the beam is under essentially constant load. Unfortunately, the unequal arm construction is fairly sensitive to variations in the temperature, which requires the zero point to be adjusted often. The capacity of an analytical balance is limited commonly to 160 or 200 g, depending on the type.

Top-loading balances, in which the total weight is read on an optical scale indicating the deflection of a beam of light caused by the weight of the object, are rapid; their accuracy range is usually 10^{-2} - 10^{-3} g. More recently, electronic balances which provide very rapid digital read-out of weight have become available. They range from 10^{-1} to 10^{-4} g accuracy. Most are extremely simple to operate, with a single push-button controlling the zero and automatic tare.

The main rules to be followed in using all balances, including triple-beam types described below are:

1. Do not weigh substances directly on the pan; use a capped weighing bottle, a watch glass, a beaker, or weighing paper (note that some chemicals may attack paper).
2. The container must be weighed empty (tared).
3. The pan and environs of the balance must be kept clean. Clean up spills immediately.
4. If the balance is not equipped with a beam arrest then some weights should be left on the beam when not in use to prevent undue wear of the knife edges caused by random movement.

8.1. TRIPLE-BEAM BALANCES

For rough weighing (10^{-1}–10^{-2} g accuracy) rugged and inexpensive triple-beam balances are used. These are so named because weighing is performed by sliding weights along three calibrated beams — one each for hundreds, tens and units. The capacity is usually 600 g with provision for extending the capacity by the addition of accessory weights. For very large masses a large double pan balance and weights usually are employed.

The triple-beam balance should be used when extreme accuracy is not needed; e.g., for weighing the starting materials in a preparation or for preparing solutions which are to be standardized later. It is much faster to use than the analytical balance.

8.2. ANALYTICAL BALANCES

If very accurate weighings are required, an analytical balance probably will be used. Newer ones may be electronic, but many are mechanical. These are delicate instruments that are easily damaged if not used properly. The beam and the pan are mounted on agate knife edges resting on flat agate plates, and the agate knife edges are subject to wear if the balance is allowed to swing freely when not in use. Hence, the arresting mechanism which lifts the knife edges from their bearing surfaces must be used. Weights or objects on the pan should not be changed unless the beam and pan have been arrested. All controls should be manipulated slowly to avoid shock which can damage knife edges, or misalign weights.

Several control knobs are found in this type of balance:

1. Arrest control — to lift the beam and pan from the knife edges when the balance is not in use, or when changing weights. This control normally has a partially released position which is used to determine if the weights added are more or less than needed, and for adding small weights, and a fully released position for final reading.
2. Zero — adjust
3. Weights — usually separate controls for 100 g, 10 g, 1 g and 0.1 g increments.
4. Some balances possess a tare knob with which the weight of a small container can be subtracted automatically from the total weight. This should be set to zero if not used.

    The following directions are general for a single-pan analytical balance but some variations can be expected with different models. Careful procedure must always be used with a precision balance.

The general procedure for using an analytical balance is summarized below:

1. Check that the balance is level (the level may be adjusted as necessary by an instructor), and that the pan is clean (a brush can be used for cleaning). As with any balance, do not put chemicals directly on the pan. Use a suitable container (e.g., capped weighing bottle); weighing paper can be used for noncorrosive substance. Volatile compounds or hygroscopic substances should only be weighed in closed, tared containers. Corrosive chemicals (e.g., concentrated acids) should not be brought near a balance except in a tightly sealed container.

2. Check the zero reading by slowly releasing the empty pan, and adjust the zero control until the illuminated scale comes to rest at zero. If the balance cannot be adjusted to zero, see the instructor. Notice that *most* of the time it does not matter what the empty pan reading is, because the weight will be determined by difference, i.e., wt (sample) = wt (sample + container) − wt (container). Obviously the zero must not be changed inadvertently between weighings, and because this always is possible zeroing is recommended.

3. Arrest the pan, place the object to be weighed in the center of the pan, and close the doors, because any air currents will cause an unsteady reading. If the load is not centered on the pan, the pan will oscillate each time it is released. Note that the object must be at room temperature to avoid errors.

4. With the pan arrested dial the 100 g weight and slowly partially release the pan. If too much weight has been added, arrest the pan and remove the 100 g weight. Be certain that the beam is *fully arrested* when adding or removing the 100 g weight or the sample. The balance must be at least partially arrested when adding smaller weights.

5. Partially arrest the pan and then slowly add weights in steps of 10 g until too much has been added. Remove the last (excess) weight and repeat the procedure with the 1 g weights and then the 0.1 g weights. *Add or remove weights slowly or the knife edges will be dulled and the weights may fall off their holders.*

6. The pan may now be released fully and the weight read by summing all the weights indicated on the dials together with the weight shown on the illuminated scale. Use the vernier scale to estimate the nearest tenth of a milligram (see sec. 8.5 on the use of the vernier scale).

7. Arrest the pan, and carefully remove the sample and weights.

8. On completion of the weighing leave the balance in perfect condition. In particular:
   a. The pan and compartment must be clean.
   b. The pan must be arrested.
   c. All weights must be removed.
   d. The doors must be closed.

9. Report any difficulties to the instructor immediately. Do not try to make repairs yourself.

Remarks

1. Finger prints do have measurable weight, so handle the samples as little as possible and wipe them clean. Use clean tongs where possible, or the finger and thumb of an old cloth or rubber glove, or commercially available finger cots. Small glassware may be handled with paper. Charges of static electricity are generated easily on wiping glassware and this can cause drifting of the readings during weighing.

2. When the chemical is to be poured into another container after weighing, determine the weight of the original container *after* the transfer. Some of the

chemical inevitably will be left behind. If the chemical is to be transferred quantitatively by washing into the other vessel, weigh the original container first.

3. Weigh objects only at room temperature. Convection currents will make warm objects "weigh" too little.

4. If all the balances are in use, time can be saved by obtaining an approximate weight on the triple-beam balance while waiting for the analytical balance to become available.

A detailed description of the structural features and theoretical principles of a large variety of balances have been described in several textbooks[1,2] of analytical chemistry.

8.3 TOP-LOADING BALANCES

This is a fast, easy to use balance that may or may not be electronic. Most have the feature of being able to tare (i.e., automatically cancel) a weight of over 100 g (e.g., that of an empty container). Some use a continuous scale, others have a digital read-out. Operational procedures vary, but the following points are generally applicable.

1. The balance must be level. If a pan arrest mechanism is present, it must be released.

2. If there is an on-off switch, be sure it is on.

3. If there is a separate tare knob, it can be used to adjust the balance to zero with an empty container present.

4. An object is weighed by first checking that the balance reads zero, then placing the object on the pan and reading the scale. Weight control knobs, if present, are adjusted as necessary to bring the reading on scale.

5. Most digital balances will have a single push button or bar control; there may or may not be a separate on-off switch. In some, a push bar turns the balance on. A second press will set the reading to zero (e.g., in taring a container). Lifting the bar will turn the balance off. The object is weighed by placing it on the pan after the balance has been turned on and zeroed.

6. As with all balances, clean up all spills after use and return all dials to their initial positions.

8.4. BUOYANCY CORRECTION

For work of the highest accuracy a buoyancy correction often is necessary. This correction normally is required when the density of the substance or materials being weighed is significantly different from that of the weights themselves. When an object is weighed in air there will be an upward force acting upon it that is equal to the weight of air displaced. Unless the density of the object being weighed is the same as the density of the weights, a different amount of air will be displaced by the object and the weights so that the upward thrusts will be different. The correction required by this buoyancy will normally be negligible for solid objects, because their densities will be close to those of the weights. For containers of gases, liquids, solutions, and low density solids, however, the effect can be quite significant.

To obtain the true weight (W_{true}) of the substance, i.e., the corrected weight in a vacuum, the following approach may be used.[3] The mass of air displaced by any object equals the volume of the object multiplied by the density of air. Therefore, to obtain the apparent mass of the object this quantity must be subtracted from the true mass, and the balance condition is given by

$$W_{true} - \frac{W_{true}}{\rho_{obj}} \cdot \rho_{air} = W_{wt} - \frac{W_{wt}}{\rho_{wt}} \cdot \rho_{air} \tag{8.1}$$

where W_{wt} is the mass of the weights *in vacuo*, ρ_{obj} the density of the object, ρ_{wt} the density of the weights, ρ_{air} the density of the air and W_{true} the true mass of the object *in vacuo*.

This equation may be rearranged to yield:

$$W_{true} = W_{wt} \left[\frac{1 - (\rho_{air}/\rho_{wt})}{1 - (\rho_{air}/\rho_{obj})} \right] \tag{8.2}$$

If the ratio ρ_{air}/ρ_{obj} is small, Eq. (8.2) can be approximated by:

$$W_{true} = W_{wt} + W_{wt} \left(\frac{\rho_{air}}{\rho_{obj}} - \frac{\rho_{air}}{\rho_{wt}} \right) \tag{8.3}$$

The density of brass weights, which are most frequently used, is 8.4, while the density of air may be assumed to be 0.0012, although the value varies slightly with the composition, temperature and pressure.

Notice that when gases or vapors are being weighed the density ratio is not small and Eq. (8.2) then must be used. Also note that when mercury is weighed the vacuum weight will be less than the air weight because its density is greater than that of the brass weights.

Examples of Buoyancy Correction

1. *Weighing 1 ℓ of water*

    For this analysis it will be assumed that the air temperature is 20°C and that the barometric pressure is 760 torr. Under these conditions the weight in air of 1 ℓ of water is 997.17 g. Hence from Eq. (8.3) the weight *in vacuo* (W_{true}) is given by

$$W_{true} = 997.17 + 997.17 \left(\frac{0.0012}{0.99823} - \frac{0.0012}{8.4} \right)$$

$$= 998.23 \text{ g}$$

where 0.0012 is the density of air at $20°C$ and 760 torr, and 0.99823 is the density of water under the same conditions. Therefore, the correction factor to be added to the observed weight is 1.06 g which is a difference of 0.1%.

2. *Weighing 2 g of potassium chloride*
   (density of KCl = 1.99)

$$W_{true} = 2.0000 + 2.0000 \left(\frac{0.0012}{1.99} - \frac{0.0012}{8.4} \right)$$

$$= 2.0009 \text{ g}$$

In this case the correction factor is 0.0009 g, which is only 0.05%.

8.5. USE OF VERNIER SCALES

To obtain maximum precision when reading a scale such as the optical scale of an analytical balance, a vernier is frequently employed. It is essentially a subsidiary scale which can be moved relative to the main scale. Two types of vernier scale exist (a) the barometer type, which is also found on the analytical balance and (b) the sextant vernier scale often found on refractometers, polarimeters, and other devices. A simple type of vernier scale of the first kind is illustrated in Fig. 8.2. Here the vernier scale is divided into 10 divisions, corresponding to nine divisions on the main scale.

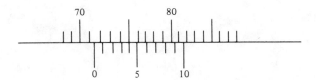

Figure 8.2. Vernier scale.

The use of the vernier scale is best explained by reference to an example such as Fig. 8.2. The zero line of the vernier (the lower scale) indicates the approximate reading. In the example, this line lies between 71 and 72 on the main scale, hence the reading is greater than 71 and less than 72. The next digit is given by the first line on the vernier scale which exactly coincides with a line on the main scale. In this case, it is line 4. Hence the actual reading is 71.4. (The 3rd digit could have been estimated from the position of the zero line — it is about one-third of the way between 71 and 72 — but the estimate is clearly less reliable than the actual vernier scale reading.)

A thorough treatment of different types of vernier scales is given by Reilley and Rae.[4]

REFERENCES TO CHAPTER 8

1. D. A. Skoog and D. M. West, "Analytical Chemistry, An Introduction," 2d ed., Holt, Rinehart, and Winston, New York, 1974.

2. J. S. Fritz and G. H. Schenk, Jr., "Quantitative Analytical Chemistry," 3d ed., Allyn and Bacon, Boston, 1974.

3. J. A. Bell (ed.), "Chemical Principles in Practice," Addison-Wesley, Reading, Mass., 1967.

4. J. Reilley and W. N. Rae, "Physico-Chemical Methods," 2d ed. (revised), Van Nostrand, New York, 1932.

Quantitative Volumetric Measurements and Titrations

Volumetric measurements are used extensively in laboratory work and it is important to emphasize at an early stage, that the establishment of good technique in the use of balances and volumetric glassware is critical for developing satisfactory laboratory practice. The discussions which follow will aid in gaining familiarity with the use of some of the most common kinds of precision volumetric glassware. Some of the important aspects in the use and care of volumetric glassware are described below. Additional information may be found in several authoritative texts.[1,2,3]

Volumetric flasks, burets, and a wide variety of pipets make up the usual types of volumetric ware. The abbreviation "TD" on the vessel indicates that it is calibrated *to deliver* the stated volume. On the other hand the letters "TC" indicate that the vessel is calibrated *to contain* the stated volume (both at specific temperatures). "TC" usually applies to volumetric flasks, but sometimes to other items, and "TD" usually applies to burets and pipets.

9.1. CLEANING VOLUMETRIC GLASSWARE

A major requirement in chemical experimentation is maintaining clean glassware. The following discussion applies to glassware such as burets, pipets, volumetric flasks, graduated cylinders, beakers, and most of the common glassware used in the laboratory.

For quantitative measurements care must be taken to remove all foreign material, both to avoid contamination and to eliminate interferences with the calibration. An important objective is the removal of grease, which prevents smooth drainage and limits accuracy. Chemical residues can cause problems if they interfere with the intended use of the solution; e.g., NaCl in a flask used to prepare $AgNO_3$ would constitute an interference, but it would not be an interference in a flask used to prepare NaOH. There are several recommended cleaning procedures and some of those which are preferred are summarized in Table 9.1. Where possible, detergent solutions should be used first, because they are less hazardous than other cleaning solutions. Care should be excercised with corrosive agents such as solutions of alkali phosphates which when left to soak can etch the glass surface. After cleaning by rinsing carefully with distilled water, inspect the glassware to see if any water streaks or droplets remain on the interior surfaces. If this is the case, the cleaning procedure should be repeated or more vigorous conditions should be tried. Water will not form beads and droplets on a well-cleaned glass surface.

Although routine use of chromic acid cleaning solution should be discouraged, it can be prepared by adding cautiously (to prevent spattering) about 400 ml of concentrated sulfuric acid to a paste made from about 20 g of sodium

TABLE 9.1

Cleaning Solutions for Volumetric Glassware

The various solutions indicated can be used for pipets, burets, volumetric flasks, beakers, graduated cylinders, among others	
Solution	Procedure
Hot 2% detergent solution (sodium lauryl sulfate) or dishwashing detergent	Add the solution until the vessel is almost full; in the case of a buret rotate it so that the inside is covered thoroughly, and use a long handled brush. After draining rinse well with distilled water.
Hot, dilute, alkaline EDTA solution (used to remove trace metal ions)	The solution should be $\sim$0.004 M with a pH of 12, Soak the glassware for no longer than 15 min to avoid etching the surface. After draining, rinse with dilute acid followed by distilled water.
Chromic acid cleaning solution **This is a strongly oxidizing solution and must be handled with care.**	The solution is more effective when warm (60°C) but if allowed to stand overnight a cold solution can be very effective. The reagent tends to decompose rapidly when heated. To clean a pipet use a rubber suction bulb to fill it (about 1/2 of the pipet bulb is sufficient) and rotate the pipet to cover the inside thoroughly. When cleaning a buret fill it with chromic acid. All chromic acid should be returned to the stock bottle provided. When green it is no longer effective. After chromic acid treatment, glassware should be rinsed with water followed by distilled water to remove chromium ions. The rinsing procedure *must* be thorough because if there is any residue of chromium remaining on the glass it can cause serious interference in EDTA titrations or spectroscopic work, for example.

dichromate in 15 ml of water, with thorough stirring between additions (technical grade chromium trioxide can also be used). Because chromic acid is a strong oxidizing agent, **use extreme caution during this process (faceshield, gloves, and lab coat are strongly recommended), and cool the container in water**. The dark red to red-brown supernatant solution should be used until it takes on a green tinge. The red precipitate may be dissolved by the cautious addition of concentrated sulfuric acid with stirring until dissolution is complete. It is more convenient to use a commercial preparation such as "Chromerge"* to avoid the necessity of mixing concentrated sulfuric acid and water.

If any of the above cleaning solutions fails to remove all the grease films, organic solvents such as acetone may be tried, followed by further washing with detergent.

While discussing the general topic of cleaning it is instructive to make some reference to *quantitative rinsing*. When rinsing a flask, beaker or other apparatus there is no need to fill the container with solvent; instead a small quantity of solvent should be added and swirled around the container before emptying it. The process should be repeated three to four times. Not only is this less wasteful but also is a more efficient process. This same method is used to transfer the remaining liquid or to wash a solid quantitatively from one container to another.

9.2. TYPES OF VOLUMETRIC GLASSWARE

9.2.1 Volumetric flasks

Volumetric flasks can be obtained in a variety of sizes from 1 ml to 5 ℓ. Their primary uses are:

1. preparation of solutions of known concentrations from accurately weighed amounts of material
2. accurate dilutions of standard solutions and analytical samples.

Prior to use it is very important to ensure that the volumetric flask is clean (see Table 9.1). If it is to contain an aqueous solution, it may be rinsed with water and used directly. Otherwise, it should be dried or rinsed with the appropriate solvent. Remember that if water is to be added to the flask there is usually no reason for it to be dry first. The substance to be dissolved or the appropriate volume of the solution to be diluted should be added carefully to the flask through a pipet or small funnel making sure to wash any surplus material from the funnel into the flask. Materials which are slow to dissolve should be dissolved by heating in a beaker or flask and then transferred quantitatively to the volumetric flask. A volumetric flask should not be heated strongly to assist solution and, if gentle warming is needed, the flask should be cooled to room temperature before

* Fisher Scientific Company

proceeding. Solvent should be added to the flask with swirling until the level is close to the calibration line. The solution should be mixed well to prevent later changes of volume on mixing, but extreme care should be exercised during this procedure to avoid bringing the solvent in contact with the ground-glass portion of the flask. Allow a few minutes for drainage and, if necessary, let the solution come to room temperature. If thermal equilibrium is not attained, expansion or contraction of the liquid caused by temperature changes will cause the final volume to be in error. The flask is then filled to the mark so that the bottom of the meniscus is in line with the mark. A long medicine dropper or small pipet or a squeeze bottle is convenient for this purpose. Finally the solution should be mixed well by inserting the stopper and inverting and shaking the flask several times. Three points should be noted:

1. There may be appreciable volume changes on mixing (volumes may not be additive on solution) so that partial mixing should be achieved before the flask is too full. This precaution is especially important with concentrated solutions.
2. If the liquid in the flask is not at room temperature, the volume will change as cooling or warming takes place. Hence the flask and its contents must be allowed to come to room temperature before the final volume is adjusted.
3. Finally, mixing in a volumetric flask is quite inefficient. To mix the material in the neck, the flask must be inverted with shaking several times, while maintaining a tight hold on the stopper.

Volumetric flasks should not be used for storage of solutions. The solution should be transferred to a suitably labeled container. This precaution is important particularly for alkaline solutions because they attack glass over long periods and subsequently would alter the volume of the volumetric flask as well as change the solution pH. Alkaline solutions attack ground joints and cause them to freeze both in flasks and standard taper Pyrex® bottles. It is preferable to use polyethylene or rubber stoppered glass containers for these solutions. Alkaline solutions also must be protected from atmospheric CO_2 which if absorbed would change the concentration of base. A disadvantage of using some kinds of plastic containers is that they are permeable to certain vapors, and some solutions slowly change concentration in them.[4]

For work of high precision the solution should be prepared at the same temperature as that used in the experimental measurements. This temperature can be attained by immersing the flask in a thermostat. Alternatively a correction can be made to take into account the volume change caused by the expansion of the glass.* The expansion of the glass is not the only factor but one must also consider

* $V_{20} = V[1 + \beta(20 - t)]$
where, V is the observed volume
V_{20} the volume at 20°C
t – the observed temperature
β – coefficient of cubical expansion of glass

the expansion of the solution, because its coefficient of cubical expansion is larger than that of glass at these higher temperatures. Generally, any vessel constructed from a borosilicate glass such as Pyrex® will change by about 0.001% per degree. The expansion of a dilute aqueous solution is about 0.025% per degree. Consequently temperature changes of about 5°C can, in some cases, be significant. Vessels constructed of soft glass have volume change of 0.003% per degree, but they are rarely encountered in modern laboratories.

9.2.2. Burets

The buret is designed to deliver variable volumes of liquid accurately, and it is used in titrations. The 50-ml buret graduated in 0.1-ml divisions, is most commonly used in the student laboratory. Readings should be estimated to ±0.01 ml, with due cognizance of errors caused by parallax.* Ten-milliliter burets, and smaller micro burets with still finer graduations may also be encountered. Because the buret normally is used for measurements of high precision it is very important to work with a clean buret. Various procedures for cleaning are described in Table 9.1. After cleaning, the buret should be rinsed thoroughly with the sample liquid. Burets are fitted with either a glass or a Teflon® stopcock. Although Teflon® is more expensive, it needs no grease and can be used with nonaqueous solvents, and also does not "freeze" after prolonged exposure to basic solutions. The main problem with Teflon® stopcocks is their tendency to deform when left screwed tightly into the barrel of the buret. Glass stopcocks are subject to leaks arising from a poor fit or improper greasing. Use of excessive amounts of grease will result in clogging of the barrel or bore of the stopcock. Special care should be observed when greasing a glass stopcock. First the old grease should be removed from the stopcock and barrel. Thick layers of grease can be removed easily with a tissue while acetone can be used to remove the last traces. A small pipe cleaner soaked in acetone often is useful for cleaning the stopcock hole, while a very fine wire can be used to dislodge any particles in the buret tip. Be sure that both ground surfaces are completely dry before the application of fresh grease.

The stopcock should be greased *lightly* with a hydrocarbon grease, taking care to avoid clogging the hole in the stopcock itself. Silicone lubricants should be avoided because they are very difficult to remove, but if they are used chloroform and/or toluene are partially effective for removing them. Insert the stopcock into the barrel and rotate it several times. If the correct amount of grease has been used, the contact area between the stopcock and barrel should appear almost transparent. The layer should be uniform with no streaks observable. After greasing the stopcock test it to see whether, in fact, it is leak proof. If it is obvious that too much grease has been used, clean all the parts again and repeat the procedure.

For proper delivery, the buret tip must be in good condition. Burets with

---

* The error caused by parallax can also be a problem with pipets and to a lesser degree with volumetric flasks.

chipped tips are not acceptable for precise work, because the drop size cannot then be controlled. Some burets have removable tips which can be replaced if damaged and are easy to clean.

Use of the buret

Before the buret is used for an actual titration it should be rinsed several times with the titrant. Close the stopcock and add some titrant from a small beaker, while holding the buret in the hand. Add 5 to 10 ml of titrant and then rotate the buret in a nearly horizontal position to wet the inside thoroughly. Repeat this treatment 2 or 3 times and each time allow the liquid to drain through the stopcock. Determine if the liquid drains uniformly and does not leave any droplets, thus indicating that the buret is clean. The buret is now filled to a level above the zero mark (a short-stemmed funnel may be helpful). It is important that there be no air bubbles in the tip of the buret. If present, these bubbles are eliminated by rapidly turning the stopcock and allowing a small quantity of liquid to flow. Also be sure that there are no leakage problems by observing if a drop of liquid forms at the tip of the buret when the stopcock is closed.

The level of the solution is now brought somewhere below the zero mark and after a minute for drainage the initial volume can be read accurately using the meniscus reader described in Fig. 9.1 with adequate precautions to avoid any error caused by parallax (see Sec. 9.2.5). As discussed previously this reading should be estimated to 0.01 ml. It is strongly recommended that time is *not* wasted trying to set the level of the liquid to any particular value.

When the buret is being used for a titration in an Erlenmeyer flask, the tip should be placed inside the neck. The preferred technique, illustrated in Fig. 9.2 is

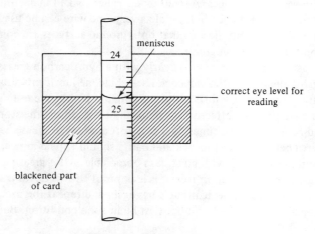

Figure 9.1.   Meniscus illuminator. (see p. 114)

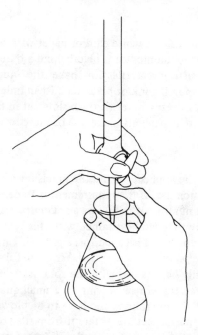

Figure 9.2. Technique for manipulation of a buret stopcock.

to swirl the flask with the right hand while manipulating the stopcock with the left hand from behind the buret. This is not particularly easy at first, especially if other methods have been employed. The advantage of this technique is that it always will apply a small pressure on the stopcock which minimizes any problems of leakage and also it is easier to control the addition of partial drops near the end of the titration. In a potentiometric titration or one using a pH meter, a magnetic stirring device normally is used and often the receiving vessel is a beaker rather than a flask.

On adding the titrant in increments, swirl the flask after each addition to ensure good mixing, unless stirring is accomplished magnetically. When the indicator begins changing color, add the titrant more slowly, approaching dropwise addition near the end point. As the end point approaches, rinse the sides of the receiving vessel with pure solvent. To attain the highest precision, drops of liquid should be split at the end point. This is achieved by allowing the smallest part of a drop to form at the tip of the buret and then touching the receiving vessel gently against the buret tip. This droplet should then be carefully washed into the bulk of the solution with a solvent wash bottle. A drop of titrant is estimated to have a volume of about 0.05 ml and because the reading is estimated to the nearest 0.01 ml the drop splitting technique is absolutely essential for high precision. About 1 min. should be allowed for drainage before the final reading is taken. After use, the titrant should be removed from the buret and discarded. The buret should then be rinsed several times with distilled water, and left filled with distilled water.

Meniscus illuminator (Fig. 9.1)

A meniscus illuminator is a white card or paper with a sharply delineated black lower half. When the illuminator is held behind a buret with the white-black dividing line at the meniscus level, reflection makes the liquid level easier to read. This device is made by applying ink or black tape to an index card, and it may be held by hand or slipped over the buret through slots cut in the top and bottom of the card. It is suggested that students make one or more of these for their own use.

9.2.3. Pipets

Depending upon the application and the precision required there are several types of pipets in common use. If the requirement is to deliver a small volume which is not an integral number of milliliters a *graduated (Mohr) pipet* often is employed. The accuracy that can be attained with this type of pipet is about 1%. When a definite volume of liquid is called for, e.g. 1 ml, 2 ml, 5 ml, 10 ml, 25 ml, 50 ml . . . , a *volumetric pipet* should be used. With careful use an accuracy of 1 part per 1000 can easily be attained with the volumetric pipet. Micropipets of less than 500 μl are employed for work involving very small quantities of liquid. These are often labeled in "λ" for microliters and have to be blown out (i.e., they are designed "TC," and in this respect are different from the volumetric or Mohr pipets). It is essential that all types of pipets be clean before being used. Cleanliness can best be ascertained by observing whether distilled water drains in a uniform manner from the pipet without leaving any streaks or droplets. Several cleaning procedures are indicated in Table 9.1.

Using the Pipet

The pipet is filled by drawing the solution into it with a rubber suction bulb (see Fig. 9.3). **Oral suction must never be used for this process.** When using the bulb make certain that it fits the stem of the pipet properly and be very careful not to draw any liquid into the rubber bulb. Do not pipet from reagent bottles.

Before filling, the pipet should be rinsed with the sample. Pour some of the sample into a clean beaker, draw some into the pipet, and then rotate the pipet horizontally to rinse the inside surfaces thoroughly. It will be necessary to tilt the pipet to clean its total length. If the pipet does not drain uniformly, further cleaning and/or rinsing will be needed. Discard the rinsing solution in the appropriate container after use.

To fill the pipet, use the rubber suction bulb until the liquid is about 1 inch above the etched line (Fig. 9.3(a)) and then quickly place the *forefinger (not* the thumb) over the upper end of the pipet to prevent the liquid from flowing (Fig. 9.3(b)). The pipet should then be tilted slightly from the vertical and the outside wiped to remove any excess liquid.

The next step requires care and often considerable practice to develop good technique. The tip of the pipet should be placed against the wall of a glass vessel and the liquid should be allowed to fall slowly by partially releasing the finger pressure,

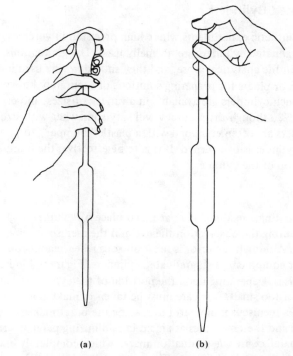

(a)   (b)

Figure 9.3.   Use of the Pipet: (a) drawing liquid above the graduation mark using a rubber bulb, (B) use of the forefinger for adjusting the liquid level to the reference mark.

until the bottom of the meniscus coincides exactly with the reference (graduation) mark on the stem of the pipet. Be sure that there are no air bubbles in the pipet and, if necessary, wipe the outside dry with a tissue, but be careful not to remove any liquid from inside the pipet when doing so. During transport of the pipet to the receiving vessel it is convenient to tilt it slightly so that solution flows back from the tip. Place the tip of the pipet well inside the receiving vessel and allow it to drain freely. Finally hold the tip of the pipet against the inside of the vessel for about 20 sec. to allow time for drainage, and then carefully withdraw the pipet with a rotating motion to remove any drops which may be held on the tip. Because the pipet has been calibrated to compensate for the small volume of liquid remaining inside the tip, this surplus should *not* be blown from the pipet. The only exceptions to this rule are micro pipets (<1 ml) which are marked "TC" and must be completely emptied.

After use, the pipet should be thoroughly rinsed with distilled water and stored carefully in the laboratory drawer or pipet stand provided. Breakage can be minimized by attaching rubber bulbs to either end of the pipet. As with burets, pipets with chipped tips will not drain properly and damaged pipets should not be used for critical work.

9.2.4. Graduated Cylinders

In many synthetic experiments where high precision in volumes of reagents is not called for, graduated cylinders are normally used. The graduations, however, are not calibrations in the analytical sense, and they should *not* be used in place of volumetric flasks or pipets for preparing solutions of accurately known concentration. Such cylinders are available in a variety of sizes, and can be read to approximately 5%, although the accuracy will vary somewhat with size. Glass graduated cylinders are often equipped with a plastic "bumper" to prevent breakage if the cylinder is overturned. If it is to be effective, the bumper must be placed near the top of the cylinder.

9.2.5. Parallax

When estimating a reading on a particular piece of volumetric glassware it is important that during the observation the eye and the meniscus being measured are in a straight line. Although the error is most obvious when reading burets it is also a problem in other equipment, e.g., graduated cylinders. Figure 9.4 indicates how the parallax error arises. Depending upon the position of the eye the reading may be either too large or too small. Due care must be taken to make observations at eye level. The use of a meniscus illuminator in reading the buret makes it much easier to see the meniscus and thereby serves as an aid in minimizing parallax errors. Another aid to avoiding parallax are the graduation marks which completely encircle the burets and other volumetric glassware. The eye is at the same level as the marks when the front and back lines merge. If the liquid level is in a position in which no marks encircle the buret, a paper loop may be used instead.

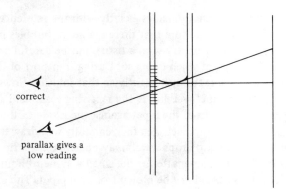

correct

parallax gives a
low reading

Figure 9.4.   Errors caused by parallax.

9.2.6. Calibration of Volumetric Glassware

One of the experiments included in Vol. 2 involves the calibration of a volumetric pipet, but some general remarks about calibration and its relationship to accuracy and precision are appropriate at this point. Specific details for the calibration of a pipet, buret and a volumetric flask have been described by Harris and Kratochvil.[5] In some cases there are significant discrepancies between the volume indicated on the glassware and the "true" volume. This "true" volume is obtained by weighing the volume of water or mercury delivered or contained by the piece of glassware and dividing the weight by the density, with due consideration being given to the buoyancy correction (see Sec. 8.4). Calibration often is necessary in accurate work for the simple reason that the measurements that can be made with a piece of glassware may be more accurate than the nominal volume stamped on it.

The volumes indicated on a pipet, buret or volumetric flask normally refer to calibration at 20°C but many experiments are performed at ambient temperature. Because the volume of glassware varies with temperature, a calibration may be required to correct for this temperature difference. The extent of this effect will depend on the type of glass used to manufacture the vessel, and can be calculated if the volume coefficient of expansion is known. Because the calibration of a buret may vary along its length, as a result of nonuniform bore, calibration data should be obtained at every fifth of its volume and, for convenience in later use, a plot should be made of the correction factor vs. the length of the buret.

9.3. PRECISION AND ACCURACY

Reference should be made to Chap. 4 to review the terms precision and accuracy. The accuracy of volumetric glassware often is indicated by a quantity known as the *tolerance*. The tolerance is a measure of the maximum error allowed in the manufacture of the apparatus. Some of the tolerances for volumetric glassware specified by the National Bureau of Standards (NBS)[6] are indicated in Table 9.2.

TABLE 9.2

NBS Tolerances for Volumetric Glassware

Maximum error of total or partial capacity, ml			
Capacity (ml) of total graduated portion less than and including	Burets	Volumetric pipets	Volumetric flasks
2		0.01	
5	0.01	0.01	0.02
10	0.02	0.02	0.02
25	0.03	0.03	0.03
50	0.05	0.05	0.05
100	0.10	0.08	0.08
500	–	–	0.15
1000	–	–	0.30

Whether or not a particular piece of volumetric glassware conforms to the NBS tolerances will depend to a great extent on the quality or grade of the glassware. Grades A and B often are used by manufactures to indicate the quality of the glassware. Grade A glassware conforms to the NBS tolerances, while the less expensive grade B has a tolerance of about twice those of NBS. Even though some of the less expensive equipment may have tolerances twice those given in Table 9.2, it is not a difficult procedure to calibrate the glassware to an accuracy at least as good as the NBS specifications.

To appreciate the significance of the tolerance of a piece of volumetric glassware and its relationship to the relative uncertainty in an obtained reading, it is instructive to examine a particular example. Consider a 50-ml buret for which the tolerance is given as 0.05 ml. Because the absolute error in the volume delivered could be as large as 0.05 ml, if 40 ml are used, the relative value of the maximum error would be

$$\frac{0.05 \text{ ml}}{40.00 \text{ ml}} \times 100 = 0.125\%$$

An estimation of the experimental uncertainty in reading a buret is somewhat subjective, but a fairly reliable estimate is ±0.01 ml. The *maximum* uncertainty in a measured volume therefore will be ±0.02 ml, because the buret is always read twice. Although the *absolute uncertainty* will be the same, i.e. 0.02 ml, for any volume delivered, the *relative uncertainty* will differ. Thus, if 40 ml are delivered from a 50-ml buret the relative uncertainty is (0.02 ml/40.00 ml) × 100 = 0.05%. This value is considerably less than the uncertainty introduced from the tolerance. If 10 ml are delivered from a 50 ml buret the relative uncertainty in the readings is (0.02 ml/10.00 ml) × 100 = 0.20%. For high precision and accuracy, titrations using a 50-ml buret should involve 35–45 ml of titrant. When smaller volumes are called for a 10-ml buret could be used, because it is divided in graduations of 0.02 ml and therefore allows a lower relative uncertainty in the reading.

To summarize, the *tolerance* represents the maximum deviation of the volume from the stated value, that is the accuracy of the glassware. One reason for calibration is to be certain glassware falls within its stated tolerance limit. A second reason would be to reduce the error introduced through this variation by knowing the true volume. On the other hand, the *uncertainty* of a reading affects the precision, that is the reproducibility of the reading, and it is not reduced by calibration. It represents the range of values that can not be visually distinguished, and is made up of the width of graduation lines and other physical limitations.

9.4. INDICATORS

Indicators are used to mark the completion of a reaction. A visual indicator reacts reversibly to change color in the titration at the point at which the sample has just fully reacted but before (appreciable) excess titrant is added. This color change signals the *end point*. Indicators for redox, complexation, acid-base, and precipitation reactions all follow the same pattern, i.e., when the concentration of a reactant reaches a critical level a change in color occurs. In addition to visual indicators, various physico-chemical techniques can be used to detect the end point. In some cases instruments are used to monitor the desired physical property directly. Examples include potentiometric titrations (emf), conductometric titrations (electrical conductivity), thermometric titrations (temperature) and acid-base titrations (pH) using a pH meter (a special case of a potentiometric titration). However, the simplest and most rapid means of detecting the end point is with a visual indicator. It should be noted that, regardless of the method used to determine the end point, the concentration is an important parameter. Very dilute solutions should be avoided, because the change in pH, emf, or other physical properties will be too small, while very concentrated solutions should be avoided because any unreacted sample (e.g., a drop on the wall of the container) represents a fairly large amount of substance. Some of the common types of indicators are discussed below.

9.4.1. Acid-Base Indicators

Acid-base indicators are highly colored acidic or basic organic molecules exhibiting the following ionization equilibria:

$$HInd + H_2O \rightleftharpoons H_3O^+ + Ind^-; \quad K_{ind} = \frac{[H_3O^+][Ind^-]}{[HInd]} \quad (9.1)$$

(acid color) $\quad$ (base color)

or

$$Ind + H_2O \rightleftharpoons IndH^+ + OH^-; \quad K_{ind} = \frac{[IndH^+][OH^-]}{[Ind]} \quad (9.2)$$

(base color) $\quad$ (acid color)

where K_{ind} represents the appropriate ionization constant.

As the pH of the solution is changed, the proportions of the acidic and basic forms of the indicator change, but normally only the color of the predominant form will be seen. For an acidic indicator* ($HIn \rightarrow H^+ + In^-$) when the pH is equal to pK_{ind}, examination of the equilibrium expression shows that the two forms will be present in the same amount and the color therefore will be intermediate. If, for example, the acidic form of such an indicator is yellow and the basic form is blue, then at pH = pK_{ind} the color will be green. If one assumes as a rough rule of thumb that when a ten-fold excess of one of the components is present then only its color will be seen, then the change from acid color to base color (or reverse) will take place roughly over the range

$$\frac{[Acid\ form]}{[Basic\ form]} \cong 10 \ to \ \frac{[Acid\ form]}{[Basic\ Form]} \cong 0.1$$

---

* Note that if the indicator is a base then an analogous discussion may be made using pOH rather than pH.

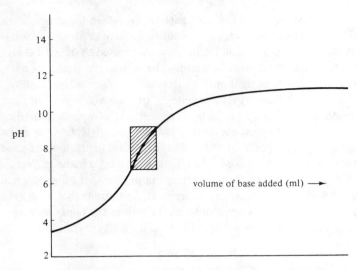

Figure 9.5. An illustration to show how an indicator changes over a range of pH and not at any *one* value of pH.▨; pH range of indicator change.

Substitution of these values into the previous expressions for the indicator equilibria show that the color change occurs over the interval from $pH = pK_{ind} - 1$ to $pH = pK_{ind} + 1$. This is illustrated in Fig. 9.5. The actual range for a particular indicator will depend on the actual color and intensities; some have a larger and some a smaller interval. In any event, the most desirable choice of indicator will be one where pK_{ind} equals the pH at the end point. If the titration involves the reaction of acid HA with base BOH, the pH at the end point or equivalence point is that of a solution of the salt BA.

When a strong acid of moderate concentration ($>10^{-3}$ M) is titrated with a strong base the pH at the end point changes very sharply with a small increment of titrant. Hence the indicator color change will be sharp. With very dilute solutions, or with weak acids or bases, the titration curve (plot of pH vs. volume of added titrant) does not show such a sharp break, so that the change in color also is less sharp. Because the change in pH is smaller in these cases, it is especially important to choose the indicator so that the pK_a of the indicator is within ±0.5 pH units of the equivalence point. With weak acids or bases, the equivalence point will not be at pH 7, because the salt of a weak acid or base is hydrolyzed. The pH at the equivalence point may be determined by calculating the pH of a solution of the salt at the appropriate concentration. The indicator should then be selected from a table or a chart, such as that given in Fig. 9.6, which gives the transition ranges of several common indicators.

As an example of how to select an indicator, consider the standardization of a 0.05 M solution of KOH with potassium hydrogen phthalate (KHP). In these calculations only the approximate pH is required so that it is not necessary to use exact volumes or concentrations. Similarly, not more than three, and generally only

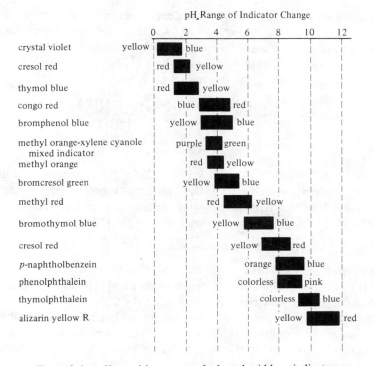

Figure 9.6. pH transition ranges of selected acid-base indicators.

two significant figures need be used in the calculation. For convenience, assume that KHP (0.204 g) was used and dissolved in 50 ml of water. The molecular weight of KHP is 204.22, so that 0.204 g corresponds to 1.00 mmol. The KOH solution contains about 0.05 mmol/ml so that about 20 ml of the KOH solution will be required. Therefore at the equivalence point the solution will contain 1 mmol of the salt of KHP in about 70 ml of water, a concentration of 1 mmol/70 ml $= 1.4 \times 10^{-2}$ M. The problem then is to determine the pH of a solution which is nominally 1.4×10^{-2} M in the salt K_2P. When the ion P^{2-} is placed in water it hydrolyzes according to the equilibrium:

$$P^{2-} + H_2O \rightleftharpoons HP^- + OH^-; \quad K_h = \frac{[HP^-][OH^-]}{[P^{2-}]} \quad (9.3)$$

Therefore the solution will be basic and the question is how basic? Such calculations are discussed in detail in any analytical or general chemistry textbook and therefore will be described only briefly here. The equilibria of concern are:

$$HP^- \rightleftharpoons P^{2-} + H^+; \quad K_a = \frac{[H^+][P^{2-}]}{[HP^-]} = 4 \times 10^{-6} \quad (9.4)$$

$$H_2O \rightleftharpoons H^+ + OH^-; \quad K_w = [H^+][OH^-] = 10^{-14} \quad (9.5)$$

The equilibrium constant for hydrolysis may be expressed in terms of the acid dissociation constant and K_w as $K_h = K_w/K_a$. This is a general expression, which applies equally to the salts of weak bases, and it may be verified by multiplying and dividing the expression for K_h by $[H^+]$ (or $[OH^-]$ in the case of a salt of a weak base). The resulting expression is then:

$$K_h = \frac{[HP^-][OH^-]}{[P^{2-}]} \frac{[H^+]}{[H^+]} = \frac{[HP^-][OH^-][H^+]}{[P^{2-}][H^+]}$$

$$= K_w/K_a \tag{9.6}$$

Thus in the case of KHP, $K_h = 10^{-14}/4 \times 10^{-6} = 2.5 \times 10^{-9}$. According to the hydrolysis equilibrium, for each mole of OH^- produced a mole of HP^- is also produced so that $[OH^-] = [HP^-]$. Hence the concentration of OH^- in a solution of $1.4 \times 10^{-2}\ M\ P^{2-}$ is given by the expression:

$$[OH^-] = \sqrt{K_h[P^{2-}]} = \sqrt{2.5 \times 10^{-9} \times 1.4 \times 10^{-2}} = 5.9 \times 10^{-6}\ M$$

Furthermore from Eq. (9.5) the concentration of H^+ must be $1.7 \times 10^{-9}\ M$, which corresponds to a pH of $8.77 \cong 8.8$ at the equivalence point. Therefore an indicator such as phenolphthalein, which changes color near pH = 9 should be used.

If due care is not exercised in selecting the correct indicator, considerable error can occur in the titration. To illustrate this effect in a qualitative manner reference should be made to Fig. 9.7, where titration curves for three acid-base titrations are given:

0.1 M HCl with 0.1 M NaOH
$10^{-3}\ M$ HCl with $10^{-3}\ M$ NaOH
0.1 M weak acid (e.g., HAc) with 0.1 M NaOH

Inspection of Fig. 9.7 shows that one cannot use the same indicator for the three titrations and expect to obtain the same accuracy or precision. For the strong acid-strong base case where moderately concentrated solutions are used (curve A) a fairly large selection of indicators is available for titrations giving an error within 1 part in 1,000. However with more dilute solutions (curve B), or in the titration of a weak acid with a strong base (curve C), the selection becomes much more critical. For instance, for curve B, if the titration using methyl orange as the indicator was stopped at pH $\simeq$ 3.5 (point a) then a considerable error would occur in the end point. Calculations show that these errors can be large. Use of phenolphthalein would similarly introduce an error. If titration using this indicator was continued to the bright red stage (pH = 10), a significant error (0.1%) would occur even for the concentrated solution.

The apparent sharpness of an indicator change depends a good deal on individual perception and experience. A white background, and the absence of reflections from other colors in the room are important. Small amounts of indicator generally give the best results and minimize errors from consumption of the titrant

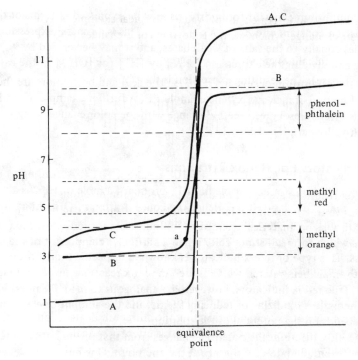

Figure 9.7. Titration curves for a strong acid-strong base and weak acid — strong base, showing the effects of different indicators.

Curve A   0.1 M HCl with 0.1 M NaOH
Curve B 10^{-3} M HCl with 10^{-3} M NaOH
Curve C   0.1 M weak acid with 0.1 M NaOH.

The bands show the pH ranges over which the indicators change color; methyl orange, 3.0–4.5; methyl red 4.8–6.2; phenolphthalein, 8.2–10.0.

by the indicator. The color changes involved with a particular end point also influence the apparent sharpness. For example, methyl orange is red at pH = 3 and yellow at pH = 4.5, while at the half neutralization point it takes on an orange color. Distinguishing these shades is not easy without experience, particularly for people lacking good color vision. Some indicators involve a colorless and a colored form, e.g., phenolphthalein. These can be used accurately only in going from colorless to colored states, when the end point is the first discernible trace of color. Thus, with phenolphthalein it is important to titrate from low pH to high pH.

To overcome some of the problems of gradual color changes, mixed indicators can be used to advantage, particularly when the need arises to have a sharp color change over a narrow and selected pH range. Generally mixed indicators are such that their pK_{ind} values are close enough so that the overlapping colors are complementary at an intermediate pH value. A typical example is xylene cyanole FF and methyl orange, which has the following color changes:

$$\text{(alkaline) green} \longrightarrow \text{grey} \longrightarrow \text{purple (acid)}$$

The grey transition occurs at pH = 3.8. The acid and base colors are the sum of those of methyl orange and xylene cyanole FF. A further example of a mixed indicator is bromocresol green-methyl orange with the color change being from blue to greenish-yellow at pH = 4.3–4.1.

9.4.2. Indicators For Redox Titrations

Redox titrations are based on oxidation-reduction reactions. A comprehensive treatment of oxidation reduction – titrations can be found in several textbooks on analytical chemistry.[2,3] Indicators for redox titrations may be substances which form a distinct color when a particular component of the reaction is in excess. This type of redox indicator is known as a specific indicator, the most common example being starch which is used in redox reactions in which iodine is produced. True redox indicators, however, are analogous to pH indicators. That is, they are themselves oxidizing or reducing agents, and the amount of each form present depends on the potential of the solution.

In a redox titration, the potential changes from that of the initial material to that of the titrant, and does so sharply at the end point. One has

$$\underset{\text{(oxidized color)}}{\text{Ind}_{(ox)}} + ne^- \rightleftharpoons \underset{\text{(reduced color)}}{\text{Ind}_{(red)}}$$

The ratio of oxidized and reduced forms is determined by the potential of the system; according to the Nernst equation which at 25° C may be written:

$$E = E_{\text{Ind}}^\circ - \frac{0.059}{n} \log\left[\frac{\text{Ind}_{(red)}}{\text{Ind}_{(ox)}} \right] \tag{9.7}$$

A color change can be seen as the potential changes, in the same way that one is seen for an acid-base indicator as the pH changes. Just as an acid-base indicator changes color over a particular range of pH determined by its pK, a redox indicator also changes color over a particular potential range, determined by its standard potential. In general the selected indicator should have a transition potential* which corresponds as closely as possible to the potential at the equivalence point in the titration. A summary of some common redox indicators, together with their

---

* The transition potential is that potential at which the color change occurs. This potential is in general different from the standard potential, where the concentrations of the oxidized and reduced forms are the same. It is the different intensities of the oxidized and reduced forms that leads to the difference in values of the transition potential and the standard potential; if the intensities were equal, the two potentials would be the same.

TABLE 9.3

Some Common Redox Indicators[a]

| | Color | | Transition Potential |
Indicator	oxidized	reduced	(V)
Ferroin (1,10-Phenanthroline iron (II) sulfate)	Pale blue	Deep red	1.11
Diphenylamine sulfonic acid	Red violet	Colorless	0.85
Methylene blue	Blue	Colorless	0.53
Indigo tetrasulfonate	Blue	Colorless	0.36

[a] For information on other indicators see, e.g., "Volumetric Analysis," I. M. Kolthoff and V. A. Stenger, Vol. 1, p. 140, Interscience Publishing Inc., New York, 2d ed., 1942.

color changes and transition potentials is given in Table 9.3. It should be noted that the transition potentials of some redox indicators are pH dependent, because a proton(s) is involved in their reduction.

As an illustration of the problem of selecting an indicator for a redox titration consider the titration of Fe(II) with Ce(IV). The Nernst equation gives

$$E = E^{\circ}_{Fe^{3+}/Fe^{2+}} + 0.059 \log \frac{Fe^{3+}}{Fe^{2+}}$$

If we assume that the end point is reached when 99.9% of the Fe(II) has been titrated then, $Fe^{3+}/Fe^{2+} = 99.9/0.1 = 999$. Using this ratio and the standard electrode potential for the redox couple Fe^{3+}/Fe^{2+} (0.771 V), the actual measured potential at the end point of the titration can be estimated. This is given by:

$$E = 0.771 + 0.059 \log 999$$
$$= +0.95 \text{ V}$$

An indicator which has a transition potential corresponding as closely as possible to the potential at the equivalence point is then selected. In this example the Fe(II) is being oxidized to Fe(III) and therefore inspection of Table 9.3 below shows that diphenylamine — sulfonic acid or Ferroin would be appropriate indicators.

The choice of a redox indicator can also be affected by the speed with which the indicator is oxidized or reduced, because it is easy to exceed the end point.

9.4.3. Indicators For Complexometric Titrations

Complexometric titrations utilize the reaction of a metal ion with one or more ligands. The ligands are generally neutral or negatively charged species which are capable of forming one or more coordinate bonds to the metal. Typical ligands used in these titrations are EDTA (ethylenediaminetetraacetic acid) and trien (triethylenetetramine). Some of the theoretical principles governing complexometric titrations will be described with the experiments using this technique. A more thorough discussion may be found in texts on analytical chemistry.[2,3]

Complexometric indicators in their simplest forms are highly colored dyes which bind to the metal ion less strongly than does the ligand employed in the titration. As long as uncomplexed metal ions are present, the indicator is bound to the metal and the solution has a color which is different from that obtained when only the free indicator is present. At the equivalence point, the bound indicator is released, and the color of the solution changes. This process is represented as follows:

$$\text{MInd} + \text{L} \longrightarrow \text{ML} + \text{Ind}$$

(Metal Indicator)	(Ligand)	(metal ligand)	(free indicator)
complex		complex	
(Color A)			(Color B)

A large variety of indicators has been used in complexometric titrations. The choice of an indicator depends upon several factors, particularly the metal ion under investigation and the pH range within which the titration is being conducted. Some of the most commonly used indicators, together with their structures, useful pH range, metal ion applications and color changes are summarized in Table 9.4.

The selection of an indicator for a complexometric titration is not as straightforward as for an acid-base or redox indicator because of the specificity of the complexometric indicator. The approach is similar but added complications caused by the various complex equilibria and their modification by pH and hydrolysis have to be considered. Kinetic effects also are important. Formalized procedures for selecting an indicator have been described (see, for example, Schwarzenbach[7]). Published procedures should be consulted to gain a full understanding of the selection and operating conditions of each indicator.[2,3,8,9] In practice an indicator is usually selected from established recipes.

In the context of complexometric titrations the metal ion concentration is normally expressed as pM, i.e., the negative logarithm of the metal ion concentration, analogous to using pH in acid-base titrations. A titration curve for a complexometric titration is then normally represented as a plot of pM vs. the percent titration or ml of added ligand. A typical curve is shown in Fig. 9.8, for the case of different metal ions with different formation constants (or stability constants) with the ligand. These curves can be constructed from calculations involving the formation constants of the metal-ligand complex and this type of information is tabulated in several textbooks and handbooks.[7,8]

TABLE 9.4

Indicators for Complexometric Titrations

Indicator (color)	Structure	Metal ions	pH range	Color of complex with ion
Eriochrome Black T (blue)	(structure of Eriochrome Black T)	Mg^{2+}, Ca^{2+}, Sr^{2+}, Mn^{2+}	6.3–11.6	red
NAS (Napthyl) Azoxine S (red violet)	(structure of Azoxine S)	Cu^{2+}, Zn^{2+}, Pb^{2+}	3–9	yellow
Xylenol Orange (yellow)	(structure of Xylenol Orange)	Bi^{3+}, Th^{4+} (often employed in cases where the metal ion forms very strong complexes with the ligand)	1.5–3.0	violet

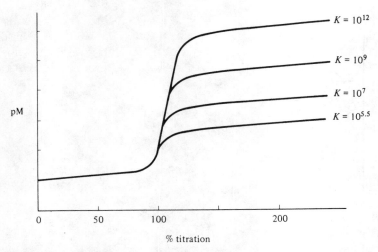

Figure 9.8. Theoretical titration curve for the titration of different metals with a given ligand for various values of K_{obs}, where K_{obs} is the observed value of the formation constant.

In addition to the problems associated with other types of indicators, complexometric indicators also are subject to a phenomenon called "blocking". When a strong or non-labile (slow reacting) metal-indicator complex is formed no end point will be observed. The metal-indicator complex does not dissociate to form the free indicator, either for kinetic or thermodynamic reasons. This problem is encountered often with metals such as Cu^{2+}, Ni^{2+}, Fe^{3+} and Al^{3+}, present either as an impurity or as the ion being titrated. If the blocking metal ion is an impurity and it forms a strong cyanide complex (e.g., Cu^{2+} or Ni^{2+}), the problem can be solved by adding cyanide, which forms an even stronger complex with the metal ion than does the indicator. If the blocking metal ion is the one being titrated, a back titration method is required. An example of this problem is illustrated in Vol. 2, exp. I, Chap. 19. Note that for some metals, e.g., Cu^{2+}, Ni^{2+}, direct titration is possible with an appropriate indicator, but for others, e.g., Fe^{3+}, Al^{3+}, indirect methods almost always are used.

9.4.4. Indicators For Precipitation Titrations

Precipitation titrations involve the formation of an insoluble product, for example, the titration of a solution of chloride ion with a silver nitrate solution.

$$AgNO_3 + Cl^- \longrightarrow NO_3^- + AgCl\downarrow$$

As the equivalence point is passed, the solution changes from one containing an excess of Cl^- ions to one containing an excess of Ag^+ ions. A suitable indicator for this titration may take the form of a colored precipitate which, because of its solubility product (K_{sp}), does not form until virtually all of the Cl^- ion has precipitated. An example of this approach is the Mohr method in which CrO_4^{2-} will react with Ag^+ to form a red precipitate of Ag_2CrO_4 at the end point. Consider the case of 0.1 M solutions of chloride ion and chromate ion. It is desired to know whether AgCl or Ag_2CrO_4 will precipitate first when Ag^+ ions are added. Because the solubility product expression is

$$[Ag^+][Cl^-] = 2.8 \times 10^{-10} = K_{sp, AgCl}$$

when $[Cl^-] = 0.1$, then

$$[Ag^+] = 2.8 \times 10^{-9} \, M$$

When this concentration of Ag^+ is reached AgCl will be precipitated. Similarly

$$[Ag^+]^2[CrO_4^{2-}] = 1.9 \times 10^{-12} = K_{sp, Ag_2CrO_4}$$

so that when $[CrO_4^{2-}] = 0.1$

$$[Ag^+] = 4.4 \times 10^{-6} \, M$$

When this concentration is reached silver will precipitate as Ag_2CrO_4. Thus silver chloride will precipitate before Ag_2CrO_4. The concentration of chloride ion when Ag_2CrO_4 first precipitates may be determined, by substituting this last concentration of Ag^+ into the expression for K_{sp} of AgCl.

$$[Ag^+][Cl^-] = 2.8 \times 10^{-10}$$

$$[4.4 \times 10^{-6}][Cl^-] = 2.8 \times 10^{-10}$$

Hence

$$[Cl^-] = 6.4 \times 10^{-5} \, M$$

This final calculation shows that most of the Cl^- has been precipitated before the Ag_2CrO_4 first precipitates. Note that using less concentrated CrO_4^{2-} would decrease even further the remaining chloride concentration although care must be taken to ensure enough precipitate will form to be detectable. If the concentration of CrO_4^{2-} is too low, the color at the end point will be very faint and difficult to see. On the other hand if the concentration of CrO_4^{2-} is too high there is an interference at the end point because of the yellow color which masks the appearance of the red precipitate.

The Mohr titration should be performed in neutral solution because in acid solutions the chromate ions will be converted to dichromate;

$$2CrO_4^{2-} + 2H^+ \rightleftharpoons Cr_2O_7^{2-} + H_2O$$

and the appearance of the end point will occur too late.

Another type of indicator used in precipitation analysis is the adsorption indicator. These indicators are adsorbed by the precipitate and change its appearance at the end point. The adsorption depends on the charge of the ions which are also adsorbed on the surface of the precipitate. In the above example, the surface ions would change from Cl^- to Ag^+ at the end point, and a suitable adsorption indicator would be a colored substance (e.g., some organic anion) which would be preferentially attracted to the positive ions (e.g., dichlorofluorescein). This process can be represented as:

$$Ag^+ + AgCl(s) + Ind \longrightarrow AgCl : Ag^+ \vdots Ind$$
$$\text{(yellow)} \qquad\qquad \text{(red)}$$

The pH is an important factor when using an adsorption indicator such as dichlorofluorescein ($C_{19}H_9O_3Cl_2COOH$). In this case the titration should be performed in the pH range 4–10. At higher pH silver will precipitate as AgOH, while in acid solutions the ionization of the indicator is repressed:

$$HIn + H_2O \rightleftharpoons H_3O^+ + In^-$$

and consequently the adsorption of the indicator is decreased.

Note also that effective dispersion of the AgCl is necessary because a large surface area is required. If the AgCl coagulates the color change at the end point will not be distinct. Colloidal AgCl is produced by stirring and the colloidal particles can be stabilized by the addition of dextrin, which is a protective colloid.

The titration of sulfate ion with Ba^{2+} can also be performed using an adsorption indicator method (alizarin red S; yellow → pink). For a more detailed discussion of the theory and practice of titrations involving precipitation and adsorption indicators consult any of several standard textbooks.[2,3,8,9]

REFERENCES TO CHAPTER 9

1. G. G. Christian, "Analytical Chemistry," Xerox College Publishing, Lexington, Mass., 1971.

2. D. A. Skoog and D. M. West. "Analytical Chemistry, An Introduction," 2d ed., Holt, Rinehart and Winston, New York, 1974.

3. J. S. Fritz and G. H. Schenk, "Quantitative Analytical Chemistry," 3d ed., Allyn and Bacon, Boston, Mass., 1974.

4. B. W. Durham, *J. Chem. Educ.* **51**, 737 (1974).

5. W. E. Harris and B. Kratochvil, "Chemical Separations and Measurements, Background Procedures for Modern Analysis," Saunders, Philadelphia, 1974.

6. J. C. Hughes, "Testing of Volumetric Glass Apparatus," National Bureau of Standards, NBS Circular 602; U.S. Department of Commerce, Washington, 1959.

7. G. Schwarzenbach, "Complexometric Titrations," Interscience, New York, 1957.

8. A. I. Vogel, "A Textbook of Quantitative Inorganic Analysis," 3d ed., Longmans, London, 1961.

9. W. J. Blaedel and V. W. Meloche, "Elementary Quantitative Analysis, Theory and Practice," 2d ed., Harper and Row, New York, 1963.

Separation and Purification Techniques

10.1. FILTRATION AND CENTRIFUGATION

10.1.1. Introduction

Filtration is a process in which a solid is separated from the liquid in which it is dispersed. The suspension is separated by using a porous medium so that the liquid may pass through it but the solid is retained. Filter paper and fritted (porous) glass are the more common media currently in use. In what follows, unless explicitly stated otherwise, the discussion centers around *qualitative* filtration procedures useful in synthesis and purification.[1] Where the contrast is informative, reference will be made to the differences between *quantitative* and *qualitative* techniques. For detailed procedures used in quantitative analysis standard works in this field should be consulted.[2]

10.1.2. Gravity Filtration

Filtration is most often effected with a glass (or plastic) funnel (see Fig. 10.1) and filter paper under the force of gravity. The various filtration methods are most effectively used when most of the supernatant liquid has been decanted first. In qualitative work the bulk of the supernatant fluid may be discarded or concentrated to yield additional product while in quantitative procedures the supernatant liquid is decanted *through* the filter to insure complete collection of the precipitate. Then the precipitate should be swirled with the remainder of the solution and then the slurry should be transferred, all at once, to the funnel. To minimize splashing add liquid to the funnel by pouring it down a glass rod touched

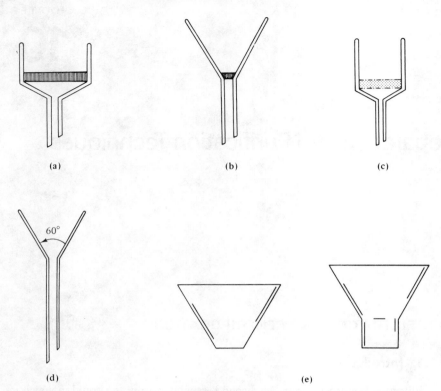

Figure 10.1. Cross-sections of different types of funnels: (a) Büchner Funnel, (b) Hirsch Funnel, (c) Fritted Glass Funnel, (d) Long-stemmed Funnel, (e) Stemless and Powder Funnel.

to the lip of the beaker or flask being emptied. The lower end of the rod should touch the inside surface of, or the surface of the liquid in, the funnel into which the liquid is being poured (see Fig. 10.2e). The liquid level should not be allowed to approach closer than $\frac{1}{4}$ in from the top of the paper, and the top of the paper should be about $\frac{1}{2}$ in below the top of the funnel.

Two methods are generally useful when folding a filter paper for gravity filtration. Figure 10.2a–d shows the simple procedure for forming a cone of paper used in simple gravity filtrations in both qualitative and quantitative work. The paper circle is folded in half, (Fig. 10.2a) and then again into quarters (Fig. 10.2b). It is opened to form a cone, with 3 layers of paper on one side and one layer on the other side, which is placed in the funnel, Fig. 10.2c. The cone of paper should be moistened and pressed against the funnel to form a seal, Fig. 10.2d. For most efficient operation the filtrate should completely fill the stem of the funnel without any air bubbles being formed.

Fluted or multiply-folded filter paper is used to provide a larger surface area for faster filtration, particularly with hot, concentrated solutions. This approach is

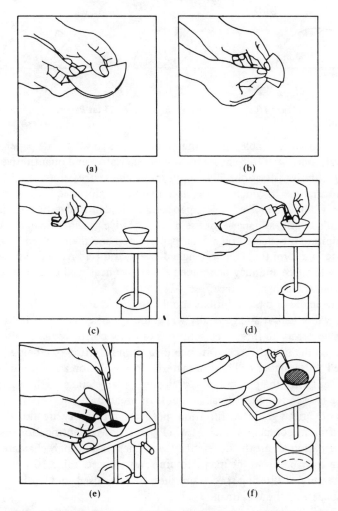

Figure 10.2 **(a)–(c)** Folding of a Filter Paper into a Cone, **(d)** Sealing a Paper Cone to the Funnel, **(e)** the Proper Method of Transferring a Suspension to a Funnel, **(f)** Washing a Precipitate.

most useful for qualitative, synthetic work with organic solvents and should be used with wide, short-stemmed or stemless funnels. To "flute" a filter paper (see Fig. 10.3a–c), place a disc of filter paper on the bench top and fold it in half and then into quarters in the conventional manner. Open the last fold as in Fig. 10.3a. Fold corners A and C to meet at B. Reopen the paper so it resembles Fig. 10.3b. Fold, in the reverse direction, corners A to D, D to B, B to E and E to C. The folded paper should look like Fig. 10.3c. When the paper is opened, a fluted cone is formed.

The retentivity of a filter paper (that is, the size of particles which it will retain) is approximately inversely proportional to the filtering speed. The types of

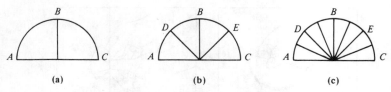

Figure 10.3. (a)–(c) Folding a Fluted Filter Paper.

filter paper which are available reflect the kinds of uses to which the paper will be put. In general, there are filter papers for use in qualitative and quantitative filtrations. Qualitative filtrations include synthetic chemistry, simple clarification of solutions, and analyses in which the precipitate either is not used further, or subsequently will be redissolved. Filter paper used in quantitative filtrations includes so-called "ashless paper" for use in quantitative gravimetric determinations where a precipitate is collected on a filter paper, dried and then ignited leaving only the precipitate in a form that can be weighed easily, and acid-hardened papers that are essentially lint-free and may be scraped to remove nearly all of a collected precipitate from the paper's surface.

Strongly acidic or basic solutions should not be filtered through paper because they will attack it and generally will cause a marked decrease in filtration speed. If it is necessary to filter such solutions or other corrosive materials (e.g., strong oxidants) the use of special filters and vacuum filtration (see below) is recommended. Filter paper will absorb material from solution and should be rinsed well with solvent after the precipitate is collected. A large circle of filter paper should not be chosen to collect small amounts of precipitate.

Papers of various grades of filtration speed and retentivity are available from a number of different companies, each of which has its own identification code. Detailed information about the different types of papers and their characteristic properties are available from the manufacturers, on request. Table 10.1 gives a qualitative measure of filtering speeds and the adjectives used to describe them for some of the papers of the Whatman Co.

10.1.3. Suction Filtration (See Fig. 10.1 for examples of different types of funnels and Fig. 10.4 for a typical suction filtration apparatus.)

In this method a Büchner funnel, a cylindrical porcelain funnel with a perforated bottom upon which a circle of filter paper is placed, is connected to a suction flask using a Neoprene adapter to make a seal. This arrangement is useful for samples of about 1 g, or larger. A Hirsch funnel fitted with the appropriate size paper, should be used for small amounts of product. The paper which is used should be large enough to cover the holes in the funnel completely, but not larger than the inner diameter of the funnel. The filter paper should be moistened with a small amount of the supernatant solution or the solvent being used. Gentle suction

TABLE 10.1

The Use of Various Types of Whatman Filter Papers

Whatman Filter Paper Chart

Filter Speed	Qualitative	Qualitative wet strengthened	Low ash	Ashless	Hardened	Hardened ashless	Retention
Fast	4	114	31	41	54	541	Coarse and gelatinous precipitates
Medium fast	1	111		43			
Medium	2		30	40	52	540	Medium crystalline
Slow	5		32	44	50	540	Crystalline
	6			42			Fine crystalline
Ash	0.06%	0.06%	0.025%	0.01%	0.025%	0.008%	

No. 3   Thick, medium speed paper with high retention
GF/A   Standard laboratory glass fiber paper
GF/B   Thicker, stronger version of GF/A
GF/C   Ultra-fine glass fiber paper

[Courtesy of Whatman Inc., 9 Bridewell Place, Clifton, N.J., 07014]

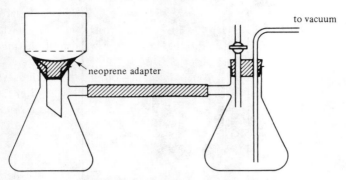

Figure 10.4. Suction filtration assembly showing a trap.

(water aspirator) is applied to the side arm of the filter flask to seal the paper to the funnel. The bulk of the material to be filtered is transferred to the funnel so that it is approximately two-thirds full and the suction is increased until a satisfactory rate of filtration is attained. Incremental additions of the material to be filtered are continued either until all the solid to be collected has been transferred, or the funnel is filled with solid. If the level of the filtrate in the filter flask should reach the tip of the funnel, interrupt the filtration, empty the flask and then resume filtration. The filtrate is the best medium to use to rinse the last quantities of precipitate into the funnel in qualitative work because it is already saturated with the product to be collected and will not dissolve more of it. To wash the precipitate, *stop the suction* and pour in the wash liquid. After a few minutes, re-apply the suction. It is helpful to stir the precipitate, but care must be taken not to disturb the paper. For more efficient washing the material may be suspended in a little fresh solvent and re-filtered.

Büchner funnels are rarely used for quantitative filtrations because of the possibility of seepage of precipitate under the paper and of deposition of material within the hollow body of the funnel. Sintered or fritted glass funnels are useful in both quantitative and qualitative filtrations. The sample may be dried to constant weight directly on these funnels, even at elevated temperatures. The fritted glass is available in porosities of coarse, medium, fine, and extra fine grades. They should not be used for very fine precipitates such as $BaSO_4$.

Consider the amount of precipitate to be collected when choosing the size of funnel to use. A fairly thick filter cake in the funnel is desirable, because thin layers may cause large losses on the filter paper, as well as more contamination from paper fibers when the filter is scraped to recover the sample. Cracks forming in a thick cake should be closed by pressing with a spatula to avoid drawing air through the cake. If air is drawn through such a crack, the solvent will evaporate leaving behind impurities to contaminate the collected precipitate. If the cracks are kept closed, the filter cake will be squeezed free of the supernatant liquid, as evidenced by the cessation of dropwise solvent flow. During filtrations requiring extended periods of time, some of the filtrate may evaporate thus cooling and concentrating the filtrate and causing the precipitation of a second "crop" of crystals. This material should be collected but it should not be combined with the initial precipitate until it has been determined (e.g., by thin layer chromatography or breadth of melting range) that the two crops are of similar levels of purity.

If the filtrate is to be kept, be very careful in stopping the suction. When a water aspirator is turned off, water may be "sucked back" into the filter flask. To prevent this, first let air slowly into the system and then stop the aspirator. A trap (Fig. 10.4) equipped with an air inlet controlled with a screw-clamp, clothes pin or stopcock is convenient for this purpose and should be connected routinely between the aspirator and filter flask.

If it is necessary to filter strongly acidic, basic or corrosive solutions, filters of modified paper (acid hardened paper for acidic materials) or inert materials (Teflon® or glass fibers) are available. These inert materials are best used in a Büchner funnel because they are not folded easily for use in gravity filtration. Funnels or filter crucibles with sintered or fritted glass bottoms (Fig. 10.1) also are useful in these instances.

Very finely divided solids are often difficult to filter, and either pass through or clog the pores of the paper or fritted glass. A finely divided mineral filter-aid, Celite, may be placed on the paper or slurried with the solution to be filtered to retain the particles. Gently scraping the surface of the cake will expose fresh surface if the filtration rate becomes too slow. This approach can be used only if the precipitate is to be discarded or dissolved in another solvent.

10.1.4. Centrifugation

Centrifugation is useful particularly as an alternate method of separating a solid from a liquid when the precipitate is finely divided and not easily filtered.[3] The suspensions to be separated should be distributed evenly among the available centrifuge tubes so that they are about half full. Insert them in the centrifuge so that the load is balanced. Failure to balance the load may cause the tubes to break and even may cause the centrifuge to move, sometimes violently, thereby creating a safety hazard. Gradually increase the speed of revolution of the centrifuge (some types are cranked manually, but they are generally driven electrically and controlled with a variable voltage source) to an empirically determined appropriate level, and allow the centrifuge to run for a short time. If, at a particular speed, the precipitate does not settle to the bottom of the tube after centrifugation for a few minutes, longer periods of centrifugation will generally *not* cause settling, either. The sample should then be spun at a higher speed, if possible. *Never* leave the centrifuge unattended. The centrifugation is interrupted by stopping the power and letting the centrifuge head come to a stop. Some centrifuges are equipped with a braking lever which can be used to slow a centrifuge. Abrupt stops may not only damage the centrifuge but will probably redisperse the precipitate. The supernatant liquid may be decanted or removed with a medicine dropper.

References to 10.1

1. A. B. Cummins and F. B. Hutts, Jr., in "Technique of Organic Chemistry," 2d ed., A. Weissberger (ed.), Interscience, New York, 1966. Vol. III, chap. 5.

2. A. I. Vogel, "A Textbook of Quantitative Inorganic Analysis," 3d ed., Longmans, London, 1961.

3. C. M. Ambler and F. W. Keith, Jr., in "Technique of Organic Chemistry," 2d ed. A. Weissberger (ed.), Interscience, New York, 1966. Vol. III, chap. 4.

10.2. THE SEPARATORY FUNNEL — EXTRACTION

The separation of two immiscible liquid phases, and the extraction of a substance from one phase into another are important steps in many analytical and synthetic procedures.[1,2]

The separatory funnel (Fig. 10.5) is used to separate immiscible liquid phases — usually an organic layer and an aqueous layer. There are several points to be noted in the proper use of a separatory funnel.

Figure 10.5. The separatory funnel.

Before filling, be sure that the stopcock is closed. The funnel is filled not more than three-quarters full and is stoppered. Grasp the funnel near the top, with two fingers over the stopper and shake the funnel vigorously (a large funnel may require the use of both hands) for a short time. Invert the funnel and cautiously open the stopcock, being careful to aim the tip of the funnel in a safe direction. This action is necessary to release pressure developed from vaporization of volatile solvents. Repeat these operations several times. Vigorous mixing is particularly important in a washing or extraction.

When the funnel is allowed to stand in an upright position, the layers should separate with the denser phase on the bottom. Often when the organic phase consists of a mixture of liquids, it becomes impossible to predict the densities of the layers. If a few milliters of one of the liquids (e.g., water) is added, it will mix with, and increase the volume of, its own phase, permitting easy identification. It is good practice to save the "unwanted" layer until the desired product actually is isolated.

If an emulsion forms, addition of a salt such as NaCl may help to disperse the emulsion and allow the layers to separate. Do not add more salt than will dissolve in the aqueous phase, because the excess may clog the stopcock. If separation is poor because of similar densities, particularly if the organic phase is itself a mixture, addition of a heavy organic component (e.g., CCl_4) may help this phase to sink, while addition of a light one (e.g., ether) may help it to rise. When the phases are separated, the stopper is removed and the lower phase is collected carefully in a suitable container by draining through the stopcock.

The separatory funnel is most often used in the process of product isolation and purification and sometimes to extract or concentrate a component during quantitative analysis. The basic principle which is applied is that of *partitioning* a mixture between two immiscible phases for which the solutes have different affinities. This principle not only is operative in extraction but also it underlies all chromatographic methods of separation (see Sec. 10.6). In the simplest and most ideal case, the two components to be separated, A and B, will have markedly different solubilities in the two phases present, so that after equilibration each component will be transferred (extracted) almost completely into a different phase. In practice, if the phases are properly chosen, the solubility of one component will be small in one phase, but not completely zero. Equilibrium between the phases can be represented as

$$(A)_1 + (B)_1 \rightleftharpoons (A)_2 + (B)_2$$
$$\text{(phase 1)} \qquad \text{(phase 2)} \tag{10.1}$$

if the states of aggregation of A and B are the same in each phase. The equilibrium constant K_D for the partition of a particular component between phase 1 and phase 2 is called the partition coefficient or the distribution coefficient and is defined as the ratio of solubilities of one component in phases 1 and 2; i.e.,

$$K_D^A = \frac{[A]_2}{[A]_1} \tag{10.2}$$

The value of K_D is characteristic for each compound and pair of solvents and also is dependent on the temperature of the system.

In most syntheses, the partitioning solvents are usually water and a water-immiscible organic solvent, such as diethyl ether, benzene or chloroform. Consider A and B with water solubilities of 10 g/100 ml and 1 g/100 ml, respectively, and benzene solubilities of 1 g/100 ml and 10 g/100 ml respectively. Thus K_D^A (for water-benzene) is

$$\frac{1 \text{ g/100 ml (benzene)}}{10 \text{ g/100 ml } (H_2O)} = 0.1$$

and K_D^B (for water-benzene) is

$$\frac{10 \text{ g/100 ml (benzene)}}{1 \text{ g/100 ml } (H_2O)} = 10$$

If 1 g each of A and B are dissolved in 100 ml of water and the resulting solution is equilibrated in a separatory funnel with 100 ml of benzene, the amount of A and B extracted into the benzene phase will be given by:

$$\frac{[A]^{benzene}}{[A]^{water}} = \frac{W_A \text{ g/100}}{(1 - W_A)\text{g/100}} = 0.1$$

and

$$\frac{[B]^{benzene}}{[B]^{water}} = \frac{W_B \text{ g/100}}{(1 - W_B)\text{ g/100}} = 10$$

where W_A and W_B represent the weights of A and B respectively which are found in the benzene. Thus 0.091 g of A and 0.909 g of B will be in the benzene layer, and 0.909 g of A and 0.091 g of B will be in the water layer. Using the same equations, it can be shown that several extractions with small volumes (e.g., the extractions with 20 ml of solvent) are more effective at extracting the components of a mixture, than one large extraction using the same total volume.

For any given partition system it becomes impractical to completely extract one component with a separatory funnel when the relative magnitude of the distribution coefficients is less than ~100. For these systems the apparatus shown in Fig. 10.6a,b is used. A solid phase may be extracted continuously with a volatile solvent in the apparatus shown in Fig. 10.6c, known as a Soxhlet extractor. The boiling solvent is condensed into the body of the extractor over the sample that is contained in a porous cellulose thimble. When the level of the solvent in the extractor exceeds the level in the side arm siphon tube, the extract is siphoned into the boiling flask.

When the K values for two compounds in a particular partitioning system are almost equal, the above principles have been used to develop a multi-step process called counter-current distribution[2] which is useful for the separation of structurally similar compounds. This process is a multiple-step extraction in which the two liquid phases are continually equilibrated, separated and each phase is then re-equilibrated with fresh immiscible phase. The behavior of a pure compound ($K = 1$) under the conditions of multiple extraction can be seen in Fig. 10.7(a). One

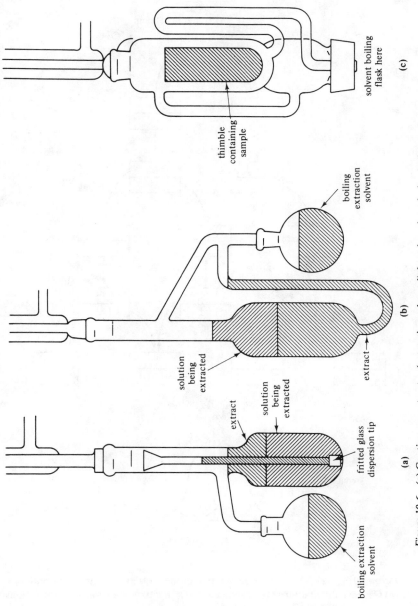

Figure 10.6 (a) Continuous extractor (extracting solvent *lighter* than the solution.), (b) continuous extractor (extracting solvent *heavier* than the solution.), (c) Soxhlet Extractor.

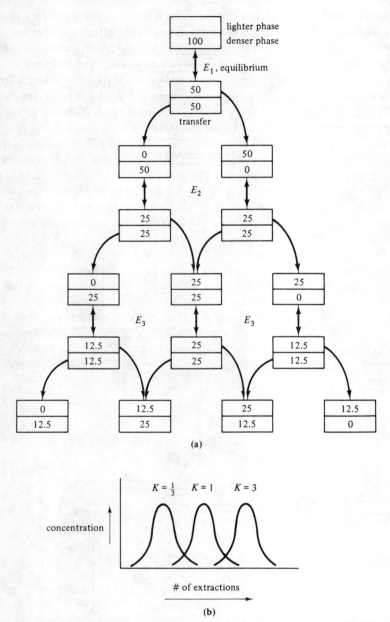

Figure 10.7. (a) Schematic representation of the steps of a multiple distribution, (b) Gaussian distributions of concentration as a function of a number of extractions for compounds with $K = \frac{1}{3}$, 1 and 3.

hundred parts of the pure compound are dissolved in the denser phase and an equivalent volume of immiscible, lighter phase is added to the separatory funnel. The contents are shaken (indicated by the symbol ↕) to effect equilibration of the phases. As specified, the compound being extracted has a $K = 1$ and after the first equilibration (E_1) 50 parts are to be found in each phase. The phases are transferred to different separatory funnels and each is re-equilibrated with fresh immiscible phase. The contents of the second set of funnels are separated and transferred to another set of funnels to be equilibrated with fresh immiscible phase. After three transfers 12.5 parts of the compound are to be found in the first and last funnels, and 37.5 parts are to be found in the second and third funnels. For a large number of transfers the concentration distribution will (in the absence of complicating factors such as complex formation) yield a bell-shaped curve (Fig. 10.7(b)). Thus, to effect separation of two substances by extraction, the values of K for each substance in the system to be used *must* be different. The smaller the difference in K values, the larger the number of transfers which must be made to effect separation.

It is sometimes possible to vary drastically the distribution coefficient of certain compounds by changing the pH of the aqueous solution. Thus, benzoic acid, which essentially is insoluble in water, can be quantitatively extracted as the benzoate ion into aqueous sodium bicarbonate solution. Similarly an amine that might be very soluble in an organic solvent but insoluble in water can be extracted into aqueous hydrochloric acid as a hydrochloride salt. Extractions of inorganic materials frequently employ complexing agents. For example, aluminum can be extracted from water into an organic phase by forming the 8-hydroxyquinoline complex. This process also is pH dependent.

References to 10.2

1. P. A. Ongley (ed.), "Organicum: Practical Handbook of Organic Chemistry," Addison-Wesley Pub. Co., Reading, Mass., 1973.
2. L. C. Craig and D. Craig, "Technique of Organic Chemistry," edited by A. Weissberger, vol. III, chap. 2, Interscience, N.Y., 2d ed., 1966.

10.3. DISTILLATION AND EVAPORATION

10.3.1. Introduction

Distillation[1,2] is the process most frequently used to separate mixtures of miscible liquids, and the effectiveness of separation depends on the boiling point differences among the components of the mixture. The composition of the vapor above a boiling liquid is richer in the component with the lower boiling point (higher vapor pressure). If the boiling point differences are large enough, *simple distillation* (i.e., condensation without fractionation) may be adequate for a separation. Otherwise, *fractional distillation* (Sec. 10.3.3) must be used. At times, it is desirable to work at lower temperatures than the normal boiling point and *vacuum distillation* (Sec. 10.3.4) may then be employed. *Steam distillation* (Sec. 10.3.5) often is effective in separating water-immiscible materials. Not all liquids, even when the components have significantly different boiling points, can be separated completely by distillation because of the formation of azeotropic mixtures with constant boiling points and constant vapor compositions at a given pressure. Ethanol and water from an azeotrope 95% in ethanol and water-ethanol mixtures initially less than 95% in ethanol cannot be concentrated to ethanol contents above 95%. The components of an azeotrope may be separated by adding a third component, or by chemically removing one of the components.

Aspects of the theory of distillation are considered in most physical chemistry textbooks. The basic principles governing liquid-vapor equilibria are discussed below.

If two liquids A and B form an ideal solution, the partial pressure of each liquid in the vapor is given by Raoult's law:

$$P_A = X_A P_A^\circ \tag{10.3}$$

$$P_B = X_B P_B^\circ \tag{10.4}$$

where P_A and P_B are the partial pressures of A and B, P_A° and P_B° are the vapor pressures of the pure liquids at the temperature in question, and X_A and X_B are the mole fractions of A and B in the solution. The total vapor pressure of the solution, P_T, is

$$P_T = P_A + P_B \tag{10.5}$$

if the vapor approximates an ideal gas.

The mole fraction composition of the vapor is given by

$$X_A^v = \frac{P_A}{P_A + P_B} = \frac{P_A}{P_T} \tag{10.6}$$

$$X_B^v = \frac{P_B}{P_A + P_B} = \frac{P_B}{P_T} \tag{10.7}$$

For an ideal solution the composition of the vapor is different from the composition of the liquid solution, as shown graphically in Fig. 10.8. The total vapor pressure of a solution of composition 1 is P, and the vapor is enriched relative to the liquid in component B, which has the higher vapor pressure.

Most liquid mixtures are not ideal; that is, the partial pressures are not linear functions of the composition. For many systems, however, the partial pressure-liquid composition plots show only relatively small curvature.

The boiling point of the solution is the temperature at which P_T equals the pressure of the surroundings. A boiling point-composition diagram for a system not too far from ideal is shown in Fig. 10.9. This form of the vapor-liquid equilibrium diagram, which is more useful for considering distillations, is obtained by plotting

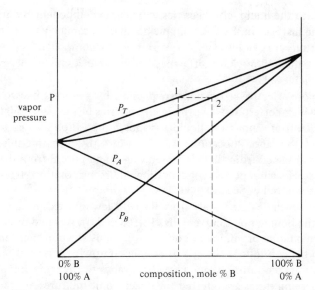

Figure 10.8. Variation of vapor pressure with composition for solution of two liquids which obey Raoult's law. $P_A^\circ < P_B^\circ$. Solution of composition 1 is in equilibrium with vapor of composition 2.

the temperature at which the total vapor pressure reaches a certain value (e.g., 1 atm) against the composition. Because the vapor is enriched in component A relative to the liquid during a distillation, the liquid left behind is gradually enriched in the higher boiling component B. The vapor composition, being in

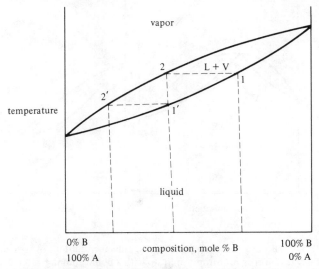

Figure 10.9. Plot of boiling point vs. composition for a solution of two liquids that form a nearly ideal solution. $P_A^\circ > P_B^\circ$ (i.e., B.P. of pure A < B.P. of pure B).

equilibrium with the liquid, also becomes enriched in component B during the progress of the distillation. Hence, a simple distillation cannot yield pure A and B, but only fractions enriched in one component. If one component is present in small amount, and if the boiling point difference is very large, a satisfactory purification may often be achieved.

Much more effective separations are possible by a process known as fractional distillation[3] which corresponds in principle to a series of simple distillations. If a very small amount of vapor from liquid 1 (vapor composition 2) is condensed to 1′ and re-distilled, the vapor 2′ becomes still richer in A. By repeating this process, pure A can be obtained from the condensed vapor, and pure B from the residual liquid. This use of such repeated condensation and re-evaporation steps in fractional distillation is discussed in Sec. 10.3.3.

Systems which deviate widely from Raoult's law may exhibit a maximum or minimum in the boiling point curves (Fig. 10.10a and b). A liquid of composition equal to the maximum or minimum boiling point boils at a constant temperature, in contrast to the continuous changes of boiling point with composition exhibited by other solution compositions, and provides a vapor of the same composition as the liquid. These mixtures are referred to as azeotropic mixtures or azeotropes. Although the boiling characteristics of an azeotrope resemble those of a pure liquid, an azeotrope is not a chemical compound, and the constant boiling composition may change with pressure.

These diagrams (Figs. 10.8–10.10) are examples of vapor-liquid phase diagrams. They obey the phase rule,

$$F = C - P + 2 \tag{10.8}$$

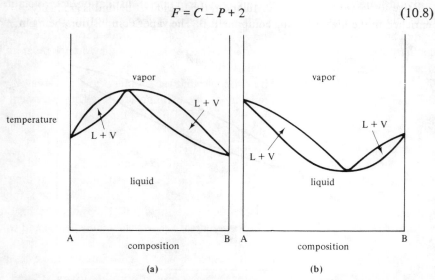

Figure 10.10. Boiling point composition diagrams for solutions of two liquids A and B, that: (a) form an azeotropic mixture with a boiling point greater than that of either component, (b) form an azeotropic mixture with a boiling point less than that of either component.

where F is the number of degrees of freedom, C is the number of components and P is the number of phases. Degrees of freedom refer to the number of parameters (temperature, pressure, composition) that can be varied independently. In the above, one variable, either T or P, is fixed, so that with this restriction,

$$F = C - P + 1 \tag{10.9}$$

Because these are two-component systems, in the single-phase regions (all liquid or all vapor), $F = 2$, but in the two-phase region, $F = 1$. That is, when vapor and liquid are in equilibrium, if the temperature is changed then the vapor and liquid composition both must change in very specific ways to maintain equilibrium.

10.3.2. Simple Distillation

An apparatus such as that shown in Fig. 10.11 is used for simple distillation. The distillation flask (or "pot") must be supported securely, either by a clamp or by a wire gauze on an iron ring. The gauze is necessary if the flask is heated by a burner flame. A burner should not be used for heating highly combustible liquids such as ether or low boiling hydrocarbons, and its use with noncombustible systems may be limited by the fact that others in the laboratory may be using flammable solvents. Alternate methods of heating include electric heating mantles, and heated water or oil baths. These baths, which may be heated by a hot plate or an immersion heater (see Sec. 7.7), have the advantage that their temperatures are easily measured and controlled to give a suitable rate of distillation. Very volatile materials with a low flash point which are ignited easily by a hot surface should be heated by a steam bath (see **Safety, Chap. 1**). Check the joints between the distillation adapter, the flask and condenser to be sure that they do not separate when liquid is added to the distillation flask.

The thermometer should be placed so that the middle of the bulb is even with the lowest part of the side-arm opening of the distilling adapter. This placement will permit an accurate reading of the temperature of the distilling vapors, i.e., the boiling point of the liquid collected. Note that this temperature is *not* the temperature of the boiling liquid (see Sec. 12.1.1).

Water should pass into the *lower* opening of the condenser jacket, and out at the upper opening. The flow of water should not be very fast. Be sure that the end of the outlet hose is in a drain and that it stays there. The hoses should be firmly attached to the condenser *before* assembling the apparatus. The rubber hosing on an unattended apparatus should be fastened with wire at all connections, and to the drain.

For high boiling liquids ($>150°$) water cooling is not necessary and even may cause thermal stresses which may crack the condenser. Blowing compressed air through the condenser jacket will increase the condenser's cooling capacity in this situation. For very volatile liquids, the receiver should be cooled in an ice-water bath. It even may be necessary to cool the condenser with a refrigerated liquid. A

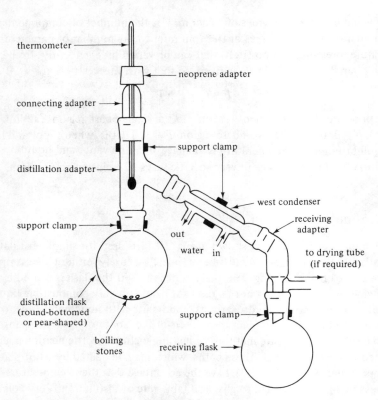

Figure 10.11. Basic apparatus for simple distillation. A drying tube may be attached to the receiving adaptor when distilling moisture sensitive materials. An inert atmosphere can be maintained by using a two-necked distillation flask and passing an inert gas through the system. A bubbler should then be attached to the receiving adaptor to prevent entrance of air. The boiling stones can be dispensed with if magnetic stirring or other form of agitation is used.

drying tube is necessary only if the distilled liquid is to be protected against moisture while special traps can be used to isolate the distillate from air. If distillations of volatile, toxic or combustible liquids such as ether cannot be conducted in a hood, a rubber hose leading to a hood should be attached to the receiver to avoid the hazard of high concentrations of such vapors in the laboratory.

To carry out a distillation, the liquid or reaction mixture is added to the distilling flask (~half-full). Two or three boiling chips are added to prevent 'bumping" (sudden evolution of vapors) which may physically carry liquid into the receiver. *Do not* add boiling chips to hot liquid, because violent boiling may result. The liquid is heated until it begins to boil and the condensate begins to drip into the receiving flask. The rate of distillation may be adjusted by controlling the amount of heat supplied to the distillation flask. *Never* distill until the distillation flask (or residue) is dry, because the material left at the end of a distillation is subjected to much higher temperatures when the distilling liquid is completely evaporated and may decompose violently!

The boiling temperature of a mixture of miscible liquids is the temperature at which the sum of the partial pressures of the components equals atmospheric pressure (*see* section 10.3.1). In a simple distillation of liquids which have boiling points less than approximately 150°C apart, the vapor will become enriched in the more volatile (lower boiling) component while the liquid will become progressively richer in the less volatile (higher boiling) component. Thus, if the vapor is continuously removed by condensation, the boiling temperature of the mixture will rise continuously, but each portion (fraction) of vapor condensed and collected in the receiving flask(s) will contain both components. As a result simple distillation is not useful for efficient purification of volatile liquids, but can be most effectively utilized for removal of a volatile liquid from a relatively nonvolatile liquid or solid (e.g., a reaction product, a drying agent, inorganic by-products, and others).

10.3.3. Fractional Distillation

Liquid-vapor equilibrium diagrams demonstrate that most often a single distillation will not accomplish an effective separation for components which differ little in boiling points. Often, to achieve high purity, a series of redistillations are needed. Fractional distillation[3] uses a column that effectively carries out a series of such multiple distillations by increasing the surface area for condensation and evaporation over which the vapors must travel. The vapors are condensed and re-evaporated as they proceed up the column and equilibrate with condensate flowing down, as if the column consisted of a series of distilling chambers connected in series. The more volatile component of the descending liquid vaporizes and the less volatile component of the ascending vapor condenses in successive stages. The vapor which eventually passes into the receiver is enriched in the more volatile component after the successive equilibration steps. The term "theoretical plate" describes the theoretical number of separate simple distillations which would be required to accomplish the same result. One of the experiments described later will deal with this concept.

The appropriate apparatus is shown in Fig. 10.12. Simply interposing a long, empty column between the distillation flask and the distillation adapter will increase the efficiency of distillation only slightly. The surface area may be increased by filling the column with materials (see Fig. 10.13) having large surface area that are sufficiently open to permit gas and liquid flow.

A simple, easily constructed, durable column for general laboratory use is the Vigreux column (Fig. 10.13(a)). The column can be prepared by heating a small area of a glass tube with a gas-oxygen flame until it softens and then pushing the hot glass in and down (~45° angle) with a thin graphite rod (or a pencil) toward the center of the tube. The indentations are made in pairs on opposite sides of the tube. Each pair of indentations is made along a plane that is perpendicular to the plane of the pair preceding it.

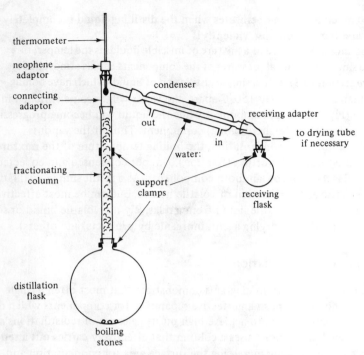

Figure 10.12. Basic apparatus for fractional distillation.

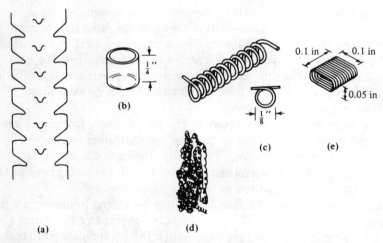

Figure 10.13. Some examples of column packings for fractional distillation, with approximate efficiencies compared to an empty tube at the same distillation rate in parenthesis: **(a)** section of a Vigreux column (5). **(b)** glass rashig ring (5), **(c)** glass helix (10), **(d)** Stainless steel scouring pad (10), **(e)** Helipak® (stainless steel) (50).

More efficient columns can be prepared by filling the annular volume of the column with the various packings shown in Fig. 10.13(b–e). The separate pieces of packing are introduced into the column (held vertically) one-at-a-time to avoid the formation of voids. The column may be filled with a liquid to avoid breakage of the glass packings. The packing is prevented from falling completely through the column by a ring of Vigreux-like indentations, or by a very loose tuft of glass wool, at the bottom of the column.

Steel sponge (commercially available as Kurly Kate®) is an efficient general-purpose packing. It cannot be used in contact with acid vapors, and it must be carefully dried by rinsing with acetone and then protected from laboratory vapors between uses, or it will rust. However, it is cheap enough to discard after one use. Columns are packed by cutting a section of stainless steel sponge of a size to give a relatively loose packing, pulling it out to a strand somewhat longer than the column to be packed and then, using a rod to insert it, pushing in portions to a uniform density. The packing may also be drawn through the condenser with a piece of wire attached to the sponge. The column can be constructed from a 30-cm condenser that has indentations at its lower end to retain a packing so that a packed column of about 20 cm is obtained. The condenser jacket is left empty, but it is best if one opening is stoppered to prevent air circulation, and the outside is wrapped with several layers of asbestos sheet, particularly if distillations above 100°C are to be conducted. Condensation is caused by cooling in the distilling adapter and along the column. Ideally, all the heat loss should occur in the head (distilling adapter) and not along the column. In practice this is difficult to achieve because the temperature varies along the column from flask to head unless the column is well insulated. When using larger diameter and longer columns it may be necessary to wrap the column with insulated heating wire powered by a variable voltage source to prevent excessive condensation in the column ("flooding"). Such columns often are enclosed in a silvered evacuated jacket which is designed to minimize radiative heat losses and thereby minimize one of the major causes of flooding.

Flooding consists of an excessive downflow of liquid in the column and often occurs if the column is not insulated adequately, but the more common cause is too rapid a distillation. Flooding is evidenced by the appearance of slugs of liquid trapped in the column. When this occurs it destroys the equilibration of liquid with vapor and may cause ejection of liquid into the receiver, contaminating the previously collected material. It may be corrected by slowing the rate of distillation (or increasing the heat to the column, if it is wrapped with a heating wire). Ideally the packing should appear to be wetted with only a thin film of condensate. The slowest distillation rate consistent with uniform heating and lack of flooding will give the highest reflux ratio (the ratio of the amount of vapor condensed on the thermometer and walls of the head and returned to the column to that collected as distillate) and, therefore, give the best separation.

If the procedures outlined above are followed, the temperature at which the liquid distills will remain relatively constant after equilibrium is attained, until most of the more volatile component has distilled and then will rapidly rise to the boiling point of the next most volatile component if the heat input to the flask (and column) are increased. Thus, to prevent contamination of purified material with less pure distillate, it is common practice to collect separate portions or fractions of the distillate as the distillation proceeds. Alternatively, the temperature of the distillate may indicate when to collect a new fraction, i.e., change receivers only when the temperature changes. Each fraction then is evaluated for purity by measuring its refractive index or subjecting it to gas chromatographic analysis or some other appropriate analysis (see Sec. 10.6.5) and fractions of similar purity may be combined. With the apparatus described to this point, the reflux ratio, if not controlled mechanically with a collection device, is controlled only by the rate of distillation.

10.3.4. Vacuum Distillation

The boiling point of a liquid is the temperature at which its vapor pressure is equal to that of the atmosphere above it. Thus, lowering the pressure permits distillation to take place at a lower temperature.[4,5] This procedure is useful with high boiling liquids which may decompose at the temperature needed to effect distillation at atmospheric pressure. The experimental apparatus is shown in Fig. 10.14 and 10.15. The boiling point of a liquid at various pressures can be estimated from the temperature-pressure nomograph shown in Fig. 10.16. The nomograph is based upon the assumption that $\Delta H_v \neq f(T)$ and is, therefore, only strictly accurate for pressures near 1 atm. As a practical matter, however, it yields a good approximate value for initial settings.

Caution! Glassware under reduced pressure is dangerous. Only round-bottom or pear-shaped flasks should be used, and these should be inspected carefully for

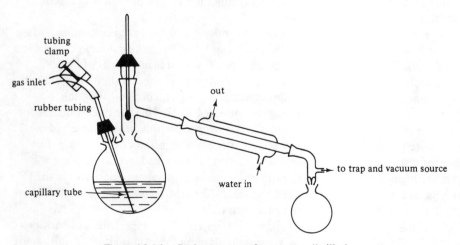

Figure 10.14. Basic apparatus for vacuum distillation.

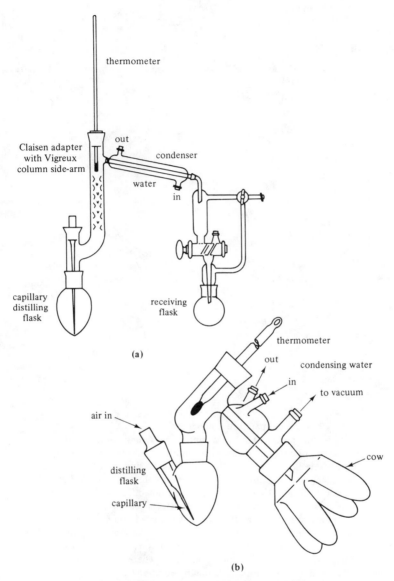

Figure 10.15. Apparatus for vacuum distillation showing two types of fraction cutters: (a) Apparatus for vacuum fractional distillation, (b) Apparatus for short path length distillation to minimize loss of material with a "cow" type fraction cutter.

cracks. The ground glass joints must be greased, preferably at the upper half only so that the product will not be contaminated with the lubricant. Be sure that the grease is distributed uniformly as evidenced by the transparency of and lack of gaps in, the resulting film.

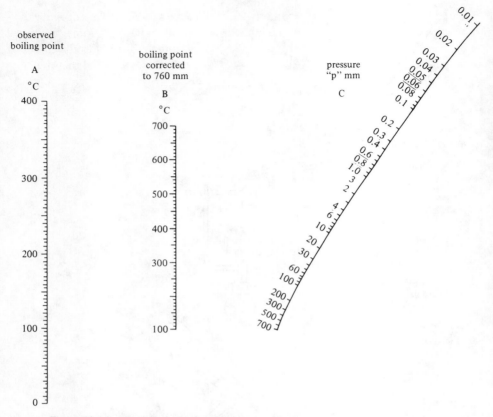

Figure 10.16. Pressure-temperature nomograph.

This nomograph can be used to compute the boiling point at a new pressure when the boiling point is known at another pressure. As an example, assume a reported boiling point of 100°C at 1 mm. To find the new boiling point at 18 mm, connect 100°C (Column A) with 1 mm (column C) and observe where this line intersects column B (about 280°). Next connect 280°C (column B) with 18 mm on column C and observe where this intersects column A. 151°C at 18 mm is the new approximate boiling point. A reported boiling point of 151°C at 18 mm may be converted similarly to one at 1 mm as follows: Connect 151°C (column A) with 18 mm (column C) and note where this intersects column B. Connect this point with 1 mm (column C) and note where this intersects column A. This will be the approximate boiling point at 1 mm.

(reprinted with the permission of MC/B Manufacturing
Chemists, Norwood, Ohio, 45212.)

At low pressure, boiling stones are ineffective because the gas they contain is rapidly exhausted. Further, if the distillation must be interrupted the pores of the chips become clogged with the liquid to be distilled. Hence, to prevent bumping, it is usual to introduce an air or nitrogen bleed into the system, below the surface of the liquid. A piece of glass tubing is pulled into a very fine capillary tip (see

Sec. 6.4) which is inserted to reach to the bottom of the flask. A stream of *fine* bubbles should be pulled through the liquid when vacuum is applied. Some control can be obtained with a screw clamp on a piece of rubber tubing fitted to the end of the capillary tube as shown in Fig. 10.14. However, it is always essential that the capillary be extremely fine.

An alternative source of agitation is a magnetic stirrer and stir-bar. Using magnetic stirring allows the lowest possible pressure to be attained and eliminates the necessity of drawing fine capillaries.

Examples of the apparatus to be placed between the distillation apparatus proper and the vacuum source are shown in Fig. 10.17. This arrangement may be modified by additional components when necessary for a particular application; for example a manostat (see below) may be inserted between the trap and manometer

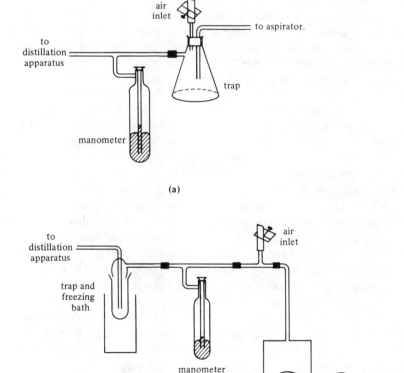

Figure 10.17. Arrangement of traps and manometers for vacuum distillations: (a) for use with a water aspirator as a source of vacuum, (b) for use with a mechanical pump as a source of vacuum.

in Fig. 10.17(b); or additional traps may be used to protect the pump. Either a water aspirator or a mechanical vacuum pump may be used as the source of vacuum, but there are minor differences in the arrangement depending on which is chosen. A water aspirator will reduce the pressure, at best, to about 12–20 torr, the vapor pressure of water at room temperature, which often is adequate. With a water aspirator, a trap is essential to protect the system from water "suck-back" (see Sec. 14.1.1).

A mechanically driven oil pump will attain lower pressures ($\sim$0.001 torr or 1 μm at the lowest). The actual pressure attained depends on the particular system (e.g., the rate of air bleed). A mechanical pump must be protected from organic vapors, and a trap (see Sec. 14.1.2) which can be immersed in a freezing bath is used before the pump.

An air inlet controlled with a stopcock or screw clamp should be provided to permit admission of air or N_2 to the system at the end of the distillation. Note that if the vacuum is released with liquid in the distilling flask, air pressure will force the liquid into the air inlet tube (Fig. 10.14), and possibly into the gas regulator if one is used. Consequently, release the vacuum slowly and open the screw clamp of the bleed to balance the pressure increase.

If the full vacuum of the pump results in a boiling point which is too low for good condensation the air inlet can be opened partially to give an intermediate pressure. A controlled leak can be produced by using the needle valve of a gas burner. This can be adjusted to give the desired pressure which, with care, can be kept relatively constant. Better pressure control is provided with a manostat (a device for maintaining constant pressure).

The Cartesian diver manostat (see Fig. 10.18a) is a common form. This device, which essentially functions as a valve, is useful in the range of approximately 5–100 torr. It consists of a bulb called the diver which moves in response to pressure changes in the system. In operation the diver floats on enough mercury to bring the diver to the point where its rubber tip just touches the capillary tip. Both stopcocks are kept open as the system is being evacuated. The upper stopcock is closed when the chosen pressure is approached. The only connection between the pump and the system will then be the capillary tip. The second stopcock is closed just before the chosen pressure is reached, setting the pressure in the diver. As the system pressure continues to decrease, the gas trapped in the diver by the mercury will expand and push the diver against the capillary sealing the system from the pump. If the pressure in the system increases it forces the diver down, opening the capillary connection to the pump.

A pressure gauge is needed, both to ensure that the system is performing well, and to make the recorded boiling temperature a meaningful observation. Some details on pressure measurements are presented in Chap. 14. A simple type of gauge is a U-tube filled with mercury, with the upper end sealed (Fig. 10.18(b)). The manometer in Fig. 10.18(d) is essentially a U-tube with concentric arms. The pressure of the system is the difference between the mercury levels when vacuum is applied to the manometer, after rotating the stopcock to connect the gauge with

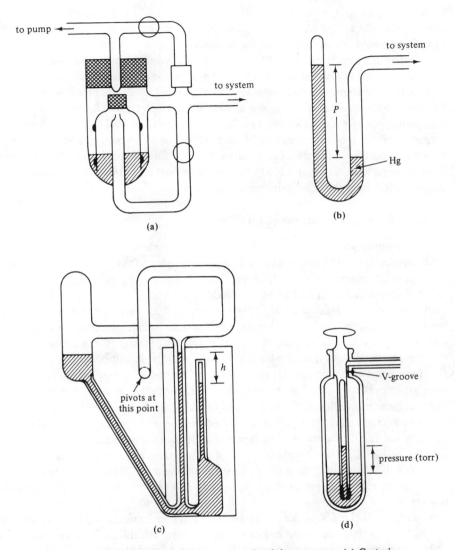

Figure 10.18. Devices for measuring and maintaining pressure: (a) Cartesian diver manostat, (b) U-tube manometer, (c) tilting McCleod gauge, (d) closed-end mercury manometer.

the system. This gauge is useful in the range above 5–10 torr. The tilting McCleod gauge, Fig. 10.18(c), is useful in the range of 5–10 torr to 0.001 torr. The gauge is rotated slowly about its axis, filling the tubes with mercury until the level in the inner tube reaches the reference mark on the attached scale. The pressure corresponding to the difference in mercury level, h, is read directly from the scale (see Chap. 14). After a reading is taken the gauge is carefully returned to the rest

157

position so that no mercury is trapped in the sealed capillary. In general, vacuum gauges should be closed to the system except when a pressure measurement is being made, to prevent contamination of the mercury.

When the apparatus in Fig. 10.14 is used, the vacuum must be released to change receivers. Allow the system to cool before *slowly* admitting air. Be especially careful that the vacuum is not re-established while the liquid is hot. This problem can be avoided with the use of special "fraction cutters", one of the more common types of which is known as a "cow" adapter (Fig. 10.15b). It has several fingers to which receiving vessels are attached. The desired vessel can be placed in the proper position to receive the distillate by rotation around the ground-glass joint. Other adapters (Fig. 10.15a) use stopcocks to permit collecting fractions without disturbing the equilibrium in the column.

10.3.5. Steam Distillation (Co-Distillation)

The above distillation processes deal with the separation of components of solutions. Steam distillation (or co-distillation with steam)[6] is a way of separating an organic compound that is insoluble in water from tarry by-products, salts, and other nonvolatile materials. For an immiscible mixture, each substance exerts its own vapor pressure independently of the others, so that the total vapor pressure is the sum of the individual vapor pressures of the pure liquids. Because the mixture will boil when the total pressure reaches atmospheric pressure, the mixture will boil at a temperature lower than the boiling point of any single component. The mole fractions of the two components in the vapor will be equal simply to the ratios of their vapor pressures. Because the molecular weight of water is small compared to many organic molecules, the actual weight of water needed to carry over several grams of a large organic molecule may not be great, even if the organic compound is not very volatile. (As a practical matter it must have a vapor pressure of at least a few torr at $100°C$.) The two phases in the distillate then are separated easily. The efficiency of steam distillation increases as the temperature is raised. Super-heated steam (water vapor at a temperature higher than $100°C$) can be generated in special heaters.

Structural characteristics are important factors in determining the separability of mixtures by steam distillation. For example, in the nitration of phenol both ortho and para nitrophenol are formed. Separation of these isomers by recrystallization is a relatively difficult process. However, the ortho isomer forms intramolecular hydrogen bonds while the para isomer forms *intermolecular*

ortho isomer   para isomer

hydrogen bonds. Thus, the ortho isomer is readily volatilized in a steam distillation, leaving the para isomer behind as a polymerically associated species.

Steam distillation usually is accomplished by passing steam into a mixture of water and the organic compound in a flask, as shown in Fig. 10.19(a). The delivery tube should reach to the bottom of the flask. To avoid "splash-over," an empty, short distilling column may be used between the condenser and the distillation flask. Water in the steam may be removed by passing it through a separatory funnel before it enters the distilling flask, as in Fig. 10.19(b).

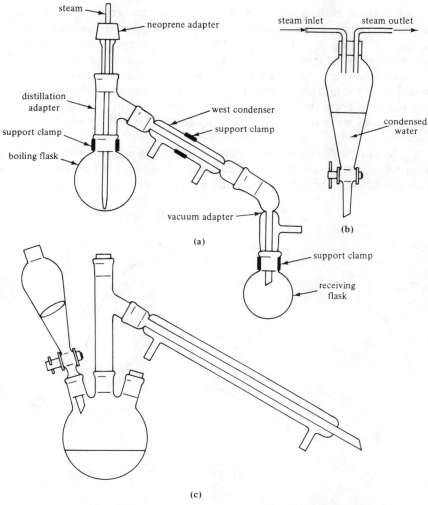

Figure 10.19. (a) Apparatus for steam distillation with externally generated steam, (b) steam condensate trap, (c) apparatus for steam distillation when steam is generated internally.

It is also possible, and often more convenient, to generate the steam simply by boiling the water-organic mixture as in a simple distillation. If this done, more water may be added constantly from a dropping funnel to maintain an adequate level (Fig. 10.19(c)).

The boiling point-composition behavior of a two-phase liquid system may be discussed in more detail. First, it should be pointed out that even though two liquids A and B may be "immiscible," usually there is still some mutual solubility, although if the chemical nature of the two liquids is greatly different, this solubility may be negligible for most purposes. However, miscibility is a function of temperature, and may increase or decrease as the temperature increases, depending on whether the heats of mixing are endothermic or exothermic. In some cases, liquids which have limited miscibility at one temperature may be totally miscible at another temperature, as illustrated in Fig. 10.20. At temperatures and compositions lying below the curve in Fig. 10.20 (e.g., point 1), two liquid phases (of compositions $1'$ and $1''$) exist in equilibrium, while above the curve, only a single liquid phase exists. The system phenol-water is one example of this behavior. Other systems show complete miscibility at low temperatures but not at higher temperatures, while still others show immiscibility over only a limited range of temperature.

Systems which show immiscibility at the boiling point are of interest in connection with co-distillation. A typical phase diagram of the boiling point as a function of overall composition is shown in Fig. 10.21. Here a and e are the

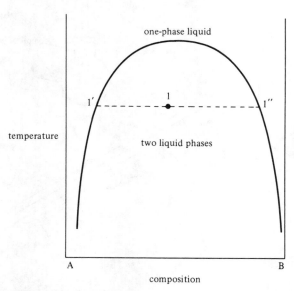

Figure 10.20. Composition-temperature plot for two liquids that are slightly soluble in one another at low temperatures but completely miscible at high temperatures.

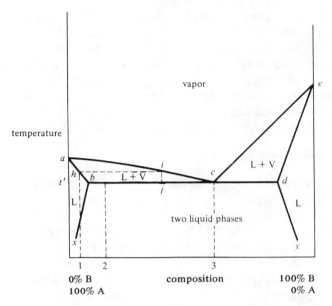

Figure 10.21. Boiling point-composition diagram for two slightly miscible liquids.

boiling points of pure A and pure B, respectively. The region "*abx*" represents a solution of B dissolved in A.

The boiling characteristics in this region are simply those described for a two-component system in Sec. 10.3.1, unless the composition of the vapor in equilibrium with the liquid lies to the right of point "*b*." Vapor with composition to the right of "*b*" (richer in B) yields a two-phase liquid when condensed, so that distillation of single-phase liquid can yield a two-phase distillate, as illustrated by path "*hij*." The compositions and relative volumes of the two phases will be determined by their final temperature, the composition of one phase falling on line '*bx*" and the composition of the second falling on line "*dy*." Because the vapor (*i*) is richer in component B than the initial one-phase liquid the residue becomes progressively richer in component A.

Any mixture with a composition to the right of "*b*" and to the left of "*d*" will consist of two liquid phases and the liquid will boil at temperature *t'* to yield a vapor of composition "*c*." Hence, a mixture of composition 2 will become progressively richer in component A, as the vapor is enriched in component B relative to the liquid, and eventually the liquid moves into the single phase region to the left of "*b*." Vapor of composition *c* condenses at temperature *t*, to yield two liquid phases of composition "*b*" and "*d*" with an overall composition of 3. As a result, distillation of a mixture such as 2 produces a distillate enriched in B and a residue enriched in A.

If '*a*" and "*d*" are very near the 100% A and 100% B axes respectively, so that miscibility is very slight, the separation may be nearly complete. Then the total vapor pressure is approximately the sum of the vapor pressures of the pure components and distillation may be performed at a temperature below the normal boiling point of either component. This principle is applied widely in steam distillation in order to isolate reaction products that are relatively immiscible with water by a co-distillation of the two phase liquid. Often the amount of water remaining in the organic phase of the distillate may be small enough to be removed with a drying agent after the layers of distillate are separated.

10.3.6. Solvent Removal

It is often necessary to remove large amounts of volatile solvents to recover a product.[7] Simple distillation may be used, but often it is more convenient to speed the process by using lower pressures.

A rotary evaporator (Fig. 10.22) is a very convenient apparatus for rapid evaporation of volatile solvents at reduced pressure. The apparatus consists of a motor which permits a round-bottomed flask to be inclined and rotated under vacuum. Evaporation takes place from the film of liquid that forms on the glass, thereby eliminating bumping. The flask may be rotated in a vessel of warm water to prevent the liquid from being chilled by evaporation, and thus to speed the process. The rate of rotation (if the evaporator has a variable rate of rotation), the pressure, and the temperature of the water should all be optimized to provide rapid

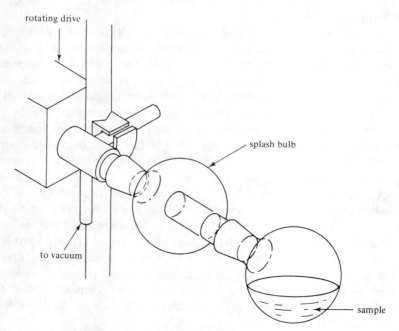

Figure 10.22. Rotary evaporator.

evaporation without violent bumping. High temperature, low pressure, and rapid rate of rotation all increase the rate of evaporation. However, the temperature must not be so high as to vaporize the desired product.

The vapors which are evaporated must be condensed and should be collected in dry ice—acetone traps. Ice or ice—salt traps are inadequate. The traps must be emptied after each use, and each person using the system should regrease the joints of the traps as necessary when finished.

If the desired product is a nonvolatile liquid, it is not always obvious when to stop the evaporation process. Normally, one stops the evaporation when the volume of the residue appears to remain constant or when the solvent ceases to distill. Because such liquid products are generally purified further by a normal distillation technique, it is not a problem if some solvent remains behind. If the desired product is a solid, the solvent may be removed completely.

As shown in Fig. 10.23, there are other ways of removing solvent which do not involve the use of the relatively expensive rotary evaporator. If the sample is stable at $80-100°C$, the solvent simply may be boiled (boiling stones) on a steam bath. The evolved solvent vapors should be entrained with an aspirator tube (see Fig. 10.23a). A thick walled test tube or a filter flask may be used (Fig. 10.23b,c, and d), if it is necessary to use reduced-pressure evaporation. Manual agitation, magnetic stirring, or a stream of air bubbles should be provided during the evaporation to avoid bumping. If the container becomes cold, immersing it in hot water will increase the rate of evaporation.

Note that these procedures are intended to be used only when the desired residue has a much higher boiling point than the solvent. There is then no likelihood of losing product if reasonable conditions are used, and there is no need to monitor temperature. If the desired product is appreciably volatile, a fractional distillation procedure should be used.

References to 10.3

1. A. Rose and E. Rose, in "Technique of Organic Chemistry" A. Weissberger (ed.), Interscience, New York, 1951. Vol. IV, chap. 1.

2. A. Rose and E. Rose, in "Technique of Organic Chemistry" A. Weissberger (ed.), Interscience, New York, 1951. Vol. IV, chap. 4.

3. A. L. Glasebrook and F. E. Williams, in "Technique of Organic Chemistry" A. Weissberger (ed.), Interscience, New York, 1951. Vol. IV, chap. 2.

4. J. R. Bowman and R. S. Tipson, in "Technique of Organic Chemistry" A. Weissbenger (ed.), Interscience, New York, 1951. Vol. IV, chap. 5.

5. E. S. Perry and J. C. Hecker, in "Technique of Organic Chemistry" A. Weissberger (ed.), Interscience, New York, 1951. Vol. IV, chap. 6.

6. C. S. Carlson, in "Technique of Organic Chemistry" A. Weissberger (ed.), Interscience, New York, 1951. Vol. IV, chap. 3.

7. G. Broughton, in "Technique of Organic Chemistry," 2d ed. A. Weissberger (ed.), Interscience, New York, 1966. Vol. III, chap. 6.

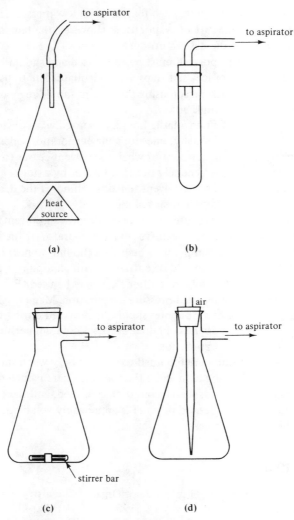

Figure 10.23. Alternate methods for concentration of solutions; (a) vapor entrainment at atmospheric pressure, (b) vacuum evaporation of small volumes, (c) vacuum evaporation of larger volumes, (d) use of an air stream in evaporation.

10.4. CRYSTALLIZATION, SUBLIMATION, AND ZONE REFINING

10.4.1. Crystallization[1]

A compound which is a solid at room temperature can be isolated from a reaction mixture by evaporation of the solution to a dry residue, or by the addition of a nonsolvent to precipitate all of the product. Under these conditions there will be no fractionation or purification of the product. When a crystal forms from solution, from the gaseous state or from the molten state, the ordered nature of the crystal may cause the exclusion of impurities from the crystal lattice, under proper conditions. This will result in the purification of the substance at hand. To permit growth at a rate that is favorable to purification, it is important to choose a solvent, or solvent pair, which allows slow crystallization.

In general, a compound's solubility in a particular solvent, or solvent pair, increases appreciably with temperature. In favorable cases, the desired material can be recovered in good yield by cooling the solution to room temperature or below. In the most satisfactory cases, a contaminated compound is dissolved in a hot solvent, the solution is filtered while hot to remove any insoluble materials, and is then allowed to cool. The desired compound crystallizes leaving most, if not all, of the impurities in solution, along with only a small amount of the desired compound.

Solubility behavior can best be understood in terms of solid-liquid equilibrium diagrams (phase diagrams). Figure 10.24 illustrates the simplest such diagram for a two-component system of A (solvent) and B (solute) in which a eutectic is formed. A mixture of composition 1, which is liquid above temperature t_1, when cooled to t_1, forms solid B. The composition of the remaining solution becomes richer in A as B is removed and the equilibrium solution composition follows curve "zy" as the temperature is lowered until the composition reaches

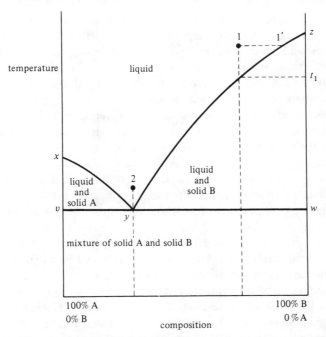

Figure 10.24. A solid-liquid phase diagram for a system of two components A and B, that form a simple eutectic mixture.

165

point "y," that corresponds to the eutectic composition. Area "yzw" represents a two-phase region in which solution is in equilibrium with solid B. At point "y" further cooling causes deposition of solid A and B at the eutectic temperature. Point "y," the eutectic point, is the only point in the diagram where three phases are in equilibrium at a fixed pressure (see Sec. 10.3.1 for an introduction to the phase rule). Below temperature "vw" a solid mixture of A and B exists as two phases. The behavior to the left of composition 2 is analogous, except that A is the solid phase which first appears.

Phase diagrams of this sort describe the phase relationships that are involved in recrystallization. It is clear from this diagram that the maximum amount of B in the liquid phase increases with temperature. The amount of B which can be recovered on cooling increases as the system is cooled to a lower temperature, but is limited by the composition at the eutectic point. Some B is always lost in a recrystallization, the amount corresponding to the solubility at the lowest temperature to which the solution is cooled (that must be above the eutectic temperature).

Curve "yz" gives the solubility of B in A as a function of temperature. A typical solubility curve, however, is a plot of the amount of B that dissolves in a fixed amount of A, while curve "yz" is a plot of the composition of the solution, usually in mole % or wt %. That is, for curve "yz," as the amount of B in solution increases, the amount of A decreases. Simple calculation can convert phase diagram data to solubility data.

If solvent A is evaporated at a constant temperature, the overall composition of the system is changed. For example, solution 1 (Fig. 10.24) will change composition toward 1′, and when the liquid composition reaches 1′, further evaporation will cause deposition of solid B.

Phase diagrams or solubility curves may be determined either by chemical analysis of the composition of the liquid phase in equilibrium with solid at a series of temperatures, or by cooling-curve measurements (thermal analysis). If a solution of overall composition 1 is allowed to cool, the temperature will decrease at a rate determined by the rate at which heat can leave the system (dependent on the thermal conductivity of the surroundings, convection, and the temperature difference between the system and the surroundings). When solid starts to form, the heat of crystallization released reduces the cooling rate, and a break appears in the plot of temperature vs. time. At the eutectic composition, the temperature must remain constant as long as any liquid is present, and a horizontal line is seen in the cooling curve. In actual practice, supercooling is likely, and the break in the curve occurs at a lower temperature than it should, often followed by a slight temperature increase. Typical cooling curves are illustrated in Fig. 10.25.

More complex phase diagrams are common; for example, a compound may be formed. Figure 10.26 illustrates the phase diagram for $LiNO_3-H_2O$, where the compound $LiNO_3 \cdot 3H_2O$ can be formed in a certain temperature range. If $LiNO_3$ is crystallized from water below 28°C, the trihydrate will be obtained, but above

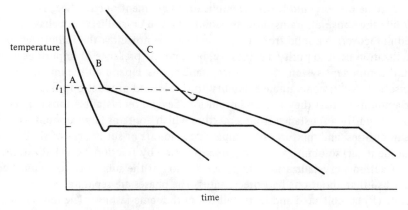

Figure 10.25. Cooling curves.

(A) Cooling curve for a single substance or a eutectic mixture such as "2" in Fig. 10.24. Supercooling is indicated.

(B) Ideal cooling curve for a mixture such as "1" in Fig. 10.24.

(C) As B, but showing supercooling. The dotted line shows how the correct freezing point is estimated.

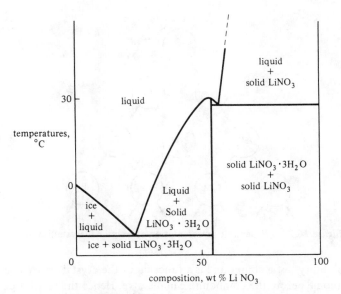

Figure 10.26. Phase diagram for the system $LiNO_3 - H_2O$, in which the compound $LiNO_3 \cdot 3H_2O$ forms. The dashed line continues upward to the 100% $LiNO_3$ axis at the melting point of $LiNO_3$, 261°C.

$28°C$, the anhydrous salt is obtained. In other systems two or more different solvates may be present, and further complications are mentioned in Sec. 10.4.3.

While the above diagrams show the solubility and crystallization behavior involved in recovering a solid from a solution, they do not show the full process of recrystallization used to purify a substance, because the presence of impurities leads to a multi-component system. Simple recrystallizations depend on the impurities being insoluble, or of comparable solubility to the desired product but present in smaller amounts so that they remain in solution. Several recrystallizations may be necessary to purify a substance from a mixture which contains large amounts of soluble impurities, and mixtures of comparable amounts of substances of similar (but not identical) solubilities may require separation by fractional crystallization.

In fractional crystallization, the concentration of the substance to be purified is adjusted so that about one half precipitates. The phases are separated and the precipitate (P_1) is collected and recrystallized in the same fashion. The mother liquor (M_1) is reduced to half its volume, giving another precipitate P_2 and mother liquor M_2. The various subfractions P_x, M_y are treated as indicated in the diagram below (Fig. 10.27). $P_x M_y$ means that P_x and M_y are combined, recrystallized, and separated to give P_{x+1} and M_{y+1}. As the process continues, the least soluble substance(s) will eventually be concentrated at the right-hand extremity of the triangle while the most soluble substance(s) will be concentrated at the left-hand side of the triangle.

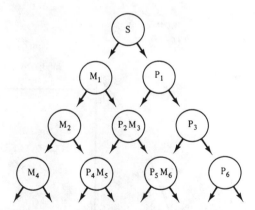

Figure 10.27. Schematic representation of fractional crystallization.

The solubility[2] of a compound is a function of the crystal lattice energy that must be overcome before the compound can dissolve. Hence the relative polarity of the solvent and solute are considerations of prime importance. A useful rule of thumb is "like dissolves like," which indicates that compounds will be most soluble in solvents which are similar in structure. Polar groups (particularly $-OH$, $-NHR$, $-CO_2H$, or $-CONH-$ which can form hydrogen bonds) increase the solubility of a compound in hydroxylic solvents such as water and alcohols, and reduce the solubility in nonpolar solvents such as hydrocarbons. Hydrocarbons are useful solvents for many compounds of low to medium polarity.

The melting point of an organic compound usually is a good indicator of the stability of its crystalline form. The higher melting an organic compound is, the more stable the crystal lattice and the less soluble it will be. This effect is demonstrated by a comparison of the melting points and relative solubilities of the isomeric nitrobenzoic acids (Table 10.2). The symmetrical para isomer can form a more stable lattice than its isomers and is, therefore, higher melting and less soluble. Inorganic salts such as NaCl are high-melting compounds but, because of the very large solvation energies of ions in a variety of polar solvents, (e.g., alcohols, acetonitrile, acetone) these salts may exhibit appreciable solubility. Inorganic compounds often are purified by recrystallization from water, ethanol, or, to a lesser extent, from the more polar organic solvents such as acetone or acetonitrile.

TABLE 10.2

Melting Points of Nitrobenzoic Acids

Isomer	Ortho	Meta	Para
Melting point	147°C	141°C	242°C
Relative solubility*			
Ether	23	28	1
Ethanol	31	37	2

* Taking the absolute solubility of the para isomer in ether as the basis for comparison.

It sometimes is difficult to find a single solvent that is effective when hot but a poor solvent when cold. In this case, the solid is dissolved in the smallest possible amount of the poorest effective solvent(hot). The resulting solution then is diluted with a hot nonsolvent for the material to be crystallized which is miscible with the solvent, until the solution becomes turbid. On standing, crystals should form. If necessary, the solution should be cooled in a refrigerator or a cold bath. Such solvent pairs may be mixtures of alcohols and water, ether and acetone, benzene and aliphatic hydrocarbons.

Preliminary tests to find a suitable solvent should be made with 10–20 mg of compound and 1–5 ml of solvent in a test tube. A solvent should give a clear solution when boiled with the compound to be purified and many crystals should form when the solution is cooled in an ice-water bath, if it is to be useful on a large scale. When insoluble impurities are present the suspension should be filtered hot. If crystals fail to form on cooling, scratch the inside of the test tube with a glass rod, or seed with a few crystals of the crude substance. Some method of purity assessment (breadth of melting range, thin layer chromatography) should be used to determine if the crystallization has removed the impurities effectively.

In some cases, the product may separate initially as an oil. This behavior will tend to concentrate impurities, and even if the oil solidifies, it is unlikely to be very pure. Note, however, that often a highly purified sample also may resist crystallization.

Seeding an oil with a very small amount of a compound structurally similar to the product being purified may induce crystallization, as may vigorously scratching the inside surface of the flask under the oil with a glass rod. Some compounds hold solvent so tenaciously, or absorb atmospheric moisture so rapidly, that they can be obtained only as crystalline solids when the most exacting drying procedures (see Sec. 10.5) are carried out. Gummy materials obtained when partial crystallization has occurred may be freed of adhering materials by rubbing (sometimes called trituration) the compound with the flattened end of a glass rod under a few milliliters of a relatively poor solvent such as pentane. This approach is often helpful with oils, as well. If all else fails, it may be necessary to purify the sample by distillation or chromatography before further attempts at crystallization will prove to be successful.

Recrystallization Procedure

1. Choose a solvent or a mixture of solvents as described above (see Table 10.3). Generally, several solvents will be found to be useful. To narrow the selection, factors such as ease of solvent removal, hazards (toxicity and flammability), and cost, among others, should be considered.

2. Boil an amount of solvent, judged to be an adequate amount to dissolve a portion of the compound to be purified, in an Erlenmeyer flask. Beakers promote evaporation and deposition of a crust of impure solid on the walls. When the solvent is boiling, cautiously add small amounts of the compound to be purified until the solution is saturated, as evidenced by the presence of undissolved solid. Add more hot solvent until the solution is again homogeneous. Continue the alternate addition of solid and solvent until the last of the solid has been added. The solution should be almost saturated. Do not confuse insoluble impurities with undissolved compound. If a residue does not seem to decrease in amount on adding more solvent, it is probably an impurity and should be filtered. Allow enough time for slowly soluble materials to dissolve.

3. If the solution is colored (and the product is expected to be colorless) it may be necessary to decolorize the solution before a pure sample may be obtained. About 10% of the weight of the compound of decolorizing (activated) charcoal should be added carefully *with the solution temperature being kept below the boiling point* of the solvent. Rapid addition of charcoal to a solution at its boiling point can lead to "bumping" and splattering. The charcoal is extremely effective at adsorbing colored compounds because of its large surface area, and its affinity for polar or conjugated molecules. The adsorption of impurities proceeds best in polar solvents, such as water or alcohols.

TABLE 10.3

Common Solvents for Crystallizations

Solvent	b.p. (°C)	Characteristics
Petroleum ether[a] (ligroin), an alkane fraction	90–100	Moderate solvent for nonpolar compounds; poor solvent for very polar compounds
Benzene	80	Moderate to good for nonpolar, moderate for polar compounds; **TOXIC**
Toluene	110	Similar to benzene, but better solvent
Carbon tetra-chloride	76	Moderate for nonpolar, compounds; **TOXIC**
Chloroform	61	Moderate to good for nonpolar, good for polar compounds; **TOXIC**
Acetone[b]	56	Moderate to good for nonpolar and polar compounds
Ethyl acetate	78	Moderate to good for nonpolar and polar compounds
Ethanol[c]	79	Moderate for nonpolar, moderate to good for polar, good for hydrogen-bonding compounds
Acetic acid	118	Moderate for nonpolar, good for polar and hydrogen-bonding compounds
Water	100	Poor for nonpolar, good for hydrogen-bonding compounds

[a] Petroleum ether fractions with different boiling ranges are available.

[b] 2-Butanone (methyl ethyl ketone), b.p. 80°C, is a slightly better solvent.

[c] Other alcohols, especially methyl alcohol (b.p. 64°C) and isopropyl alcohol (b.p. 82°C) have slightly different solubility characteristics and often are useful.

4. Prepare a fluted filter paper (see Sec. 10.1) of appropriate size and place it in a suitable stemless (or short-stem, large-bore) funnel. Heat a small amount of the solvent in another Erlenmeyer flask and place the funnel with filter paper in the mouth of this flask, so that the refluxing solvent condenses on the tip of the funnel, keeping it warm. When the funnel has been adequately warmed (to prevent premature crystallization) remove the flask and funnel from the heat and filter the *hot* solution into the warmed flask, as rapidly as possible. If some product crystallizes in the funnel it should be assayed for purity before combining it with the bulk of the crystals which should be obtained on allowing the filtrate to cool.

5. Allow a reasonable time for crystals to form, and, if necessary, cool the solution in an ice bath. Scratching the walls of the flask with a glass rod may initiate crystallization. In stubborn cases the addition of a tiny crystal of impure product to the cold solution may be required. (Always save a little crude product for this purpose.)

6. Filter the cold solution by suction. To remove all solvent, the filter cake may be pressed with a clean flat object. The cake should be washed with cold solvent either by stopping the suction, adding the solvent, and letting it stand briefly before reapplying suction, or by removing the material, suspending it in a little cold solvent, and refiltering.

7. Once the solid is freed from solvent as outlined in the section on suction filtration, the compound may be spread on a clean porous plate or filter paper to evaporate any traces of solvent. Blotting the solid with dry filter paper will help. Oven drying is possible if *no* flammable solvents have been used, if the oven temperature is at least 30°C below the melting point of the compound, and if the compound is not sensitive to oxidation.

    Moisture-sensitive compounds should be taken at once from the filter and placed in an oven or in a desiccator (see Sec. 10.5.2).

10.4.2. Sublimation

Sublimation[3] is a process whereby a solid, when heated below its triple point (see Fig. 10.28), vaporizes and is condensed directly from the vapor on a cold surface. Some apparatus which may be used for sublimation is shown in Fig. 10.29. Sublimation will proceed faster if the pressure is lowered and the temperature raised. It is also faster if the material is finely divided and the condenser surface close to the solid. Too rapid sublimation should be avoided, because it may be accompanied by spattering of solid particles that carry impurities to the sublimate.

Sublimation often is useful for the purification of solids that are too thermally sensitive for distillation and it is especially effective for the purification of small amounts of material. For sublimation to be effective as a means of purification, the sample should be appreciably more volatile than the contaminants. The sublimation temperature can be lowered by conducting the sublimation at reduced pressures (Fig. 10.29b,c). The sublimed sample can either be scraped from

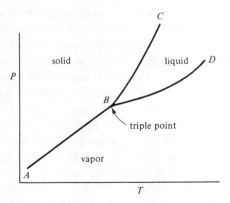

Figure 10.28.   Phase diagram for one-component system (triple-point diagram).

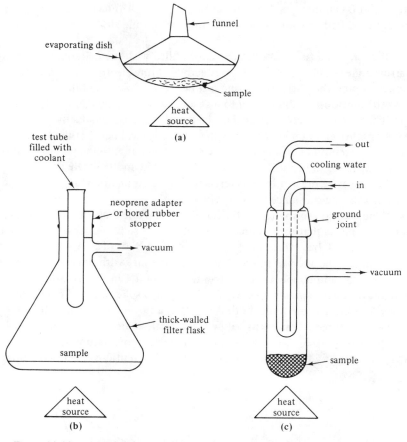

Figure 10.29.   Apparatus for sublimation.
(a) A watch glass covered with a funnel can be used for sublimation at atmospheric pressure. The sublimate will condense inside the funnel.
(b) Simple vacuum sublimation apparatus.
(c) A more sophisticated vacuum sublimation apparatus.

the condenser surface or dissolved by rinsing the condenser with an appropriate solvent.

If the desired compound and the impurity have similar volatilities, it may be necessary to perform a fractional sublimation by interrupting the sublimation periodically to collect the sublimate. Each fraction should be assayed for purity by thin layer chromatography or other appropriate means.

10.4.3. Zone Refining

Zone refining[4] is another purification process based on a multi-component phase diagram. Using Fig. 10.24 for a 2-component system as an example, it is clear that if solid B containing a small amount of impurity A is melted and the liquid is partially solidified, the solid will consist of pure B while all A remains in the liquid phase.

If a long glass tube sealed at one end is filled with the material to be purified, a narrow zone of substance may be melted by heating electrically a band or wire wrapped around the glass. If the band is now moved down the tube at a rate that allows the previously molten band to just solidify as the new band melts, the impurities present are concentrated in the liquid phase and gradually move down the column. After a number of cycles the impurities will be all at the bottom of the column. This method is effective for the preparation of extremely pure compounds, but the substance to be purified must be stable in the molten state and the impurities must be present in relatively small amounts and must not form solid solutions with the substance to be purified.

Solid solution formation is illustrated in the phase diagram of Fig. 10.30. This diagram resembles that for a simple eutectic, but the solid phases which form are not pure A and B, bur rather solid solutions of A in B, or of B in A. The equilibrium composition of the solid solution may vary with temperature. A system of composition and temperature represented by point 1 in Fig. 10.30 consists of a liquid phase of composition 1′ in equilibrium with a solid phase of composition 1″ This solid, although containing both A and B, is a single phase, rather than the two-phase mixture such as exists below the eutectic temperature in Fig. 10.24. If mixture 1 in Fig. 10.30 is cooled below the eutectic temperature, t, the system consists of a two-phase mixture of two solid solutions. Clearly, compounds

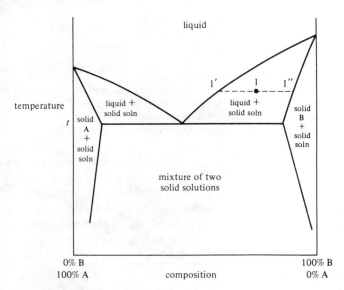

Figure 10.30. Melting point-composition diagram for a system of two components A and B that form solid solutions in one another over a limited composition range.

that form solid solutions with one another are not separated by zone refining, although enrichment is possible. Solid solutions are most commonly formed by compounds of similar crystal structure and chemical nature.

More elaborate solid solution phase diagrams are possible, and are discussed in many physical chemistry texts.

References to 10.4

1. R. S. Tipson, in "Technique of Organic Chemistry," 2d ed. A. Weissberger (ed.), Interscience, New York, 1966. Vol. III, chap. 3.
2. H. Stephen and T. Stephen (eds.), "Solubilities of Inorganic and Organic Compounds," Pergamon Press, New York, 1963. Vol. II, part 1, 2, 3.
3. R. S. Tipson, in "Technique of Organic Chemistry," A. Weissberger (ed.), Interscience, New York, 1951. Vol. IV, chap. 7.
4. W. G. Pfann, "Zone Melting," 2d ed., Wiley, New York, 1966.

10.5. DRYING

10.5.1. Drying Agents

After extractions with (or from) aqueous solutions, organic substances will be contaminated with water that it may be necessary to remove.[1] Solvents often must be freed of water before they are used in a reaction or an analysis. A variety of materials are available for use as drying agents. The key points to consider in the selection of a drying agent are, efficiency, formation of by-products (e.g., $LiAlH_4 \xrightarrow{\text{H}_2\text{O}} H_2$), possibility of chemical reaction with the substances present other than water, regeneration, and the relative cost. Some of these factors will be considered below for specific cases. The surface nature of the drying agent, which depends upon the method of preparation and/or pretreatment, often is a critical factor in determining the efficiency of the desiccant. Both liquids and solids may be used to combine with water by the formation of azeotropes or hydrates, respectively. The formation of azeotropes also is used effectively to force chemical reactions to completion, e.g., in the esterification of acids or in the formation of ketals. Among the many compounds which form azeotropes with water are benzene, toluene, and carbon tetrachloride. Thus, these solvents can be dried by simply distilling, and discarding, the first 10–20% of the solvent and then collecting the next 75% as essentially anhydrous solvent. These solvents also may be used effectively to dry solvents which boil at much higher temperatures than the drying solvent. The drying solvent is added in sufficient quantity to entrain the estimated amount of water present and the azeotrope is distilled, followed by a few percent of an intermediate fraction. After the purity of the distillate is checked (e.g., refractive index), the bulk of the dried solvent is collected.

In addition to the materials described below which are useful for organic materials, barium and magnesium perchlorates are useful for drying inert gases and inorganic materials (see Sec. 5.3). Among the anhydrous solids which are useful as drying agents for organic materials by virtue of the fact that they form hydrates are such desiccants as $MgSO_4$, Na_2SO_4, $CaCl_2$, K_2CO_3, $CuSO_4$, $CaSO_4$, aluminum oxide, silica gel, and molecular sieves. Copper sulfate, magnesium sulfate, and sodium sulfate are commonly used, neutral drying agents. Sodium sulfate is the most used of the three because of its high capacity and low cost (see Table 10.4) and because it may be obtained in granular form that is easily filtered. All three have relatively high capacities and absorb water at a rapid rate, but they do not bind water tightly. Thus, if they are used to dry a solvent, they should be filtered *before* the solvent is distilled or they will release the water they have absorbed when the temperature of the suspension is raised. This behavior is advantageous because the exhausted desiccant can be regenerated by heating after the solvent has evaporated.

Granular calcium chloride binds water tightly and rapidly, and has a relatively high capacity. Its major drawback is that it forms complexes with many nitrogen- or oxygen-containing solvents. It is, therefore, generally useful only for the drying of hydrocarbons or halogenated solvents. Potassium or sodium carbonate (as well as potassium or sodium hydroxide) are useful for drying basic compounds but are relatively less efficient than the others listed. Calcium sulfate is an extremely efficient and rapid drying agent, but it suffers from low capacity (it can only absorb about 7% of its weight of water). It is helpful to dry a solvent (or a solution) first with one of the other reagents, followed by calcium sulfate for final drying. Anhydrous calcium sulfate is a powder and is relatively inexpensive. A granular form, called Drierite®, is available commercially, at a slightly higher cost. "Indicating Drierite"® is Drierite® impregnated with $CoCl_2$ which is blue when dry but pink when wet. Thus, it gives an indication of when the desiccant is saturated with water. It is approximately twice as expensive as Drierite® and should therefore only be used to indicate the progressive exhaustion of other drying agents (e.g., by adding it judiciously to $CaCl_2$ when filling a drying tube or a gas drying tower). Indicating Drierite® should not be added to organic solvents or solutions because they may be contaminated by the cobalt salt.

Aluminum oxide and silica gel, available in plain and indicating forms, are rapid, efficient drying agents, but are relatively expensive. They also have the advantage of being able to absorb (or destroy) other impurities (peroxides by aluminum oxide). Many workers allow highly purified materials to percolate through a short column of aluminum oxide before distilling them, immediately prior to use.

Molecular sieves are crystalline, hydrated metal aluminosilicates with a three dimensional interconnecting network structure of silica and alumina tetrahedra. Various types of molecular sieves are available based upon the different arrangements of the tetrahedra. Common forms are Type A molecular sieves, which can be represented chemically as

$$Na_{12}[(AlO_2)_{12}(SiO_2)_{12}] \cdot xH_2O$$

and Type X molecular sieves, which can be represented as

$$Na_{86}[(AlO_2)_{86}(SiO_2)_{106}] \cdot xH_2O$$

The crystalline molecular sieves consist of interconnecting cavities of uniform size, separated by narrow openings or pores. Initially the structural network is full of water, but with moderate heating the moisture escapes from the cavities without changing the crystalline structure. The cavities remaining have large surface areas and thus are readily available for adsorption of water or other materials. The process of evacuating and refilling the cavities may be repeated indefinitely, under favorable conditions. Molecular sieves can also be used for separation purposes utilizing pore uniformity to differentiate substances on the basis of molecular size and configuration. They may be used directly from the bottle but for maximum drying efficiency they should be activated by heating at $\sim 180°C$ at ~ 1 torr for a few hours and then flushing with dry nitrogen.

Aluminum oxide, silica gel, and molecular sieves can catalyze chemical reactions and care should be used when subjecting reactive materials to their influence. Table 10.4 surveys some common drying agents, their applicability to various systems, their relative cost, their capacities, and some pertinent comments. Additional information on the efficiency of desiccants is given in Blaedel and Meloche.[2]

Drying agents which combine chemically with water include phosphorus pentoxide (P_2O_5), barium and calcium oxide (BaO, CaO), calcium hydride (CaH_2), lithium aluminum hydride ($LiAlH_4$), sodium hydride (NaH), sodium, potassium, and sodium-potassium alloy. The first is a potent dehydrating agent that yields phosphoric acid as the end product of hydration. Thus, easily dehydrated or acid sensitive liquids such as alcohols, or amides, cannot in general be dried by treatment with P_2O_5. A further difficulty that is encountered is the tendency of the powdery oxide to form a dense impermeable crust that shields the lower layers of desiccant from contact with water. To be most effective, the mass of material should be stirred periodically. All the other desiccants listed are strongly basic compounds that react with water (but also with any acidic functional group) to form the corresponding hydroxide (calcium and barium oxides) or hydrogen (all the others). In addition, lithium aluminum hydride, sodium hydride and sodium can also react with carbonyl groups and other activated multiple bonds, to limit further their applicability. Elemental sodium is not a very efficient drying agent because it becomes rapidly coated with a layer of NaOH which impedes drastically the further reaction of water with sodium. If the solvent to be dried boils at a temperature higher than the melting point of sodium (97°C) or if the **very dangerous** liquid Na-K alloy (formed by cautiously pressing equal amounts of sodium and potassium together under an inert solvent such as xylene) is used, new reactive surface is constantly reformed and the removal of water therefore is much more efficient than with the solid metal(s). **Because these reagents liberate large amounts of H_2 in extremely vigorous and exothermic reactions, it is important to remove the bulk of the water first by treating the solvent with some of the drying agents which**

TABLE 10.4

A Survey of Useful Drying Agents

Desiccant	Most useful for	Capacity[a]	Economy[b]	Regenerate	Hazard (if any)
Aluminum Oxide	Hydrocarbons	0.2	10	175°C	–
Magnesium Perchlorate, anhydrous (Anhydrone®)	Inert gas streams	0.24	30	250°C vacuum	May form explosive mixtures on contact with organic materials
Calcium Chloride	Inert organics	0.3	8	250°C	Forms complexes with O- and N-containing compounds
Calcium oxide, barium oxide	Ethers, esters, alcohols, amines	0.3	6	–	–
Calcium sulfate (Drierite®)	Most organic materials	0.07	55	200°C	–
Barium perchlorate, anhydrous (Desicchlora®)	Inert gas streams	0.17	30	140°C	May form explosive mixtures on contact with organic materials
Sodium-lead alloy (Dri-Na®)	Hydrocarbons, ethers	0.08	31	–	Hydrogen is generated
Granular phosphorus pentoxide (Granusic®)	Gas streams	0.5	9	–	Phosphoric acid is formed
Magnesium sulfate, anhydrous	Most organic	0.2	14	–	–
Silica gel	Most organic	0.2	14	250°C	–
Sodium sulfate, anhydrous	Most organic	1.25	1	150°C	–

[a] Grams of water combined per gram of desiccant
[b] Approximate relative cost per 100 grams of water combined.

function by physical combination, or by the formation of crystalline hydrates. Final drying is then accomplished. Caution also should be exercised when destroying unreacted drying agent with methanol (in a hood) so that the hydrogen liberated (as well as any flammable solvent present) will not be ignited. Chlorinated solvents should *never* be exposed to any of the strongly basic reagents, particularly the active hydrides or the alkali metals which can react explosively when so treated.

10.5.2. The Desiccator

Solids can sometimes be effectively dried by spreading the finely divided material in a thin layer on absorbent material (filter paper) and allowing the residual solvent or water to evaporate. However, in general, this is a less than satisfactory procedure because solvent and/or water cannot be removed completely in this way. A desiccator is a glass (or plastic) chamber (Fig. 10.31) in which a controlled, dry atmosphere can be maintained and a sample can be dried effectively. It is used for storage of materials which may absorb water from the air as well as for final drying of products. In analysis, substances that have been dried by heating should be placed in a desiccator to cool before weighing. If the item to be cooled is large with respect to the desiccator or is very hot, such as a freshly

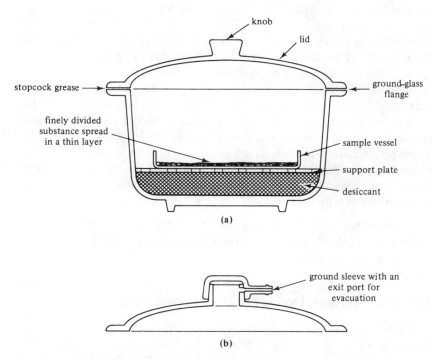

Figure 10.31. (a) Simple desiccator, (b) top for the conversion of (a) to a vacuum desiccator.

heated Pt crucible, pressure may develop because the gas in the desiccator will expand. To prevent the lid from "popping-off," a desiccator with a vacuum outlet can be used, so that if pressure develops, it may be equalized by opening the valve briefly. Alternatively, the cover can be left *slightly* ajar, until the object has cooled briefly.

The usual desiccator is a heavy glass vessel with a lid having a ground-glass lip fitting a ground-glass flange. The selected desiccant (commonly Drierite®, silica gel, magnesium perchlorate, H_2SO_4, or P_2O_5) is placed in the bottom of the desiccator. A porcelain plate or wire screen is put above this to support containers holding the materials to be dried. The ground flange is greased with petroleum jelly or stopcock lubricant (the use of the much more expensive high vacuum grease is wasteful and because this material is stiffer, it tends to bind), the lid is put on and rotated until the grease forms a uniform transparent coating. To remove the lid, slide it to one side; do not attempt to lift it. Note that the desiccator should not be left open longer than necessary, because this will exhaust the desiccant. Residual organic solvent in a sample often may be absorbed by fine shavings of paraffin wax placed in a tray in the desiccator. If the compound is sensitive to acid vapors (or if the substance to be dried contains a volatile acidic contaminant) a tray of sodium hydroxide should be placed in the desiccator.

The drying process can be made much more rapid and effective by using a cover modified for connection to a vacuum source. Vacuum desiccators should be lubricated with a vacuum grade grease. The samples for vacuum desiccation should be placed in shallow, open containers to facilitate the evaporation of water into the open volume of the desiccator where it may be trapped by the drying agents being used. The samples should be freed largely of solvent or water by simple evaporation before putting them in the vacuum desiccator or the sample may splatter when the system is evacuated. The vacuum pump that is used should be protected from contamination by volatile material with a dry ice − acetone trap. The pump should not be *continuously* connected to the desiccator. Usually it is sufficient periodically to re-evacuate the desiccator until the weight of the sample no longer changes. When admitting air to the desiccator, care should be taken to do it slowly and gradually so that the sample will not be blown from its container or be contaminated with desiccant.

Be aware that an evacuated glass vessel may implode violently and is a potential bomb. It should never be carried about when evacuated and should always be kept in a protective shield.

10.5.3. Drying Ovens, Drying Pistols, and Heat Lamps

Drying ovens are heated cabinets of varying size (some types can be evacuated) which are used to free a sample from water or other nonflammable, volatile solvents. In practice the sample must be thermally stable and not subject to oxidative degradation. Flammable solvents should *never* be evaporated in an oven, because of the potential hazard of fire and/or explosion. See also Sec. 7.7.

A small-scale apparatus useful for the rigorous drying of reagents is called the Abderhalden drying pistol (Fig. 10.32). The sample chamber is heated either with refluxing solvent or by electrical means, after it has been evacuated with a vacuum pump and closed with the vacuum stopcock. The vapors evolved during heating will be trapped by the desiccant (usually P_2O_5) in the drying chamber.

Infrared lamps are useful for the rapid drying of thermally and oxidatively stable compounds. They are discussed in Sec. 7.7.

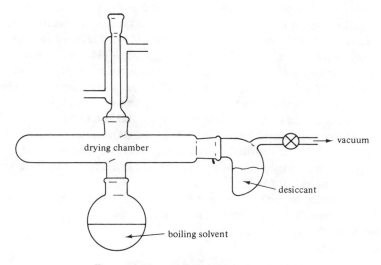

Figure 10.32. Abderhalden drying pistol.

References to 10.5

1. G. Broughton, in "Technique of Organic Chemistry," 2d ed. A. Weissberger (ed.), Interscience, New York 1966, vol. III, chap. 6.
2. W. J. Blaedel and V. W. Meloche; "Elementary Quantitative Analysis," Row, Peterson, White Plains, N.Y., 1957, p. 157.

10.6. CHROMATOGRAPHY

10.6.1. Introduction

Chromatography is a general term which describes a number of techniques for the separation of complex mixtures for both analytical and preparative purposes. As a preparative tool, chromatography is used to isolate and purify reaction products for characterization, and for further use. As an analytical technique, chromatography is used for separation, identification, and quantitative determination of the components of complex samples. The basis of chromatography is the selective distribution (partitioning) of the constituents of a sample between a stationary phase in the form of a sheet or a column and a mobile phase which flows continuously past the stationary phase. In the usual mode of operation, which will be treated here, each component travels as a band, and the velocity of the band relative to the mobile phase is determined by the distribution of that component. The separation process is termed development or elution, and the mobile phase is often referred to as the eluent.

The stationary phase may be either a solid or a liquid immobilized on a solid support, and the mobile phase may be either a liquid or a gas. Two of the four possible combinations of mobile and stationary phase types will be considered here. The first is gas-liquid chromatography (GLC), which is often referred to as gas chromatography (GC) or as vapor phase chromatography (VPC), and the second is liquid-solid chromatography (LSC), which is often referred to simply as liquid chromatography (LC). In GLC, the mobile phase is a gas and the stationary phase is a liquid on a suitable substrate. In LSC, the mobile phase is a liquid, and the stationary phase is a solid or a liquid adsorbed on a solid. In GLC, the distribution of each component is governed by its solubility in the stationary phase and by its vapor pressure. In LSC, the distribution of a component may be controlled by its adsorption on the stationary phase or by its solubility in a liquid phase held by the stationary phase, or by both factors.

The distribution of each component between the stationary phase and the mobile phase is governed by an equilibrium constant known as the partition coefficient, K, which is defined by Eq. (10.10),

$$K = C_s/C_m \qquad (10.10)$$

where C_s represents the concentration of the component in the stationary phase and C_m represents its concentration in the mobile phase. Distribution in chromatography is analogous to the distribution of a solute between two immiscible phases, which is discussed in detail in Sec. 10.2 in connection with extraction of organic products from reaction mixtures. It is shown in standard texts[1-6] that each component is distributed in such a manner that f, the fraction of the time it spends in the mobile phase, is given by Eq. (10.11)

$$f = \frac{1}{\dfrac{KV_s}{V_m} + 1} \qquad (10.11)$$

where V_s and V_m represent respectively the volumes of the stationary phase and the mobile phase. The larger the value of K, the less time a component spends in the mobile phase and the more slowly it moves. Each component is thus retarded with respect to the mobile phase, and this selective retardation is the basis of chromatographic separations. Separation of two components requires that the partition coefficients of the two components differ, and the greater the difference between the partition coefficients, the easier the separation will be.

As the chromatographic band moves ahead in the elution process most of the molecules move forward at approximately the same velocity, but some molecules move forward more rapidly, and some move ahead more slowly. The result is a gradual broadening of the band as it moves ahead. Peak broadening reduces the effectiveness of the separation process, and careful attention to the preparation and use of chromatographic columns and sheets is necessary to obtain efficient separations.

In discussion of the ability of a chromatographic column or sheet to effect a separation, it often is usual to consider the column or sheet as being equivalent to a certain number of stages, in each of which the partition equilibrium is established in an ideal manner. That is, the actual chromatographic process separates the sample as effectively as a hypothetical process made up of a certain number of ideal equilibrium stages. These ideal equilibrium stages are known as theoretical plates, and the separating power of a chromatographic column or sheet is often defined in terms of the equivalent number of theoretical plates. Because the equivalent number of theoretical plates is proportional to the length of the column or sheet, the separating efficiency of columns or sheets of different lengths is commonly defined in terms of the length of column or sheet that is equivalent to a single theoretical plate. Because the theoretical plate concept was first applied to distillation columns, it is customary to refer to the height equivalent to a theoretical plate, and to abbreviate this term as HETP, or more simply H. (See Exp. 4, chap. 22).

Each of the various types of chromatography that has been developed is particularly suitable for certain applications. A brief summary of the major types of chromatography that will be discussed is given below. More detailed discussions are given by Dean,[1] Berg,[2] Bobbit et al.,[3] and Heftmann.[4]

Liquid-solid chromatography can be based on columns or sheets of stationary phase. The nomenclature of the various modifications is not entirely systematic, and sometimes it can be misleading. LSC on alumina (Al_2O_3) or silica (SiO_2) columns, which is often referred to as *column chromatography*, is a powerful and widely used method for separation of organic mixtures both for preparative and analytical purposes. In addition to SiO_2 and Al_2O_3, other adsorbents are often used in column chromatography. A recent modification of traditional column LSC, *high pressure liquid chromatography*, abbreviated as HPLC, has vastly improved the efficiency of this process through the use of very fine particle sizes. *Ion exchange* chromatography is a form of column chromatography based on the selective adsorption of anions or cations on a stationary phase that is usually a synthetic polymer with specific ion exchange sites. Ion exchange chromatography is useful for separation of many organic mixtures as well as inorganic mixtures, and it is applicable for both preparative and analytical purposes. *Paper chromatography* is performed using a sheet of paper, and because the sample size is usually limited to an amount of about 1 mg, it is most useful for analytical purposes. *Thin layer chromatography* is related to column chromatography, but the stationary phase is in the form of a thin sheet on a supporting substrate. Thin layer chromatography is often used to monitor organic reactions or to determine the optimum conditions for column chromatography. The sample size is of the order of 1 mg in the usual TLC experiment, but much larger samples can be accommodated by increasing the thickness of the stationary phase ("thick layer" or "preparative thin layer chromatography").

10.6.2. Adsorption Chromatography on Columns

In general, the adsorption affinity of a compound is determined by its polarity. Nonpolar molecules are adsorbed weakly, and polar molecules are adsorbed strongly. The sample components to be separated compete with the mobile phase for a limited number of adsorption sites on the stationary phase. The extent of adsorption of each component is determined by its polarity relative to that of the mobile phase and the activity or adsorbent strength of the absorbent. The order of elution in adsorption chromatography is thus determined primarily by the differences of polarity among the sample constituents. The polarity of the mobile phase and the activity of the absorbent influence the elution properties of all sample constituents in approximately the same manner. Hence, these factors do not change substantially the order of elution. The usual order of elution of common types of organic compounds is shown in Table 10.5, along with a number of common mobile phases ranked in order of increasing eluent strength, and representative adsorbents arranged in order of increasing activity.

Selection of an appropriate adsorbent and mobile phase is guided by consideration of the polarity of the sample constituents. The susceptibility of the sample constituents to chemical changes which may be induced by the adsorbent must also be considered. Strongly adsorbed materials require a relatively weak adsorbent and a relatively strong eluent, while weakly adsorbed materials require a strong adsorbent and a weak eluent to be moved on the column at a suitable rate. In general, materials which are adsorbed moderately are better separated using a relatively weak adsorbent and a relatively weak eluent than with a strong adsorbent and a strong eluent. Use of a strong adsorbent and a strong eluent usually leads to more band spreading and poorer resolution than use of a weak adsorbent and a weak eluent.

The acid-base properties of the adsorbent should be considered, both as a factor in solute retention, and because they may cause solute decomposition. Thus, for example, strongly basic solutes, and solutes which are decomposed under acidic conditions should not be separated on an acidic adsorbent. Silica is an acidic adsorbent, and alumina is available with acidic, neutral, or basic characteristics. Magnesium silicate (Florisil) is essentially neutral as are cellulose and starch. In general, only materials labelled "chromatographic quality" or otherwise designated as suitable for chromatography should be used as adsorbents.

TABLE 10.5

Summary of Adsorptive Properties of Compound Types, Eluents and Adsorbents

Elution order of common functional groups in adsorption chromatography	Eluent strengths of commonly used mobile phases in adsorption chromatography	Order of adsorbent strengths or common adsorbents
Aliphatic hydrocarbons (eluted first)	Fluorocarbons (weakest eluent)	Cellulose (weakly adsorbing)
Olefins	Petroleum ether	Starch
Ethers	Carbon tetrachloride	Magnesium silicate (Florisil)
Halogenated hydrocarbons	Benzene	Silicic acid $SiO_2 \cdot X\ H_2O$ (silica gel)
Aromatic hydrocarbons	Ethyl ether	Alumina, Al_2O_3 (strongly adsorbing)
Aldehydes and ketones	Chloroform	
Esters	Ethyl acetate	
Alcohols and amines	Pyridine	
Carboxylic acids (eluted last)	Ethanol	
	Methanol	
	Water	
	Acetic acid (strongest eluent)	

The chromatographic activity of alumina and silica adsorbents can be moderated by addition of small quantities of water to the fully activated materials Adsorbent activities are usually defined by the Brockmann scale, which ranges from I for a fully activated adsorbent to V for a nearly deactivated adsorbent. Details of the preparation of adsorbents are given by Dean,[1] by Lederer and Lederer,[5] and by Snyder.[6]

Chromatographic columns must be packed carefully to obtain optimum separations, and it is especially important to avoid entrained air and uneven distribution of the adsorbent. A column suitable for the routine separation of small amounts of organic reaction products can be made in a 50-ml buret (Fig. 10.33). A small plug of glass wool is pushed to the bottom of the buret with a glass rod, and enough dry sand is added to form a layer approximately 5 mm thick. The buret is clamped in a vertical position with a single clamp near the middle. The column is half filled with hexane or petroleum ether, and the adsorbent, in the form of a powder or a slurry in the same solvent, is slowly added to the column through a powder funnel. The stopcock is opened to allow the solvent to drip out at such a rate that the height of the liquid always remains approximately 10–15 cm above the height of the adsorbent. The column is tapped gently and continuously as the adsorbent is added to ensure that the column is packed evenly. A short piece of rubber tubing is useful for this purpose. It is essential that the packing be free of

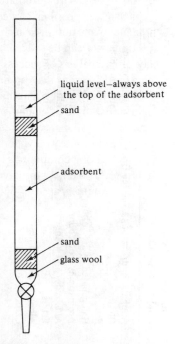

Figure 10.33. Details of an adsorption chromatography column made using a buret. Note the different layers of material.

voids or channels to prevent band spreading. When the column is packed to the desired height, enough sand is added to form a 5-mm layer above the adsorbent. The sand will prevent disturbing the adsorbent when eluent or sample is added during operation of the column.

Because the capacity of a column made from a 50-ml buret is limited, it is often necessary to construct columns with a larger capacity. Glass tubing of the proper diameter fitted with a one-hole stopper is convenient, if a commercial column of the proper size is not available. A length of 7-mm glass tubing connected to a short length of rubber tubing is inserted in the stopper and the flow rate is controlled with an adjustable clamp on the rubber tubing. A rule of thumb useful for judging the proper amount of packing is that approximately 30 g of adsorbent should be used for a 1-g sample. In general, the column dimensions should be such that the ratio of the packing height to the column diameter is approximately 10.

The solvent level must never be allowed to fall below the top of the adsorbent bed. The sample is added by lowering the solvent level to the top of the upper layer of sand, and adding the sample so that it runs down the interior wall of the column without disturbing the adsorbent bed. The sample is applied to the bed as a solution in a nonpolar solvent, and the same solvent is usually selected as the mobile phase to begin the elution. Elution should be continued with this mobile phase as long as the bands move at a reasonable rate; i.e., a rate that will permit the separation to be finished in a period of a few hours or less. With automated equipment, unattended chromatographic separations are possible, and elution times can be as long as the order of a day. With manual operation, it is desirable that elution be complete within several hours. Elution rates can be judged by estimating the progress of a colored band visually or by analysis of successive eluted fractions by an appropriate method. Although a single solvent may be suitable for the entire elution, it is often necessary to increase the eluent strength one or more times during the elution process. The appropriate rate of change of eluent strength is determined by the nature of the compounds being separated. For closely related compounds, changing the solvent composition by a few per cent each 50–100 ml may be adequate, whereas for widely differing compounds a much more rapid increase of eluent power may be necessary to elute strongly adsorbed constituents in a reasonable time. Thin layer chromatography (see below) often is useful in selecting approximate eluent compositions.

10.6.3. Thin Layer Chromatography

The adsorbent in TLC is in the form of a sheet approximately 0.25 mm thick on a glass, plastic or aluminum support. TLC is applicable, in general, to the same types of compounds as column chromatography, and it has the advantages of speed and convenience. The sample is applied to the adsorbent near one end of the plate as a spot and the end of the plate is immersed in a shallow pool of the mobile phase (often called the elution solvent). The sample spot must be above the liquid level. The solvent is pulled upward through the adsorbent by capillary action and each solute moves upward as a spot at its characteristic rate. Separated solutes may be recovered from the plate by scraping the spot from the plate and extracting the solute. Solutes are often detected by fluorescence or by use of a visualizing reagent that yields a characteristic color with the solute. Detailed discussions can be found in Refs. 1–6 and in the texts by Stahl[7] and Kirchner.[8]

The TLC method is used principally for rapid qualitative analysis, as for example, in monitoring the progress of a reaction, and for optimizing conditions for column chromatography. The resolving power is limited because in general the TLC plates are equivalent to no more than 300 theoretical plates. The capacity of conventional 0.25-mm thick plates is restricted to microgram-level samples, but larger samples can be separated by increasing the thickness of the adsorbent. TLC is applicable only to relatively nonvolatile substances, because very small amounts of material are exposed on the open surface of the adsorbent. In this respect, it is complementary to GLC. Quantitative determinations based on measurement of the intensity of spots on the plates are difficult, but good results can be obtained by collecting the spot and determining the quantity of solute recovered by an appropriate instrumental method such as spectrophotometry.

A solute can be characterized by its R_f value, which is defined by Eq. (10.12).

$$R_f = \frac{\text{Distance Traveled by Solute Spot}}{\text{Distance Traveled by Solvent Front}} \qquad (10.12)$$

The distance is measured from the starting point to the center of the most densely colored portion of the spot. It should be recognized that equality of the R_f values of a standard and an unknown compound is a necessary, but not sufficient, condition for identification of the unknown compound. For detection of a given compound in a mixture, the TLC behavior of the unknown and the TLC behavior of the authentic material should be determined by placing a spot of each side by side *on the same TLC plate* and developing the plate in the usual manner. Determination of the TLC behavior of unknowns and reference materials on separate plates is unreliable, because differences in the characteristics of the plates can influence R_f values significantly. The absolute R_f value also is sensitive to experimental conditions, for example whether or not the atmosphere of the chromatography chamber is saturated with eluent vapor.

Plates for TLC can be obtained commercially or prepared in the laboratory. Small TLC plates for routine laboratory work can be prepared readily as needed from microscope slides, using one of the two techniques described below. Larger TLC plates, up to 20 cm in length and width, can be purchased or they can be prepared by using a special coating device.

Dipping Method for Microscope Slides

Plates are dipped in an adsorbent slurry of 60 g of alumina or 35 g of silica in 100 ml of 2 : 1 (v/v) chloroform : methanol contained in a screw-cap, wide-mouth jar. Be certain that TLC grade adsorbent is used, because this contains a binder necessary for an adherent coating. The slurry should be approximately 3 in deep and the level must be maintained by periodic addition of slurry. Solvent must also be added to replace evaporation losses, that are minimized by closing the container tightly when it is not in use.

Shake the slurry to mix it thoroughly before use. Hold two clean and dry microscope slides together by one end, immerse them in the slurry to within $\frac{1}{4}$ in from the upper end, and withdraw them slowly and steadily. Touch a bottom corner of the slides to the lip of the slurry container to remove excess coating, separate the slides, and allow them to dry at room temperature, coated side up. They can be used as soon as they are uniformly white. Streaked or unevenly coated plates are unsuitable and indicate incomplete mixing of the slurry. A thin or grainy coating after thorough mixing of the slurry indicates that the slurry contains excess solvent.

Coating Method for Microscope Slides

Prepare a slurry of 1 : 2 (w/w) TLC grade alumina or silica gel : water by shaking the proper amounts of materials in a small stoppered bottle or Erlenmeyer flask, making certain that the gel is thoroughly wet. Prepare only as much material as can be used at the moment, because the adsorbent contains a plaster binder which will begin to harden within a few minutes. Spread the gel on the slides using a medicine dropper. Do not attempt to use disposable pipettes, which have narrow tips and are often clogged by the slurry. Form a bead of gel around the slide, as shown in Fig. 10.34a, leaving $\frac{1}{4}$-in space at one end for holding the slide. Fill in the rectangular space by running beads of gel across the width of the plate as shown in Fig. 10.34b, and immediately tap the plate gently from below to distribute the gel evenly. Prepare as many slides as possible before the binder hardens and the gel

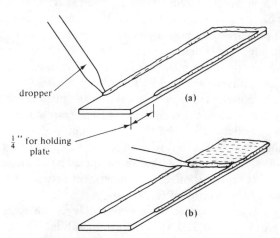

dropper

$\frac{1}{4}''$ for holding plate

(a)

(b)

Figure 10.34. Preparation of TLC plates by coating. Figure 10.22a illustrates the placing of the initial bead of gel around the edges of the plate. Figure 10.22b illustrates the manner in which the remainder of the plate is coated.

becomes too stiff to spread easily. The useable life of the gel can be extended somewhat by adding a small amount of water when the gel first stiffens.

Place the coated slides on a level surface and allow them to stand undisturbed until the initially glossy surface of the gel becomes dull. Place the slides in an oven or on a hot plate at 90–100°C to complete the drying.

TLC Separations

The sample is applied as a 5–10% solution in a volatile solvent such as acetone or methylene chloride. As a rule of thumb, a few crystals of sample are dissolved in a few drops of the solvent. Apply the sample with a very fine capillary prepared by drawing out a melting point capillary in a microburner flame until the diameter is approximately one tenth the original value. The capillary must be open at both ends. Dip the capillary in the sample and touch the capillary to the adsorbent plate approximately 1 cm from the end, being careful not to disturb the adsorbent. The spot should be no larger than 1 mm in diameter, and it should be 5 mm from the edge of the plate. If it is necessary to apply additional sample, let the spot dry and repeat the application at the same point. Two common errors are allowing the spot to become too large and applying too much sample. Slides prepared by dipping usually require a smaller amount of sample than do coated slides. With care, three samples can be placed side by side on a single microscope slide. In preliminary studies, it is useful to place spots made up of one, two or three applications of the same sample on a single slide to determine the proper quantity of sample.

The choice of a developing solvent in TLC is based on the fact that optimum resolution and maximum precision in estimating R_f values usually can be obtained when the R_f values are between 0.3 and 0.7. A brief list of developing solvents is given in Table 10.5, and more extensive lists are given in Refs. 1–8. Although solvent mixtures can be used profitably in TLC to attain an intermediate polarity, the use of several solvents of successively increasing polarity, which is common in column chromatography, is seldom encountered in TLC. Chloroform is a useful solvent for a wide variety of compounds, and if the R_f values are too low, addition of a few percent of methanol will increase the migration rates.

Solvents may be evaluated rapidly in a systematic manner by the following technique. Apply test spots of the sample about 1 cm apart along a plate and let the spots dry. Touch a capillary filled with the solvent to be evaluated to the center of a test spot and allow the solvent to flow until it forms a circle about 1 cm in diameter. Mark the solvent front and observe the spreading of the sample spot either directly or after visualization. Sample movement between $\frac{1}{3}$ and $\frac{2}{3}$ that of the solvent indicates a solvent worthy of further study.

Chromatograms are developed in a covered, wide-mouth container or jar filled to a depth of approximately 0.5 cm with the eluting solvent. Be certain that the solvent level is below the sample spots. Evaporation of the developing solvent from the surface of the TLC plate can be prevented by ensuring that the atmosphere in the developing vessel is saturated with the developing solvent. Filter paper wicks immersed in the developer and extending upwards along the walls of the vessel often are useful for this purpose. Stand the TLC plate in the developing solvent with the spotted end down and replace the cover of the developing chamber. When the solvent front has risen to within approximately 1 cm of the upper end of the plate, remove the plate, mark the solvent front, and allow the plate to dry for a few minutes. Mark any visible spots by pricking an outline in the adsorbent using a fine capillary or a sharp pencil.

Colorless spots can be visualized by using a number of color-forming reagents. One of the most useful reagents is iodine vapor, which forms colored complexes with nearly all compounds except saturated hydrocarbons and mercaptans. Mercaptans reduce iodine and give a white spot on a yellow background, but saturated hydrocarbons are not detected with iodine. To develop the color, the slide is placed for a few minutes in a closed container containing a few iodine crystals.

Samples which fluoresce can be visualized by exposing the plate to ultraviolet light. Commercial TLC plates containing fluorescent zinc sulfide can also be used to advantage. Colorless components are readily located by their ability to quench the fluorescence of the zinc sulfide when the plate is exposed to ultraviolet light.

10.6.4. Paper Chromatography

The weak adsorptive power of cellulose (Table 10.5) renders paper an ideal stationary phase for chromatography of highly polar substances such as polyols and amino acids that cannot be analyzed readily using more active adsorbents. Operationally, paper chromatography is similar in all important respects to TLC except that downward development as well as upward development is possible in paper chromatography. Another difference is that the size of the sheet in paper chromatography is larger than the usual TLC plate. Paper suitable for chromatography is available in rolls, sheets, and strips. For special applications, chemically modified papers are available. For routine work, Whatman No. 1 or 2 paper is suitable. A detailed discussion of grades of paper and useful developing solvents for a large range of compounds is given by Lederer and Lederer.[5]

Selection of a suitable solvent system for paper chromatography is very important, but few generalizations can be made concerning selection of the solvent. Fortunately, a vast amount of information concerning solvents suitable for chromatography of many kinds of compounds has been collected and organized. Useful summaries are given in the general texts by Lederer and Lederer,[5] Berg,[2] and Heftmann,[4] and in the specialized texts on paper chromatography by Block et al.[9] and by Feinberg and Smith.[10]

10.6.5. Gas-Liquid Chromatography

GLC is characterized by extremely high resolving power that permits separation of complex mixtures in short times. The high resolving power is possible because highly selective liquid phases can be used in columns equivalent to many thousands of theoretical plates. Detection limits for single components are as low as 10^{-12} g with sophisticated detectors, and detection limits of 10^{-6} g are readily achieved with simple detectors, with the result that only small samples are necessary. A recorder tracing of the chromatogram is obtained for reference and analysis. Quantitative determinations are performed readily by measuring peak heights or peak areas and using a calibration curve. The major limitation of GLC is the requirement that the sample be sufficiently volatile. In principle, a compound having a vapor pressure as low as a few torr at the operating temperature can be analyzed, but for reasonably rapid elution from a typical GLC column, a compound should possess a vapor pressure of the order of 30–300 torr at the operating temperature. In terms of boiling points, compounds with boiling points of 250°C or less are, in general, readily analyzed by gas chromatography using equipment of the type normally found in undergraduate laboratories.

The major components of a gas chromatograph are illustrated in Fig. 10.35. A high pressure cylinder of gas (usually He or N_2) with a regulator to reduce the pressure to the 10–100 psi range, is the source of carrier gas (mobile phase). The carrier gas passes first through a heated block that is fitted with a septum through which the sample is injected into the flowing gas stream using a microliter syringe. Liquids can be injected directly, and solids are generally introduced as a solution in a volatile solvent. The carrier gas carrying the volatilized sample next enters the chromatographic column that is enclosed in an oven maintained at a precisely controlled temperature. An introductory treatment of GLC is given by Jones,[11] and comprehensive discussions of the theory and practice are given by Dal Nogare and Juvet,[12] Littlewood,[13] and Purnell.[14] Additional information is given by Refs. 1–6, which have sections on GLC.

Two general types of GLC columns are used, packed columns and capillary columns. Packed columns are typically 3–6 mm in diameter and are packed with a

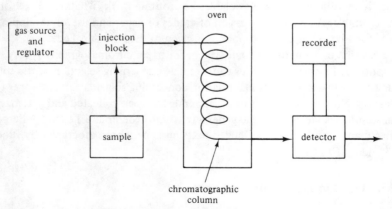

Figure 10.35. Simplified diagram of essential components of a gas chromatograph showing the functional relationships.

porous solid which is coated with the stationary liquid phase. Packed columns range in length from approximately 1 ft for easily separated mixtures to perhaps 20 ft for difficult separations. Capillary columns range in diameter from 0.2 to 0.5 mm, and may be as long as 100 m. The liquid phase is coated directly on the capillary walls as a thin film, which gives HETP values as low as 0.3 mm. The result is that columns equivalent to as many as one million theoretical plates can be used for difficult separations. Because the volume of the stationary phase in capillary columns is low, the sample size must be reduced to the 10-μg range. This reduction is usually achieved by splitting the carrier gas stream just before it enters the column, and allowing only a small portion of the carrier gas (and sample) to enter the column. The low sample capacity of the capillary column requires use of a highly sensitive detector such as the flame ionization detector.

The gas stream emerges from the column and enters the detector, which senses the separated components by monitoring some physical property of the gas mixture. Although many types of detectors have been proposed, nearly all gas chromatography is performed using one of three detectors; the thermal conductivity detector (TCD), the flame ionization detector (FID), and the electron capture detector (ECD).

The sensitivity of a detector depends in general on the identity of the sample component, and to obtain reliable quantitative results it is necessary to prepare a separate calibration plot for each component to be determined. The calibration plot must be obtained under exactly the same operating conditions as are used in the analysis of unknowns. It is wise to check the calibration periodically by interspersing standards among the unknown samples.

Thermal conductivity detectors, which are probably the most common type of detector, use a thermistor or a heated resistance wire to sense changes in the thermal conductivity of the carrier gas induced by sample peaks. Two transducers are used, the first located in the carrier gas stream upstream of the sample injection port to act as a reference, and the second located at the end of the chromatographic column. Each transducer, which is placed in a thermostated metal block and heated above the temperature of the surroundings by an electric current, loses heat at a rate governed by the thermal conductivity of the surrounding gas stream. The sensitivity of the thermal conductivity detector is proportional to the difference in thermal conductivity between the carrier gas and the sample, and to obtain maximum sensitivity, helium is used as the carrier gas because it has very high thermal conductivity. When no sample peak is passing through the detector, both transducers are at the same temperature and exhibit the same resistance, but when a sample peak passes through the detector, the temperature of the detector transducer is higher than that of the reference transducer, and the resistances are no longer equal. The transducers form part of a Wheatstone bridge circuit that converts the difference of resistance to an electrical potential that is recorded.

The flame ionization detector senses organic compounds in the effluent gas by detecting the ions which are formed when the organic compound is burned in a hydrogen flame. Part of the carrier gas emerging from the column is fed into a small hydrogen flame, the ions formed by combustion of the organic compound are collected by electrodes near the flame, and the ion signal is amplified and recorded. The main advantage of the flame ionization detector is its high sensitivity, but it is more complex than the thermal conductivity detector and it does not respond to some compounds such as CO_2, CO, O_2, and H_2O.

The electron capture detector is a selective detector that is very sensitive to compounds with high electron affinities such as highly chlorinated compounds and aromatic compounds with electronegative substituents. A stream of low energy electrons passed through the carrier gas is collected by a charged electrode and the current is amplified and recorded. Electronegative compounds passing through the detector attenuate the electron beam by capturing electrons and thus reduce the current generated by the detector. Because of the high sensitivity and selectivity of this detector for chlorinated compounds and phosphorus esters, it is widely used for detection of traces of pesticides.

The output of the detector is displayed on a recorder as a series of peaks as illustrated in Fig. 10.36. An air peak is obtained if air is present in the syringe along with the sample to be injected. It is sometimes convenient to introduce air deliberately as a reference peak. The detector signal can be attenuated by factors ranging from 1 to 1000 so that measurable peaks can be obtained for both major and minor components of a mixture.

Because the resolving power of GLC is related directly to the number of theoretical plates, determination of the theoretical plate number of a GLC column is often necessary. The number of theoretical plates can be calculated readily from the properties of a single peak using Eq. (10.13) (see Refs. 11–14),

$$r = 16 \left(\frac{t'_R}{t_w} \right)^2 \tag{10.13}$$

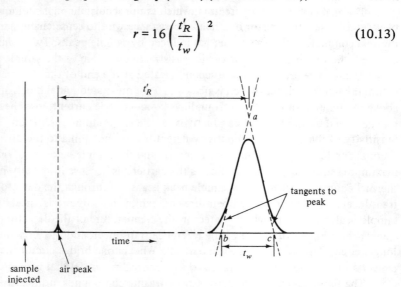

Figure 10.36. Gas chromatogram of a single component mixture showing the adjusted retention time, t'_R, and the peak width, t_w.

where

r is the number of theoretical plates equivalent to the GLC column

t'_R, the adjusted retention time, is the retention time measured from the air peak as shown in Fig. 10.36

t_w is the width of the peak at the baseline as shown in Fig. 10.36

Effective use of GLC requires careful selection of the stationary liquid phase and the operating temperature. The carrier gas flow rate also exerts significant effects on the efficiency of the separation. Detailed discussions concerning selection of optimum operating conditions are presented in Refs. 11–13, and these sources should be consulted for further details.

In practice, most if not all, of the samples that will be encountered in an undergraduate laboratory program can be resolved using one of the liquid phases listed in Table 10.6. It should be noted that many liquid phases are nearly equal in separating power and can be used interchangeably, if the maximum operating temperature of the liquid phase is not exceeded.

The operating temperature has a strong effect on the resolution and the retention times in GLC. Raising the temperature reduces retention times sharply, and thus shortens the analysis, but almost always at the expense of decreased resolution. In general, the operating temperature should be as low as is consistent with reasonably short retention times. For samples with a wide range of boiling points, the temperature can be programmed to increase continuously in order to reduce the time required for analysis without compromising the resolution. Temperature programming is not available, however, on simple instruments.

The flow rate of the carrier gas influences the efficiency of the GLC column, as shown by Fig. 10.37, that illustrates the variation of the height equivalent

TABLE 10.6

Commonly Used GLC Liquid Phases

Common abbreviation	Composition	Polarity	Max. Temp. (°C)
Apiezon M	High mol. wt. hydrocarbons	N	275
Squalane	High mol. wt. hydrocarbon	N	150
DC-200	Methyl silicone fluid	M	225
DIDP	Diisodecyl phthalate	M	175
DEGS	Diethylene glycol succinate	M-S	225
TCEP	1,2,3-tris (2-cyanoethyl) propane	S	180
Carbowax 20 M	Poly (ethylene glycol)	S	250

Polarity
N = Nonpolar
M = Moderate Polarity
S = Strongly Polar

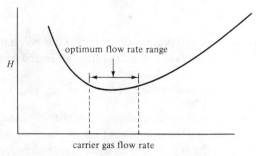

Figure 10.37. Effect of carrier gas flow rate on efficiency of a GLC column.

theoretical plate, H, with the carrier gas flow rate. The carrier gas flow rate is *not* a critical variable, and reasonably high efficiency can be achieved over a relatively broad range of flow rates.

Operation of Gas Chromatographs

The operating instructions for gas chromatographs vary somewhat from one model to another, but a number of general rules should be observed:

1. Never exceed the maximum operating temperature of the liquid phase. Failure to observe this precaution will result in excessive volatilization of the liquid phase from the column. This volatilization will cause the baseline to drift and, once the liquid phase condenses in the detector, it is difficult to remove. Prolonged operation at too high a temperature eventually will destroy the column and foul the detector.
2. When using a thermal conductivity detector, be sure that the carrier gas is flowing before applying current to the detector. Failure to do so will result in damage to the detector. The detector oven and the column oven can be heated safely without carrier gas flowing.
3. Before attempting to analyze a sample, permit the chromatograph to reach thermal equilibrium. This will require from half an hour to perhaps 2 hr, depending on the temperature chosen. Thermal equilibration is indicated by a stable baseline with the detector attenuator on the most sensitive scale.
4. A noisy or unstable baseline which persists after adequate time for thermal equilibration usually means that the injection septum has become porous and should be replaced. Consult the instructor before attempting *any* repairs, such as replacing the septum.
5. Particular care is necessary in injecting samples to avoid damaging the expensive microliter syringe and to obtain optimum peak separation. Push the needle through the septum quickly, but be careful not to strike the metal plate which is behind the injection port in some chromatographs. Forcing the needle against the plate will damage the needle. Inject the sample by pushing the plunger quickly and smoothly, and withdraw the needle. A common error

is failure to push the needle through the septum, with the result that the sample never reaches the chromatographic column. Some practice will be necessary to obtain sharp peaks.

6. Determine the proper attenuation factors necessary to obtain peaks of optimal size for each component of interest. The attenuator may be adjusted during the progress of the chromatogram to accommodate major and minor components. Note that the smallest number on the attenuator scale corresponds to the highest sensitivity. It is important to recognize when the chromatogram is sufficiently detailed for your purposes. For example, if only two components of a complex mixture are of interest, the attenuator should be set so that the peaks corresponding to those components are recorded accurately.

7. A useful method of obtaining an air peak for reference and for avoiding premature volatilization of the sample from the syringe needle is as follows. Draw a portion of sample into the syringe, hold the needle upwards and gently tap the syringe to dislodge the air bubble above the plunger and move it toward the needle. Empty the syringe and again draw in a portion of the sample in excess of that needed. Typically, $1-2$ μl is suitable for a simple mixture, whereas larger sample sizes are needed for complex mixtures or detection of minor components. Eject the excess sample, remove any pendent drops from the needle, and then pull back the plunger nearly to fill the *needle* with air.

8. The extreme sensitivity of GLC makes it *absolutely necessary* to avoid contaminating the sample accidentally. Rinse the syringe several times with the sample prior to taking the sample for analysis, and when work is completed rinse the syringe several times with a volatile solvent such as acetone. Dry the syringe by drawing air into and out of it alternately. Check to be certain that the solvent has been removed from the syringe by drawing a few microliters of air into the syringe and injecting it into the chromatograph.

Quantitative Analysis by GLC

Quantitative analysis by GLC is based on measurement of the height or the area of the peak of interest and comparison with a calibration curve prepared using known amounts of the compound. In general, a separate calibration curve is necessary for each component to be analyzed, and the calibration curve must span the concentration range of the samples. To obtain reliable quantitative results by GLC, it is necessary to control the operating conditions rigorously. In particular, the column temperature, the carrier gas flow rate, and factors that influence the sensitivity of the detector must be controlled very carefully. The calibration curve must be obtained on the same chromatograph that is used for the analysis of the samples, and it is desirable that the calibration be performed just before or just after the analysis.

Measurement of peak height is a simple although relatively imprecise basis for quantitative analysis. The peak height is measured from the baseline, as illustrated

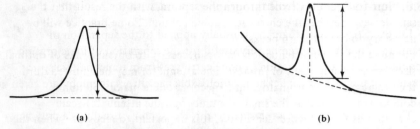

(a)   (b)

Figure 10.38. Baseline location and peak height determination for quantitative analysis by GLC: (a) constant baseline, (b) varying baseline.

in Fig. 10.38. Location of the baseline is straightforward, unless the baseline varies significantly in the neighborhood of the peak, as illustrated in Fig. 10.38b. The accuracy of locating the baseline under these conditions is usually poor, and the precision of the results is degraded accordingly.

Measurement of peak area is more time-consuming than measurement of peak height, but usually it is significantly more precise. The baseline is estimated in the same manner as described for measurement of peak height, and the area under the peak is measured using one of the following methods.

If the peak is relatively symmetrical, and the baseline is level, the peak area is readily measured by triangulation. Tangents are drawn to the peak and the peak area is taken as the area of the resulting triangle *abc*, as illustrated in Fig. 10.36. Triangulation is not readily applied if the peak is unsymmetrical or if the baseline is not linear as shown in Fig. 10.38(b). In such cases, it is preferable to determine the peak area by using a planimeter or by cutting out the peak and weighing the paper. If available, recorders which automatically integrate peaks can be used.

In using either peak height or peak area for quantitative analysis, it is necessary to correct for changes in the attenuator setting. The peak height or peak area is normalized by multiplying it by the attenuation factor. Be sure to record the attenuation factor on the recorder chart beside each peak. With practice in sample injection and attention to careful quantitative technique in preparation of samples and standards, it is possible to obtain accuracy and precision of the order of 2–5% in quantitative analysis by GLC.

Sometimes it is possible to reduce the degree of control required somewhat by the use of an internal standard, a fixed quantity of which is added to each sample. In the internal standard method, the ratio of the height or area of the peak of interest to that of the peak of the internal standard is measured, thus avoiding the need for precise control of sample volume, and relaxing somewhat the need to control the instrumental parameters. The internal standard should appear in the chromatogram near the peak of interest, its level of concentration should be such that both peaks can be recorded at the same attenuator setting, and the standard must not react with any of the other components of the sample. When these conditions are met, the internal standard method is a useful technique for quantitative determination.

10.6.6. Ion-Exchange Chromatography

Ion-exchange chromatography is usually applied to the separation of inorganic ions, but it can often be very useful for the separation of ionic organic compounds, and for the separation of nonelectrolytes from ionic materials. Discussions of the various applications of ion exchange chromatography are given in Refs. 1–6.

The stationary phase in ion exchange chromatography usually is a polymeric resin which contains fixed charged groups and mobile counter ions of opposite charge. Ion exchange resins are designated as anionic if the mobile counter ions are anions and cationic if the mobile counter ions are cations. Cation exchangers are subdivided into strongly acidic types and weakly acidic types, according to the affinity of the exchange site for hydrogen ion. Anion exchangers are subdivided into strongly basic types and weakly basic types according to the affinity of the exchange site for hydroxide ion. A list of representative ion exchange resins is given in Table 10.7. In addition to the characteristics of the ion exchange sites, the degree of cross-linking of the polymeric matrix and the particle size of the resin beads also influence ion exchange processes to some extent. For most applications, the degree

TABLE 10.7

Representative Commercial Ion Exchange Resins

Trade Name	Manufacturer
Strong acid cation exchangers (sulfonic acid)	
Amberlite IR-120	Rohm and Haas Co., Philadelphia, PA
Dowex 50 W	Dow Chemical Co., Midland, MI
Duolite C-25	Chemical Process Co., Redwood City, CA
Lewatit S-100	Farbenfabriken Bayer, Leverkusen, West Germany
Ion Exchanger I	E. Merck AG, Darmstadt, West Germany
Weak acid cation exchangers (carboxylic acid)	
Amberlite IRC-150	Rohm and Haas Co., Philadelphia, PA
Duolite CS-101	Chemical Process Co., Redwood City, CA
Ion exchanger IV	E. Merck AG, Darmstadt, West Germany
Strong base anion exchangers (quaternary ammonium salt)	
Amberlite IRA-410	Rohm and Haas Co., Philadelphia, PA
Dowex 1 and Dowex 2	Dow Chemical Co., Midland, MI
Duolite A-40	Chemical Process Co., Redwood City, CA
Lewatit M 500	Farbenfabriken Bayer, Leverkusen, West Germany
Ion Exchanger III	E. Merck AG, Darmstadt, West Germany
Weak base anion exchangers (amine salt)	
Amberlite IR-4	Rohm and Haas Co., Philadelphia, PA
Dowex 4	Dow Chemical Co., Midland, MI
Ion Exchanger II	E. Merck AG, Darmstadt, West Germany

of cross-linking is not a critical factor, and resins with 4–12% cross-linking can be used. The particle size also influences the ion exchange kinetics. The rate of attainment of the ion exchange equilibrium, and the effectiveness of the separation process, are increased by decreasing the particle size, but at the expense of decreasing the permeability of the ion exchange column. For most chromatographic purposes resins with particle size in the 50–100 mesh range or 100–200 mesh range are suitable.

The experimental techniques associated with ion exchange chromatography are similar to those associated with column chromatography. Conditions for separating a given mixture of ions may be calculated, in principle, from the ion exchange properties of the particular ions in a given sample, as described by Dean.[1] An alternative approach is to use conditions which have already proved successful for the separation in question. The "Handbook of Analytical Chemistry"[15] contains an extensive tabulation of such data, and additional information can be obtained from journals devoted to chromatographic techniques, such as the *Journal of Chromatography*.

10.6.7 High Pressure Liquid Chromatography

By reducing the particle size of the stationary phase to the order of 20 *u*m, the efficiency of column chromatography is increased to a remarkable degree. The use of such small particles decreases the permeability of the column drastically, so that it is often necessary to use pressures of 1000 psi or more to obtain reasonable flow rates. However, several thousand theoretical plates are easily obtained and elution times of the order of minutes are common. Because of the excellent resolution and rapid separations, the term HPLC, which can mean high pressure liquid chromatography, is now commonly interpreted as high performance liquid chromatography. This technique is the solution complement to gas chromatography[16] and is widely applied using stationary phases which function by partition, adsorption, or ion exchange. The stationary phases are specially prepared, in some cases by bonding the active material to an inert support. Examples are reverse phase columns, in which a hydrocarbon chain is bonded to a solid surface. Reverse phase columns reverse the usual polarity relations of the stationary and mobile phases such that non-polar materials are more strongly held than polar ones. Consequently, polar impurities in the usual organic eluting solvents, e.g., water, do not compete. Preparative scale columns also are available.

High pressure liquid chromatographs typically have the following components. The solvent is drawn from a solvent reservoir by a special pump which generates the pressure needed to force the solvent through the column. Some instruments permit the composition of the eluting solvent to be changed continuously during a separation. There is a control to set the desired pressure, and usually a pulse dampener that smooths out the pressure pulses generated by each stroke of the pump. A bubble trap often precedes the pump to ensure that air is not accidentally pumped into the column; such traps may require bleeding from time to time. Some less sophisticated instruments operate under the pressure provided by a cylinder of compressed gas such as nitrogen.

The sample solution is injected onto the column either through a septum or, more conveniently, through a special sampling valve.

The eluent passes through a sensitive detector that is commonly a U.V. detector. Routinely, absorption at wavelengths of either 254 nm or 280 nm is monitored, because these can be obtained at high intensity from a mercury light source, and many organic compounds absorb in these regions. (A substance that does not absorb at the detector wavelength will not be detected, and the eluting solvent must not absorb at this wavelength.) More elaborate detectors provide a wide range of wavelengths, while others are available which measure such properties as refractive index, electrical conductivity, or electrochemical parameters.

Because HPLC columns are easily plugged, it is essential that all eluting solvents and sample solutions be clean and free of solid materials; samples especially must be carefully filtered.

References to 10.6

1. J. A. Dean, "Chemical Separation Methods," Van Nostrand, New York, 1969.

2. E. W. Berg, "Physical and Chemical Methods of Separation," McGraw-Hill, New York, 1963.

3. J. M. Bobbitt, A. E. Schwarting, and R. J. Gutler," Introduction to Chromatography," Reinhold, New York, 1968.

4. E. Heftmann (ed.), "Chromatography," Reinhold, New York, 1967.

5. E. Lederer and M. Lederer, "Chromatography," 2d ed., Elsevier, New York, 1967.

6. L. R. Snyder, "Principles of Adsorption Chromatography," Marcel Dekker, New York, 1968.

7. E. Stahl, "Thin Layer Chromatography," Academic Press, New York, 1964.

8. J. G. Kirchner, "Thin-Layer Chromatography," Interscience, New York, 1967.

9. R. J. Block, E. L. Durrum, and G. Zweig, "A Manual of Paper Chromatography and Electrophoresis," 2d ed., Academic Press, New York, 1955.

10. J. G. Feinberg and I. Smith, "Chromatography and Electrophoresis on Paper," Shandon, London, 1962.

11. R. A. Jones, "An Introduction to Gas-Liquid Chromatography," Academic Press, New York, 1970.

12. S. Dal Nogare and R. S. Juvet, Jr., "Gas-Liquid Chromatography," Interscience, New York, 1962.

13. A. B. Littlewood, "Gas Chromatography, Techniques and Applications," 2d ed., Academic Press, New York, 1970.

14. H. Purnell, "Gas Chromatography," Wiley-Interscience, New York, 1968.

15. L. Meites (ed.), "Handbook of Analytical Chemistry," McGraw-Hill, New York, 1962.

16. N. A. Parris, "Instrumental Liquid Chromatography," Elsevier, New York, 1976.

Basic Synthetic Techniques

11.1. INTRODUCTION

This section deals with the basic operations which must be performed when conducting routine syntheses.[1] The variety of approaches cannot be treated exhaustively, but rather an attempt is made here to present the more common types of equipment and to describe their correct use. This limited outline cannot encompass all situations but it should provide a framework within which experience, common sense, and necessity can be combined to produce a practical solution to any problem that may arise.

11.2. ASSEMBLY AND SUPPORT OF APPARATUS

To be most effective, the glassware and mechanical components found in the laboratory must be properly connected and supported to avoid breakage or loosening of connections during a preparation.

Standard taper joints are rigid and require careful alignment when assembled. Keep the following points in mind:

1. The jaws of metal clamps should have rubber or asbestos (if the clamp will be heated) covers. Clamps and clamp holders may be obtained in a large number of sizes, shapes, and materials of construction too numerous to describe here in any detail. The simplest sources of information on these items are the catalogues of equipment published by chemical equipment supply houses.
2. Use care in tightening clamps, so that the glassware is not strained. Clamp jaws should be tightened firmly but *not* excessively. It is usually best to tighten

clamp jaws before the clamp is fully tightened on the ring stand or support framework, but care must be taken not to introduce stress or to cause misalignment when final tightening is performed. After chemicals are added to an assembled apparatus, the increase in weight may cause some movement and separation of joints. The occurrence of such misalignment or separation should be carefully checked, and corrected.

3. All parts of an apparatus should be supported. An addition funnel may not seem to require a clamp, but it can exert considerable torque on the neck of the reaction flask if the top of the funnel is struck accidentally.

4. Plan the assembly with a view to the manipulations required in the experiment. Can delivery flasks be changed quickly and easily? Is there adequate room for heating or cooling? Can heating equipment be removed or a cooling bath put in place rapidly if the reaction mixture becomes too hot? When heating mantles are used, be sure to have room below the flask for rapid removal of the heat source.

5. Place the apparatus near the necessary facilities (water, gas, steam, electricity, vacuum). Do not try to move an assembled apparatus *in toto*. It is safer to disassemble the apparatus and move the individual pieces, unless it has been assembled on one ring stand.

11.3. CHOICE OF APPARATUS

1. Round-bottomed flasks are best when heating with a burner (use a wire gauze-asbestos pad to provide better heat distribution). When using heating mantles be sure the flask and mantle are of the same size. An over-sized mantle can become overheated if it is not in contact with the flask. The temperature of the mantle is controlled by using a variable transformer as a voltage source. For further information on heating mantles see Sec. 7.7(f). Stirring in round-bottomed flasks can be accomplished conveniently using mechanical or magnetic stirring, even through a soft heating mantle (but not through those encased in metals other than aluminum), or through vessels used as heating baths. Magnetic stirring bars must be short enough to allow the bar to turn without hitting the walls of the flask. Egg-shaped bars should be used if available.

2. **Only round-bottomed or pear-shaped flasks should be used for experiments conducted at reduced pressures. The only exception is a heavy-walled suction flask. Evacuated apparatus can implode with considerable violence and flat-bottomed flasks are not uniformly strong over their surfaces.**

3. **Standard laboratory ware is unsafe for use when any significant internal positive pressure may develop. Unless specifically instructed otherwise, always be sure the apparatus is open to the atmosphere (unless evacuated), through a drying tube, if appropriate. Very modest pressures above 760 torr (~100 torr) may be maintained with a pressure release system such as that described in Chap. 14.**

4. Condensers of a variety of types and uses are shown in Fig. 11.1. Use of these pieces of equipment to best advantage requires a knowledge of the major design points and disadvantages of each type. A condenser is a heat exchanger which conducts the heat of condensation from a vapor passing through it and over its cooling surface. To be most effective, contact with the cooling surface should be as long as possible, the cooling surface should have a large area and the temperature differential between the vapor and coolant should be as large as is practical. The simplest type of condenser is an air condenser (Fig. 11.1a) which is a glass tube that transfers heat to the surroundings through the glass wall. Because of its very low condensing capacity this type of condenser is useful only for liquids boiling above 150°C. The efficiency of condensation can be increased slightly by enlarging the surface area of the condenser and, at the same time, creating some turbulence in the vapor path by introducing bulbs as in Fig. 11.1b. The introduction of a jacket through which water (or other suitable coolant) may be passed (Fig. 11.1c−f, i−k) is a definite improvement over air-cooled types. The West (11.1c), Liebig (11.1d), Allihn (11.1e), and Graham (11.1f) condensers increase in condensation capacity in the order given. However, the Allihn and Graham condensers may only be used in a vertical position because the bulbs will partially fill with liquid and the coil will become blocked with liquid if these condensers are tilted. The Graham condenser is used for condensing volatile distillates *only*. It is not effective as a reflux condenser because the coils do not drain well and when rapid evolution of vapors occurs, liquid may be ejected from the condenser. A further disadvantage of water-cooled condensers is that materials boiling higher than about 150°C should not be condensed with them because this large temperature differential may cause sufficient thermal strain to cause cracking where the water jacket is sealed to the condenser. On humid days moisture may condense on the cooled exterior of the condenser creating the possibility of contaminating the contents of the reaction flask with water when the condenser is removed. The Dimroth condenser (11.1i) and the Friedrichs condenser (11.1k) are designed to overcome most of these difficulties. They may both be used for reflux or distillation, both are more efficient than condensers 11.1a−f and both have a minimum of cooled external surface upon which moisture can condense. Also the joint between the condenser and the water jacket is far removed from the condensing zone, so that these types may be used up to about 160°C. However, very volatile liquids such as ether, over extended periods of reflux, can work their way gradually up the uncooled exterior wall and evaporate even from these very efficient condensers. Another drawback of the Dimroth condenser is the relative fragility of the pendent spiral. A much sturdier and very efficient, but also more expensive, condenser is the so-called intensive condenser (11.1j) that is essentially a combination of the Liebig and Dimroth condensers. This variant is especially useful for condensing very volatile vapors. A water-cooled

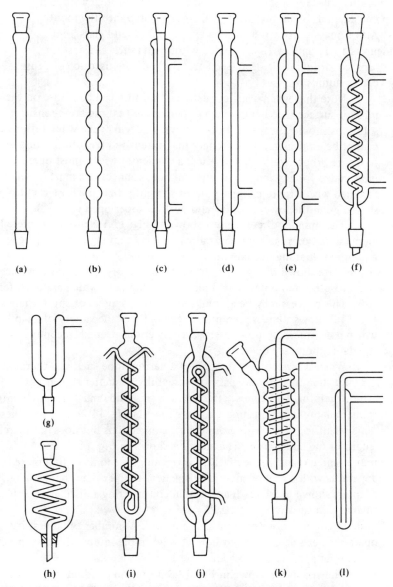

Figure 11.1. (a) Simple air condenser, (b) bulbed air condenser, (c) West condenser, (d) Liebig condenser, (e) Allihn condenser, (f) Graham condenser, (g) simple dry ice condenser, (h) spiral dry ice condenser, (i) Dimroth condenser, (j) intensive condenser, (k) Friedrichs condenser, (l) cold-finger condenser.

condenser of a different type is the cold-finger condenser (11.1l). This piece of apparatus is usually inserted into a vapor stream, through a stopper or Neoprene adapter, or simply suspended from a clamp into the mouth of a flask. Also it may be constructed with an integral ground glass joint. Condensers 11.1g and 11.1h are cooled with agents such as ice, slush baths, dry ice, or liquid nitrogen (see Sec. 13.2) and are used to condense very volatile liquids or gases.

Note that when using condensers, moisture can condense on the interior surface through the upper opening, so that moisture-sensitive solutions should always be protected with a drying tube when refluxed. It cannot be emphasized too strongly that extensive precautions must be taken to insure proper water flow through a condenser which must operate unattended for any length of time. All hose connections must be securely fastened with wire or clamps, but not so tightly that the hose is cut. Water should go in the lower opening and out the upper opening.

The number of condensers shown in Fig. 11.1 is not a complete list of all possible types but represents a reasonable sampling of those pieces of equipment that may be commonly encountered.

5. Adapters are shown in Fig. 11.2. In many laboratory situations it is necessary to fit together temporarily two pieces of equipment which are of different sizes. This process may be accomplished with adapters of appropriate sizes. Fig. 11.2 shows some representative examples of the variety of adapters which are commonly used to effect connections needed to conduct some standard laboratory operations.

When it is necessary to connect a narrow tube such as a thermometer or gas inlet tube to a ground glass joint the tubing adapter shown in Fig. 11.2a is most helpful. To be most effective, the glass tube must fit snugly into the rubber adapter. The bushing adapter shown in Fig. 11.2b accomplishes the same result for tubes with ground glass connections but does not increase the length of the connection. The adapters shown in Fig. 11.2c and 11.2d reduce and expand, respectively, the opening of a joint to fit a piece of apparatus with a ground glass joint smaller or larger than the opening. The adapters shown in 11.2e–h are used in constructing a distillation apparatus such as that shown in Fig. 10.11.

6. Common combinations of apparatus are used. A number of laboratory operations can be carried out in relatively simple setups which exemplify the correct approach. These are described below.

The apparatus shown in Fig. 11.3 is used to conduct the simple operation of boiling or refluxing a reaction mixture. The drying tube need only be used when moisture would interfere with the reaction. In certain cases a selective absorbent (e.g., Ascarite for CO_2) can be used in a drying tube to protect the reaction mixture from contamination.

When it is necessary to conduct a reaction under an inert gas or to introduce a reactive gas into a reaction mixture, the apparatus shown in Fig.

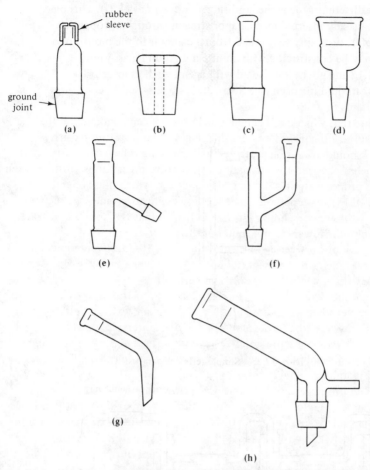

Figure 11.2. Adapters: (a) tubing adapter, (b) bushing adapter, (c) reducing adapter, (d) expanding adapter, (e) distillation adapter, (f) Claisen adapter, (g) take-off adapter, (h) vacuum take-off adapter.

11.4 is used. A bubble counter connected to the condenser outlet serves not only to indicate the flow of gases leaving the vessel but also to prevent atmospheric contaminants (air, oxygen, carbon dioxide, water) from entering the condenser. The bubbler may be filled with a reactive substance (e.g., $Ba(OH)_2$ solution for the detection of CO_2) or with paraffin or silicone oil. The bubbler should be replaced with the trap shown in Fig. 11.5 when large amounts of evolved gases must be dealt with. Traps and bubblers should be preceded and followed by an empty trap to prevent mixing should suck-back or blow-over occur.

When conducting chemical reactions it is often necessary to add reagents to the reaction mixture. One arrangement for achieving this is shown

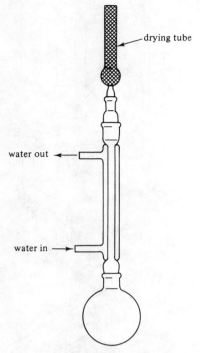

Figure 11.3. Simple reflux apparatus.

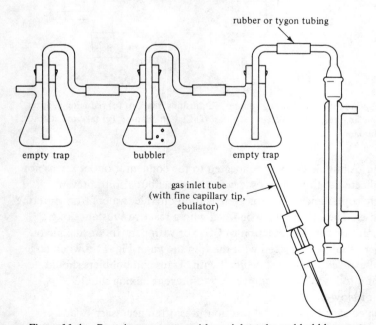

Figure 11.4. Reaction apparatus with gas inlet tube and bubble counter.

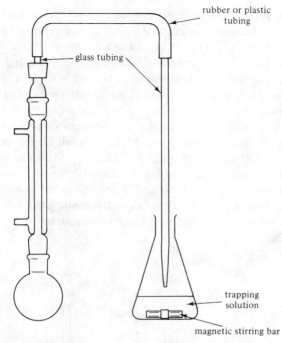

Figure 11.5. Reaction apparatus with trap for evolved gases.

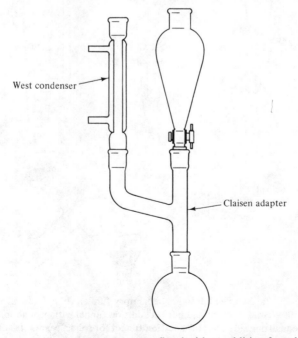

Figure 11.6. Reaction apparatus fitted with an addition funnel.

in Fig. 11.6. An alternate arrangement using a two-necked flask and eliminating the Claisen adapter can be used. Several types of addition funnels are shown in Fig. 11.7; those in 11.7a and b are the simplest. To allow proper flow, they must be open to the atmosphere, although a drying tube may be used to protect the contents from contact with air. The vented funnel shown in Fig. 11.7c is more convenient, because it can be used with the stopper in place.

To add a stable, moisture and air-insensitive solid to a reaction vessel in increments, the funnel shown in Fig. 11.7d is useful. Raising and lowering the rubber stopper allows the controlled addition of the reagent. If the reagent is sensitive or must be added to a closed system it should be placed in the apparatus shown in Fig. 11.7e and connected to the reaction vessel. Rotating the bulb around the joint axis to the position indicated by the dotted line will allow the solid to fall into the reaction vessel.

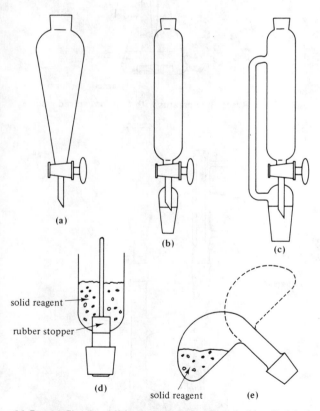

Figure 11.7. (a) Simple addition or separatory funnel, (b) cylindrical addition funnel with ground joint, (c) cylindrical addition funnel with ground joint, and pressure-equalizing side arm, (d) addition funnel for solids, (e) addition funnel for adding solids to a closed system.

Other commonly used sets of apparatus are those used for distillations; they have been described in Chap. 10.

REFERENCE TO CHAPTER 11

1. A. I. Vogel, "Elementary Practical Organic Chemistry," 2d ed., Wiley, New York, 1966.

Physical Properties

12.1. PURE LIQUIDS AND SOLUTIONS

Knowledge of the physical properties of liquids is always necessary when choosing a solvent for a particular reaction or measurement. In some cases it is necessary actually to determine the value of one or more of these properties before the solution data can be evaluated. Physical properties also are used in identifying and characterizing compounds. Some of the important properties of liquids and/or solutions are boiling point, freezing point, vapor pressure, dielectric constant, density, viscosity, surface tension, refractive index, heat of vaporization, and, if the sample is optically active, optical rotation. For pure liquids, the boiling point, density, and refractive index are the most common properties which are determined regularly and easily. The other properties mentioned above are somewhat more specific but nevertheless are important in certain situations. This section describes the common techniques used for measuring these properties. Sources for locating information on the physical properties of solvents are discussed in Chap. 2.

12.1.1. Boiling Point

The boiling point of a liquid is the temperature at which the vapor pressure equals the pressure above the liquid. Normal boiling points refer to a pressure of 1 atm. A volatile liquid product should be characterized by its boiling point, just as a solid is characterized by its melting point. This property also is useful for purposes of identification. The boiling point depends on the purity, because the vapor pressure varies according to Raoult's law ($P_A = X_A P_A^o$, where P_A is the vapor pressure of A in equilibrium with the solution, P_A^o is the vapor pressure of pure A and X_A is the mole fraction of A in the solution). The molecular weight of a nonvolatile substance B can be obtained by observing the elevation of the boiling point which it produces when dissolved in a liquid and using the relationship

$$\Delta T_b = K_b m = \left[\frac{\text{wt of B}}{\text{mol. wt of B}}\right] \bigg/ \left[\frac{\text{wt of solvent}}{1000}\right] K_b \qquad (12.1)$$

where ΔT_b is the boiling point elevation, K_b is the boiling point elevation constant characteristic of the solvent used and m is the molality of the solution.

The Cottrell apparatus shown in Fig. 12.1 is designed for the determination of molecular weights of nonvolatile solutes by the method of boiling point elevation.[1] The solvent is placed in the large outer bulb with a few boiling chips and heated gently. When boiling commences a stream of bubbles containing liquid rises up the pump, or lift, and bathes the thermometer bulb continuously with vapor and boiling liquid. A guard mantle is used to protect the thermometer bulb from cold condensate. This process is repeated with the solution. Temperature changes may be recorded with high sensitivity using a Beckmann thermometer. With this apparatus errors caused by superheating are minimized because a thin layer of solvent covering the thermometer bulb is in equilibrium with the vapor at atmospheric pressure.

If the boiling point of a pure liquid is determined during purification by

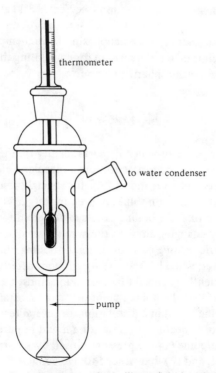

Figure 12.1. Cottrell apparatus for boiling point elevation measurements.

distillation, no further determination is necessary. If the boiling point must be determined separately, a simple distillation set-up is adequate. For smaller samples (~0.5 ml), the liquid and a boiling chip are placed in a test tube into which a thermometer is suspended with the bulb a little above the liquid surface. A cork with a slot (to allow pressure equalization with the atmosphere) should be used to hold the thermometer. The tube is heated gently until a continuous reflux of liquid from the thermometer bulb and the test tube wall is attained. If care is taken that the sample is not overheated, the temperature attained at reflux will be the boiling point. It may be helpful to cool the upper part of the test tube to avoid losing the sample.

For micro-samples (2 or 3 drops), the liquid is placed in a micro test tube (i.e., a 6-mm tube with the bottom sealed) and a fine capillary with one end sealed is placed (open end down) in the sample. The test tube is attached to a thermometer and is partially immersed in an oil or water bath that is heated until a continuous stream of bubbles comes from the capillary. The bath is then cooled and is stirred continuously until bubbling just stops. The temperature at which this occurs is the boiling point; the vapor pressure of the liquid is equal to the pressure in the capillary, that, in turn, is equal to atmospheric pressure. This procedure should be repeated until a reproducible reading is obtained.

Comparison of measured boiling points with literature values requires a thermometer calibration and stem correction (see Sect. 13.1) and a correction for the effect of pressure on boiling point. The pressure variation may be expressed in terms of the Clapeyron equation, but an approximate semi-empirical equation that arises from this approach is often more useful for calculating the effect on the boiling point of changes in atmospheric pressure:[2]

$$T_{\text{normal b.p., K}} = T_{\text{obs. b.p}}\left(1 + n \log \frac{760}{p_{\text{torr}}}\right) \qquad (12.2)$$

where n is a constant the value of which depends slightly on the structure of the liquid. A value of $n = 0.23$ is appropriate for low molecular weight hydrocarbons, of 0.20 for halides, esters, ketones, and of 0.17 for strongly associated liquids such as ethanol. A more detailed list of n values is given in Ref. 2. The above equation is useful for estimating approximate boiling points expected in vacuum distillations. While changes caused by changing ambient pressure are small except at high temperature, a barometric reading or a correction of boiling point should be a part of precise work. However, when laboratory work is conducted at high altitudes boiling points are sufficiently changed that corrections must be made routinely.

It should be noted that the boiling point of a pure substance is not the temperature of the boiling liquid in a distillation for several reasons. The process of forming bubbles of vapor beneath the surface of a liquid requires a higher pressure inside the bubble to overcome (1) the pressure caused by the liquid head, i.e., the depth below the surface, and (2) that needed to overcome surface tension, that is very large for a very small bubble. A second reason for higher pressure is that vapor must be transferred, in a distillation, from the area just above the liquid to a condenser surface; this process requires a pressure difference which may be considerable if the distillation rate is high, or if the path for vapor flow is not free. Finally, the effect of nonvolatile impurities, Eq. (12.1), is eliminated by measuring the temperature of the vapor. Thus, the temperature of the condensing vapor is measured to determine the boiling point of a pure substance. On the other hand, the boiling point of a solution must be determined by measuring the temperature of the boiling liquid, but precautions must be taken to prevent superheating because of the factors mentioned above.

12.1.2. Density

The density of a pure liquid may be used for identification and the density of a solvent or solution often is required in the treatment of data, such as in the evaluation of partial molal volumes. Density is defined as a mass per unit volume (i.e., g/ml). Its measurement is basically simple, but several precautions should be observed if precise results are to be obtained. The basic techniques for density determination are:

pycnometric (determination of the mass required to fill a container of known
    volume).
dilatometric (displacement method; normally used to measure density as a function
    of temperature, e.g., molten salts).
Archimedean buoyancy method (Westphal balance).

There are several designs of pycnometers, the most common being the Ostwald-Sprengel type and the Weld pycnometer, that are illustrated in Fig. 12.2. The Ostwald-Sprengel type, which contains a larger volume, should be used when high precision is required (e.g., five significant figures), while the Weld design is easier and quicker to use, although less precise ($\sim$0.1%).

Procedures for Using a Pycnometer

Initially, make sure that the pycnometer is clean, both inside and outside, and also take care to avoid touching it directly during the measurement because skin oil will lead to significant errors in weighing.

After it is cleaned and rinsed with the liquid being investigated, the pycnometer should be filled with liquid, at a temperature *below* that of the determination. The filling of a Weld pycnometer can be achieved directly by pouring, while the Ostwald-Sprengel type is filled by placing one arm into the liquid and applying suction to the other arm. It is important to avoid getting air bubbles into the capillary section of the Ostwald-Sprengel pycnometer. If bubble formation does occur, the pycnometer should be tilted to release them. The pycnometer is then suspended in a constant temperature bath at the required temperature until thermal equilibrium has been reached (about 15 min). In the case of the Weld pycnometer

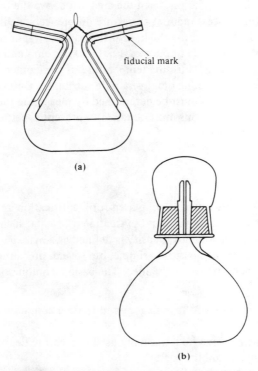

fiducial mark

(a)

(b)

Figure 12.2 **(a)** Ostwald-Sprengel pycnometer, **(b)** Weld pycnometer.

the ground glass plunger is inserted into the mouth of the pycnometer, the excess liquid which is ejected through the capillary is removed with filter paper, and when the neck is completely dry the cap is placed on top of the plunger. In the case of the Ostwald-Sprengel type the two menisci are adjusted to the fiducial marks with the aid of filter paper. The whole pycnometer then is removed from the thermostat and dried completely before weighing. A loop of copper wire is useful for suspending the pycnometer in the bath. If the temperature of the bath is appreciably different from room temperature, the pycnometer then must be brought to room temperature before weighing. If this step is necessary care must be taken not to lose any liquid from the pycnometer.

In either case, the pycnometers must also be weighed when empty and dry and then when filled with a calibrating liquid (e.g., water). It is more efficient to perform the empty weighing first. When using volatile liquids it is essential to use a cap for the pycnometer to prevent evaporation during weighing. For very precise work a buoyancy correction for weights in vacuum should be made.

The density of the unknown liquid (ρ_a) is obtained from the relationship

$$\rho_a = \frac{m_a}{V} = \frac{m_a\rho_b}{m_b} \tag{12.3}$$

where m_a is the mass of the unknown liquid, V is the volume of the pycnometer, m_b the mass of the reference liquid (H_2O) and ρ_b is the density of the reference liquid.

Density is often reported as the specific gravity or relative density. The specific gravity of a substance is the ratio of the density of that substance to that of water at some specificed temperature, usually $4°C$, that is:

$$\text{sp. gr.} = \frac{\rho}{\rho_{H_2O}} = \frac{\text{mass of a given volume of liquid}}{\text{mass of an equal volume of water}} \qquad (12.4)$$

The specific gravity is often denoted by the symbol d_4^t, where the t represents the temperature at which the density of the substance was determined and the 4 indicates the density is being taken relative to that of water at $4°C$. Notice that specific gravity and relative density are dimensionless quantities, while the units of density are normally g/ml.

The buoyancy method of determining the density or the specific gravity of a liquid uses Archimedes' principle which states that the buoyancy effect is directly proportional to the weight of liquid displaced.

In practice a Westphal balance, which consists of a sinker or plummet suspended by a fine platinum wire, as shown schematically in Fig. 12.3, is used. The gravitational force on the riders opposes the buoyant force, and when the beam is at equilibrium in the horizontal position the forces are equal. The sinker is immersed in the liquid and weighed. For direct reading of specific gravities the balance is calibrated by suspending the sinker in pure water with the weights adjusted to a value of unity. Adjustments are made with a counterpoise to restore equilibrium (zero reading). On immersing the dried sinker in the liquid under investigation the weights are readjusted until balance is restored. The position or value of the weights indicates the density of the liquid. Some designs of this balance are capable of a precision of 1 in 10,000.

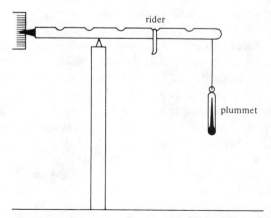

Figure 12.3.   Schematic diagram of a Westphal balance.

12.1.3. Refractive Index

The refractive index of a substance generally is denoted by the symbol n and is defined as the ratio of the velocity of light in vacuum to that in the medium being investigated. The refractive index of a liquid is a property which can be determined simply and precisely, usually to five or more significant figures. Accordingly it is one of the more important physical constants to indicate the purity of a compound. It also is used to determine the composition of binary mixtures, for identifying unknown compounds and also can be used to measure reaction kinetics in appropriate situations. Measurements of refractive index are needed in the evaluation of dipole moments, and the molar refraction,

$$R = \frac{(n^2 - 1)M}{(n^2 + 2)\rho}$$

where M and ρ are the molecular weight and density, respectively, is a useful function in discussing molecular structure.

The refractive index of a liquid is measured with an instrument known as a refractometer. The Abbé refractometer, which is the most common design, consists of two glass prisms with the liquid sandwiched between them as shown in Fig. 12.4. (Immersion refractometers are also available.) When a light ray passes from a medium of one refractive index to one of a lower refractive index, it is refracted. If the angle of incidence exceeds a critical value (θ_{crit}) the ray is reflected totally. This critical angle is related to the difference in refractive indices of the prism and the liquid. Incoming light is scattered in all directions from the rough upper surface of the illuminating prism. Ray 1 in Fig. 12.4 is reflected totally, while Ray 2 has the critical angle of incidence. The total reflection causes the dark region in the field of the telescope. The telescope is rotated to vary the critical angle until the dividing line between the light and dark field is coincident with the cross-hairs of the telescope. The instrument is calibrated so that measurements of θ_{crit} are directly related to the refractive index of the liquid. Some instruments are used directly, while others require the use of conversion tables.

The index of refraction is a function of both wavelength and temperature. In many cases a sodium lamp is used as the light source and the prism housing is surrounded with water circulating from a constant temperature bath. In this case the index of refraction is represented usually as n_D^t, where t is the temperature in degrees centigrade (normally 20 or 25°C) and D represents the sodium D line (actually a doublet at wavelengths 5890–5896 Å). For a wide variety of organic liquids the refractive index decreases by 3.5×10^{-4} to 5.5×10^{-4} per degree increase in temperature.[3] In most cases refractive indices are referred to air rather than to vacuum; the conversion factor is 1.0027. Most Abbé refractometers are built with a compensator consisting of two Amici prisms (composite prism of two different types of glass) which allows the use of white light with this instrument, and produces the same result as that obtained with a sodium lamp. (Most of the recent designs of refractometer do not use a sodium lamp.) The precision attainable with the Abbé refractometer is about ±0.0001.

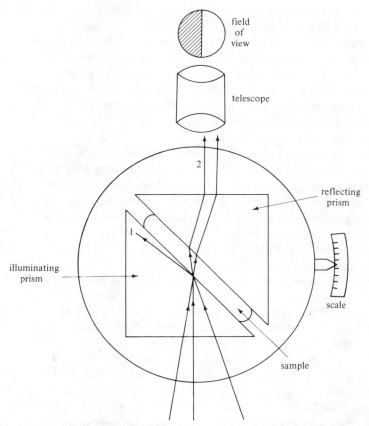

Figure 12.4. A schematic diagram illustrating the principle of a refractometer. The system consists of light rays reflected from a mirror and passing into the illuminating prism, a layer of liquid and a refracting prism. Ray 1 is totally internally reflected while Ray 2 passes through.

General Operating Procedures for Refractometers

As discussed above some designs of refractometer use a sodium lamp, while others use ordinary light. If a sodium lamp is used it should be turned on about $\frac{1}{2}$ hr before the measurements are made. Sodium lamps have a limited lifetime and if the lamp does not emit an intense yellow light it should be checked or replaced by the instructor. The lifetime of the lamp is prolonged if it is turned on and off as little as possible. The lamp should be placed in an appropriate position close to the refractometer. Normally a movable mirror is attached to the refractometer to assist in directing the light onto the prism.

If a circulating bath is used to control temperature it should be turned on at least $\frac{1}{2}$ hr before use. The thermometer in the prism housing should be checked to determine when the instrument has attained the required temperature. Necessary adjustments should be made by regulating the constant temperature bath.

The prisms should be examined and cleaned if necessary, before making any measurements. Depending upon the type of refractometer the prisms may be located at the back or at the front of the instrument. The prisms can be opened by unlocking a cam or releasing a lever and lifting or pulling. Cleaning of the prisms should be carried out with a fresh cotton swab moistened with *small* amounts of reagent grade alcohol. The prisms should be closed as soon as they are clean and dry to prevent dust particles from settling on them. Allow the prisms to dry by evaporation; do not wipe them dry. Avoid touching the face of the prism with anything other than cotton or lens paper.

A few drops of sample should be introduced between the prisms using an eyedropper through the appropriate entrance hole (in some designs the sample is applied directly to the prisms before they are closed). When using very volatile solvents more sample may be needed, because considerable evaporation can take place during the measurement. Avoid scratching the prism surfaces with the eyedropper. It is recommended that a preliminary measurement be made with distilled water ($n_D^{20} = 1.33299$). This measurement is useful to gain experience with the apparatus and to check the calibration of the instrument.

After the sample is introduced, the control should be rotated until the demarcation line between light and dark appears in the field of view of the telescope. To obtain the maximum illumination it may be necessary to adjust the position of the light source or mirror; if an Amici prism compensator is present this should be adjusted to eliminate any color fringes.

When a distinct boundary between light and dark is obtained, the position of the boundary should be brought to coincide with the intersection of the cross hairs in the telescope using the fine adjustment knob. The refractive index or scale reading can then be read through the appropriate eyepiece. (Sometimes one eyepiece serves both purposes and it may be necessary only to move your eye position or push a switch to read the scale.)

Measurements should be repeated several times until reproducible readings are obtained. On completion of the measurement, the prism should be cleaned with a cotton swab lightly moistened with alcohol and then closed. At no time should the whole prism assembly be bathed in acetone or other solvents, because this may dissolve the resin holding the prisms in place.

12.1.4. Viscosity

The viscosity of a fluid is a measure of the resistance to flow when a shearing force is applied. The resisting force is proportional to the velocity gradient, defined below. The basic equation for the coefficient of viscosity (η) of a liquid is given by

$$f = \eta A \frac{dv}{dx} \tag{12.5}$$

where f = the applied force needed to maintain a constant velocity gradient between liquid layers.

A = the area over which the force is exerted.

dv/dx = the velocity gradient between adjacent layers of fluid.

Viscosity is a molecular property and the viscosities of liquids reflect differences in molecular size and shape and of intermolecular and/or intramolecular attractions. Knowledge of absolute or relative viscosities is necessary for the interpretation of a number of physical processes in solution and for theoretical understanding of liquid structure. Viscosity measurements also are important in the study of polymer solutions and in particular an average molecular weight can be determined for a polymer sample from a determination of the intrinsic viscosity $[\eta]$, where $[\eta] = KM^a$; the constants K and a, which are functions of the solvent, solute and temperature, must be determined by independent, absolute methods. Some values of K and a for particular systems may be found in the literature. The variation of viscosity of a liquid with temperature may be described according to an Arrhenius equation:

$$\eta = Be^{E_\eta / RT} \tag{12.6}$$

where B is a constant and E_η is the energy of activation for viscous flow, a quantity which is often used in discussing transport mechanisms in liquids. The value of E_η can be obtained by measuring the viscosity as a function of temperature.

The absolute methods of determining liquid viscosities fall into three categories: (1) the capillary method, in which the liquid flow through a capillary tube is timed; (2) the oscillational technique,[4] in which a cylinder or pendulum is allowed to rotate or oscillate in the medium and the period is measured; and (3) the falling-ball technique,[4] which is used for very viscous liquids, and in which the rate of descent of a ball falling through the liquid is measured. Relative methods (employing any of the techniques above) in which the viscosity is compared to some standard liquid usually are quicker and easier but ultimately must be referred to some precise absolute determination of the viscosity of the reference liquid. Common liquids that are available commercially as calibration standards include n-hexane and methylcyclohexane. For routine measurements water is often used as the calibrant.

Capillary methods are most commonly used for room temperature measurements and are based upon the Poiseuille equation:

$$\eta = \frac{\pi r^4 Pt}{8Vl} \tag{12.7}$$

where P is the pressure head, or hydrostatic pressure, under which flow occurs, r and l are the radius and length of the capillary tube, respectively, and t is the time it takes for the volume V to flow through the capillary.

Absolute measurements require the capillary radius to be measured especially carefully because it appears in the equation raised to the fourth power. Determination of the geometric parameters of the viscometer is avoided when relative measurements are made.

One of the most common types of viscometer found in the laboratory is an Ostwald capillary viscometer, that is based on the Poiseuille equation and is shown in Fig. 12.5. It can be shown that the ratio of the viscosities of two liquids is given by the expression:

$$\frac{\eta_1}{\eta_2} = \frac{t_1 \rho_1}{t_2 \rho_2} \tag{12.8}$$

where t_1 and t_2 are the efflux times (i.e., the time for the liquid level to fall from point a to b in Fig. 12.5) for liquids 1 and 2; and ρ_1 and ρ_2 are the corresponding densities. The Ubbelhode design (a modified capillary viscometer) is useful for experiments involving several dilutions.

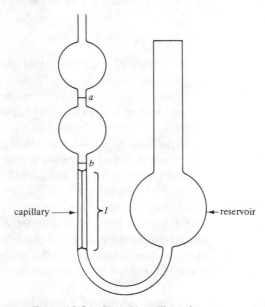

Figure 12.5. Ostwald capillary viscometer.

Use of a Capillary Viscometer

A common problem in using a capillary viscometer is that unless the viscometer is clean it is difficult to obtain reproducible readings. Grease or solid particles in the capillary tube are obvious sources of error. Hot chromic acid (*Caution!*) should be used if the viscometer obviously is contaminated, but care must be taken not to plug the capillary with solid material. The precautions for using this cleaning agent (Sect. 9.1) should be observed. For most experiments described in this text, other cleaning solutions such as detergents or nitric acid can be used, followed with several rinses of distilled water. A water aspirator can be used to draw distilled water through the capillary. For measurements on organic liquids, acetone is suitable for cleaning. Before a new sample is placed in the viscometer, the viscometer should be rinsed several times with small portions of the test sample.

Some viscosity measurements can be made at room temperature without any temperature regulation, while others will require more accurate temperature control. The temperature coefficient of viscosity for most liquids is about 1% per degree. It is, therefore, important to consider the precision needed in the measurement. If a constant temperature bath is used it is important to make sure that the viscometer is sufficiently immersed in the water so that the upper mark "a" (see Fig. 12.5) is below the water level. Care should be exercised in clamping the viscometer in the bath to avoid breakage.

Capillary viscometers are manufactured in a variety of designs and sizes. Each viscometer has a number engraved on it which is related to the diameter of the capillary tube; the larger the number the larger is the diameter of the capillary. Further information indicating the viscosity range appropriate to a given number can be found by consulting a supplier's catalog. It is desirable to select a viscometer for which the efflux time is at least 100 sec to minimize errors involved in timing, but to avoid an excessively long time interval.

A suitable quantity of the test liquid should be transferred to the reservoir bulb of the viscometer with a pipet, making sure that there will still be some liquid in the reservoir after the liquid has been drawn above point a in the capillary arm of the viscometer (Fig. 12.5). Once this volume has been determined it is important to use approximately the same volume for all subsequent measurements to minimize errors caused by the varying pressure head of the liquid in the reservoir. Dilution viscometers are so designed that volume changes do not introduce significant errors.

After the viscometer is mounted vertically, the liquid level should be raised by suction, using a rubber bulb or water aspirator. A dust-free rubber tube attached to the small tube of the viscometer is useful for this procedure. The time for the liquid to fall between points a and b on the viscometer then is determined using an accurate stopclock. Several readings should be recorded with each solvent or concentration until the desired precision (generally within 0.1 sec) is obtained.

After use, the viscometer should be cleaned thoroughly and dried. Filtered air should be blown through the reservoir to avoid drawing dust into the capillary.

Rotational, oscillational, and falling ball viscometers are used for certain liquids, e.g., oils and molten salts. For a detailed discussion of these methods the reader is referred elsewhere.[4]

12.1.5. Surface Tension

The surface tension (γ) of a liquid is defined as the force acting parallel to a unit length of the surface. To increase the surface area of a liquid at constant temperature and pressure, work must be done against the intermolecular forces in the bulk of the liquid to bring more molecules to the surface. Surface tension can be related to surface free energy according to the relation:

$$\gamma = \left(\frac{\partial G}{\partial A}\right)_{T,P} = G^s \qquad (12.9)$$

where G^s is the surface free energy per unit area, ∂A is the change in surface area, and γ is the surface tension in dyn/cm. A tension may also exist at the interface between two immiscible liquids and is called the interfacial tension.

Several methods are available for determining the surface tension of a liquid:

the capillary-rise method (in which the height of the liquid rise in a capillary tube is measured)

detachment methods (in which the force required to pull a ring or plate from the surface is determined)

the maximum-bubble-pressure method (determination of the maximum gas pressure obtained in forming a gas bubble at the end of a tube immersed in the liquid)

the drop-weight method (determination of the weight of a drop which falls from a tube of known radius)

The most commonly used methods in the laboratory for liquids at ambient temperature are the capillary-rise method and the ring-detachment method (Du Nouy method).

(a) Capillary-rise Method

This method is based upon the fact that when a capillary is immersed in a liquid that wets the walls of the tube,* the liquid will rise in the tube until the upward force caused by surface tension is balanced by the downward hydrostatic force. At equilibrium this condition can be represented as

$$\gamma = \frac{h\rho g r}{2\cos\theta} \qquad (12.10)$$

where h is the height of the liquid in the capillary tube above the level in the container, ρ is the density of the liquid, g is the acceleration due to gravity, r is the radius of the capillary tube and θ is the angle of contact between the meniscus and the capillary wall that for many liquids is approximately zero.

A typical experimental arrangement for this measurement is shown in Fig. 12.6. The radius of the capillary tube should be determined by weighing the mercury needed to fill a known length of the tube. This determination should be repeated for several parts of the tube to check the uniformity of the bore.

---

* For liquids such as mercury that do not wet the walls of the tube, the liquid is depressed in the capillary tube relative to the liquid level in the container.

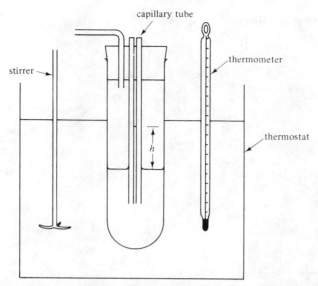

Figure 12.6. Apparatus required for the capillary-rise method of surface tension measurement.

Alternatively the tube can be calibrated with a liquid of known surface tension and density.

It is very important to use a clean capillary tube because surface forces are being investigated. The tube should be cleaned with hot nitric acid (*Caution!*) followed with several rinses with distilled water. The capillary is placed vertically in the tube containing the test liquid and then both are immersed in a constant temperature bath. When thermal equilibrium has been reached, the height of the liquid in the capillary tube above the level of the liquid in the outer vessel, is measured with a cathetometer.

(b) The Ring-Detachment or Du Nouy Method

The principle governing this method is based upon a determination of the maximum force required to detach a ring from the surface of a liquid. The relevant equation is

$$F = l\gamma \tag{12.11}$$

where, F is the maximum force, l is the circumference of the ring and γ is the surface tension of the liquid. However, because the meniscus interacts with both the outside and the inside of the ring the operating equation is

$$F = 4\pi R\gamma \tag{12.12}$$

where R is the radius of the ring.

The Du Nouy tensiometer employs a platinum-iridium ring and can be used to measure both surface tension and interfacial tension between liquids. A

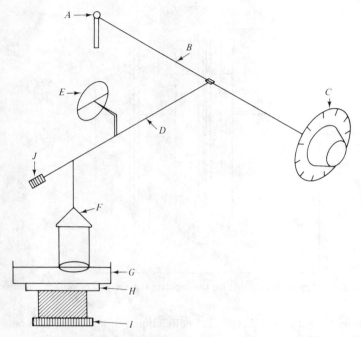

Figure 12.7.   Diagram of a Du Nouy tensiometer. *A*, torsion clamp; *B*, torsion
wire; *C*, knob for applying torsion with graduated scale and vernier; *D*, movable
lever arm; *E*, mirror with an index-line pointer on a movable arm;
F, platinum-iridium ring; *G*, large dish; *H*, movable table; *I*, adjustment screw for
raising or lowering the table; *J*, adjustable nuts for altering the length of the arm
during calibration.

schematic diagram of the Du Nouy tensiometer is given in Fig. 12.7. A fine torsion
wire *B* held horizontally between the torsion head and a fixed clamp is used to
withdraw the ring *F* from the surface with a lever arm *D*. A vernier scale is
connected to the torsion head, thus permitting direct readings of surface tension
after the instrument is calibrated.

It is essential to use a clean ring and clean glassware to obtain the highest
precision. The ring can be cleaned with an oxidizing flame but special care must be
taken because the ring is fragile. It is also important to avoid making kinks in the
ring, otherwise the measurements will not be valid.

The tensiometer can be calibrated by using one or more small weights. First
the lever arm should be allowed to swing freely from its initial clamped position
with the ring attached. The knurled knob should be turned until the index pointer,
its image and the reference line in the mirror are colinear. Then the dial should be
adjusted until the vernier reads zero using the fine adjustment. A mass between 500
and 800 mg then should be attached to the cross-bar of the ring and the knurled

knob readjusted until the index coincides with the reference line in the mirror. The true dial reading can be calculated from the relation

$$P = \frac{Mg}{2\ell} \qquad (12.13)$$

where P = the true dial reading, M = the mass, g = the acceleration caused by gravity, and ℓ = mean circumference of the ring (supplied with the ring). If the observed dial reading is greater than the calculated value, the nut in the lever arm should be adjusted to shorten the arm. If the reading is less than the calculated value the arm should be lengthened. The calibration procedure described above should be repeated until the dial reading agrees with the calculated value. Be sure that the zero adjustment is repeated each time.

For measurements of surface tension the ring should be immersed below the surface of the liquid which is contained in a suitable glass vessel (small beaker or Petri dish) placed on the movable platform. Initially, it is most convenient to set the platform in its highest position. Then the beam should be leveled, as described for the calibration procedure, with the ring positioned as accurately as possible in the surface of the liquid when the index is on zero.

The torsion in the wire then should be slowly increased and simultaneously the platform should be lowered to keep the index on zero throughout the measurement. These adjustments should be continued until the film of liquid breaks away from the ring, when the dial position should be noted. This particular reading will be the apparent surface tension. Each measurement should be repeated until the results are reproducible. To obtain the true surface tension the reading should be multiplied by a correction factor. This is necessary because the equations used above are based on the assumption that the liquid drawn from the surface by the ring forms a perfect cylinder. This assumption, however, is not strictly the case because at the breaking point, the surface becomes distorted. The actual shape is a function of R^3/V and R/r where V is the volume of liquid held up and r is the radius of the wire of the ring. The correction factor can be found in tables or calculated from the appropriate equation.[5] The correction term is a function of the circumference of the ring, the size of the wire in the ring, the total downward pull on the ring, and the density of the liquid. In some cases the correction factor can be as large as 0.75.

It is suggested that a trial measurement be made with distilled water, for which the value of the surface tension is well known.

12.1.6. Dielectric Constant

(a) Theory

When a substance is placed between the charged plates of a condenser, the molecules, because of permanent or induced dipoles, align themselves with the electric field. The total polarization consists of two parts. The first of these (P_μ) is caused by the orientation of any permanent dipoles that may exist in the molecules, while the second (P_d) is caused by induced (or distortion) dipoles produced because the applied field distorts the nuclear and electronic charge distributions in the molecules. The total polarization can be determined by measuring the dielectric constant (ϵ) of the substance, which is defined as the ratio of the capacitance (C) observed when the substance is between the plates of the condenser to that observed when there is a vacuum between them, (C_0); that is,

$$\epsilon = C/C_0 \qquad (12.14)$$

The total molar polarization[6] of the compound is then given by:

$$P = P_\mu + P_d = (4\pi N/3)(\alpha + \mu^2/3kT) = (M/\rho)(\epsilon - 1)/(\epsilon + 2) \qquad (12.15)$$

where N is Avogadro's number, k is Boltzmann's constant, T is the absolute temperature, M is the molecular weight, ρ is the density, α is the polarizability, and μ is the permanent dipole moment of the compound. From Eq. (12.15) it is apparent that both α and μ may be obtained from the temperature dependence observed for the dielectric constant. In practice, however, this method generally is useful only for gases, because of the presence in liquids of intermolecular interactions and the small temperature range that is accessible between the freezing and boiling points.

Although the above discussion implied a static (dc) field, the equations are applicable to an alternating (ac) field changing with sufficiently low frequency to enable the molecules with permanent dipoles to orient themselves in response to it. Alternating electric fields are associated with electromagnetic radiation, and at very low frequencies (radio frequencies and below) both the induced and permanent polarizations contribute to P. However, at higher frequencies, e.g., visible light, the permanent dipoles are unable to follow the changing field and only the induced polarization is present. In this case the dielectric constant is equal to the square of the index of refraction (n), i.e.,

$$\epsilon = n^2 \qquad (12.16)$$

Because the refractive index provides a measure of the induced polarization (e.g., 12.15), it is possible to obtain the dipole moment from a measurement of the dielectric constant at a single temperature plus a measurement of the index of refraction at that temperature.

The derivation of Eq. (12.15) was based on the assumption that the molecules were oriented randomly in the absence of an externally applied field. This condition is met strictly in the gas phase only. However, satisfactory results may be obtained for dilute solutions in which the polar molecules are widely separated from one another by nonpolar solvent molecules such as benzene, or heptane. In the absence of intermolecular interactions molar polarizations are additive, hence the expression for solutions is

$$P = X_1 P_1 + X_2 P_2 = \frac{(\epsilon - 1)}{(\epsilon + 2)} \frac{(M_1 X_1 + M_2 X_2)}{\rho} \tag{12.17}$$

where X represents the mole fraction. The subscripts follow the commonly used convention of denoting the solvent as 1 and the solute as 2, and the unsubscripted quantities refer to the solution. Because the solvent is nonpolar and therefore only has distortion polarization that is not greatly affected by intermolecular interactions, P_1 can be assumed to have the same value in solution as in the pure solvent,

$$P_1 = \frac{(\epsilon_1 - 1)(M_1)}{(\epsilon_1 + 2)(\rho_1)} \tag{12.18}$$

Thus, P_2 can be calculated by substitution of Eq. (12.18) into Eq. (12.17). Unfortunately, P_2 calculated in this way is found usually to vary with X_2, because of intermolecular interactions between the solute molecules, and so the value at infinite dilution, P_2^o, is used in the calculations. The usual method of obtaining P_2^o is that given by Hedestrand[7] in which a linear dependence of ϵ and ρ on X_2 is assumed,

$$\epsilon = \epsilon_1 + aX_2 \qquad \rho = \rho_1 + bX_2 \tag{12.19}$$

The final expression for P_2^o is obtained by substituting Eq. (12.19) and Eq. (12.18) into Eq. (12.17) solving for P_2, and taking the limit as $X_2 \to 0$. The result is

$$P_2^o = \frac{3M_1 a}{(\epsilon_1 + 2)^2 \rho_1} + \frac{\epsilon_1 - 1}{(\epsilon_1 + 2)\rho_1} \left(\frac{M_2 \rho_1 - M_1 b}{\rho_1} \right) \tag{12.20}$$

To obtain the dipole moment, the contribution from the induced dipoles must be subtracted. As discussed previously, this contribution is found from a measurement of the index of refraction, that is,

$$P_{2_d}^o = \left(\frac{M_2}{\rho_2} \right) (n_2^2 - 1)/(n_2^2 + 2) \tag{12.21}$$

The quantity $P_{2_d}^o$ is also known as the molar refraction, and the equation is often referred to as the Lorenz-Lorentz formula. The dipole moment is then obtained from Eq. (12.15) as

$$\mu = [9kT(P_2^o - P_{2_d}^o)/4\pi N]^{1/2} = 0.0128_1 T^{1/2}(P_2^o - P_{2_d}^o)^{1/2} \times 10^{-18} \text{ esu-cm} \tag{12.22}$$

The quantity 10^{-18} esu-cm is defined as one Debye.

An alternate method due to Guggenheim[8] does not require knowing the densities of the solutions, but rather the indices of refraction. In this method the quantity Δ, defined as

$$\Delta = (\epsilon - n^2) - (\epsilon_1 - n_1^2) \tag{12.23}$$

is plotted against the concentration in moles/milliliter, and the value of the slope at $c = 0$ is used to find the dipole moment by substituting into Eq. (12.24)

$$\mu = \left[\frac{27kT}{4\pi N(\epsilon_1 + 2)(n_1^2 + 2)} \left(\frac{\Delta}{c} \right)_0 \right]^{1/2} \tag{12.24}$$

where $(\Delta/c)_0$ is the slope at $c = 0$.

It should be noted that the above plot will not necessarily be linear, but that the slope at $c = 0$ is obtained easily. Also note that because $\epsilon_1 - n_1^2$ is a constant, one could plot $\epsilon - n^2$ vs. c to obtain the same slope. (A plot which should be linear[14] for small c is the quantity $\epsilon - n^2/(\epsilon + 2)(n^2 + 2)$ vs. c; the slope of this plot is $4\pi N\mu^2/27kT$.)

(b) Dipole Moment and Molecular Structure

Dipole moments may be treated as vectors. The dipole moment of a molecule is generally regarded as the resultant of the individual vectors associated with each of the bonds in the molecule. These individual dipole moments often are referred to as bond moments. For many applications, it is convenient to assign dipole moments to a functional group as a whole. Logically, these are known as group moments. Bond or group moments then may be used to gain information about molecular structure by calculating the expected dipole moment for the various structures under consideration. The observed dipole moment will be the root mean square dipole moment of the various species, weighted by the mole fraction of that species; that is,

$$\mu = \left(\sum_{i=1}^{n} f_i \mu_i^2 \right)^{1/2} \tag{12.25}$$

where f_i and μ_i are the fraction and dipole moment, respectively, of the i-th species in solution. If, as is often the case, only two conformations are being considered, then the fraction of each present in a sample may be determined from Eq. (12.25), because then $f_1 + f_2 = 1$. Furthermore, if the two species are in equilibrium with each other, a temperature study of the dipole moment will yield ΔH for the change in conformation in the reaction. A determination of the temperature dependence (or lack thereof) of the dipole moment also yields information as to whether the sample contains a mixture of the two conformers, or whether the molecules are being converted quickly from one conformer to another, or whether the molecules merely exist in some form intermediate between those being considered.[9] In these comparisons, however, the limitations of the method must be taken into consideration. For example, often the bond moments used for the calculation are obtained from measurements in the gas phase (or another solvent), while the experiment is done in solution, and the contribution of the C–H bonds to the dipole moment is usually ignored. This problem could, in principle, be remedied easily, because it is agreed generally that the magnitude of the C–H dipole moment is 0.4D. Unfortunately there is little agreement as to its direction. The problem of using different solvents can be minimized by use of an empirical formula such as the one given by Müller[10] that relates the gas phase dipole moment with that obtained in solution:

$$\mu(\text{soln})/\mu(\text{gas}) = 1 - 0.038 \, (\epsilon_1 - 1)^2 \qquad (12.26)$$

Ordinarily, however, these corrections, which rarely exceed 10% and usually are much less, are not made if dipole moments of a series of similar compounds in the same solvent are to be compared.

(c) Dielectric Constant Measurements

There are three major methods for determining the dielectric constant of a substance by measurement of the capacitance of a cell; the bridge method, the resonance method, and the heterodyne-beat method. The first two methods work well with liquids or solutions with high electrical conductance, but for solutions with low conductance the last method gives the best results. Because the beat method is used most widely, it will be discussed briefly. For a more thorough discussion of this and the other methods see Ref. 11.

A simplified block diagram of a heterodyne-beat apparatus is shown in Fig. 12.8. The signal from an oscillator of fixed frequency, f_0, and that from a variable-frequency oscillator are fed into a mixer, where they are combined to give a "beat" frequency, that is the difference $f - f_0$. (This process of mixing two or more waveforms to generate sum and difference frequencies is also known as heterodyning; hence the name for the method.) Although the individual frequencies are fairly high (about 2MHz in this case), it is possible to detect differences

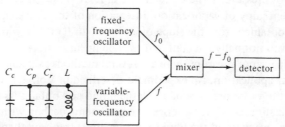

Figure 12.8. Block diagram of the heterodyne-beat apparatus for measuring the capacitance of a dielectric cell. C_c, capacitance of the cell, C_p, capacitance of a precision measuring condenser, C_r, capacitance of a rough tuning condenser, L, inductance of the circuit, and f_0, f, signals from fixed frequency and variable-frequency oscillators, respectively.

231

between the frequencies of much less than 1 Hz. Because the beats are in the audible range it is possible to use earphones as the detector, but a "magic tuning eye" or oscilloscope is more precise. The variable frequency oscillator is adjusted until its frequency exactly matches that of the fixed oscillator, that is, zero beat is obtained. Because the frequency of the variable oscillator depends on the values of the inductance and capacitance of the circuit according to the equation

$$f = \frac{1}{2\pi(LC)^{1/2}}$$

and because the capacitance is the sum of C_c, and C_p, changes in the cell capacitance (C_c) caused by changes in the solutions or substances between the plates can be determined by reading the changes in the precision capacitor (C_p) that are needed to reproduce zero beat.

In most commercial instruments it is possible to construct a calibration curve of known dielectric constant against a scale reading at zero beat. The dielectric constant of an unknown then can be measured directly by referring to this calibration curve.

Various designs of dielectric cell are available. Basically two separated conductors, flat or cylindrical metal plates, are placed opposite each other at a fixed distance. The form of the cell is determined by the nature of the sample, the type of measurement to be made, and the range involved. Some cells are water-jacketed so that temperature can be controlled.

12.1.7. Vapor Pressure and Heat of Vaporization

The equilibrium vapor pressure of a liquid at a given temperature is the pressure of the vapor in equilibrium with the liquid at that temperature. The enthalpy of vaporization of a liquid can be evaluated by determining the vapor pressure of a liquid as a function of temperature and then applying the Clapeyron-Clausius equation:

$$\frac{d \ln P}{dT} = \frac{\Delta H_v}{RT^2} \tag{12.27}$$

where P is the vapor pressure, T the temperature in K, R is the gas constant and ΔH_v is the molar enthalpy of vaporization. Integration of the above equation shows that ΔH_v may be obtained from the slope of a plot of $\ln P$ vs. $1/T$ which, for the temperature intervals normally encountered, will be linear.

For measuring vapor pressure there are several methods available depending upon the particular application. An experiment described in Vol. 2, Chap. 41 directly measures the vapor pressure of a volatile liquid and obtains its molar enthalpy of vaporization as part of a series of experiments dealing with a high vacuum system. In this method the liquid is frozen into a trap, and then the system is evacuated and sealed. The liquid is allowed to warm to the desired temperature and then the vapor pressure is read directly on a suitable manometer. For accurate measurements of vapor pressure it is important to remove any volatile impurities from the liquid, because these will tend to increase the observed pressure (nonvolatile soluble impurities will lower the vapor pressure). Although this method is general, a restriction is that the liquid must be at a lower temperature than any other part of the system. If this restriction is not met, the liquid will condense in the coldest area and equilibrium pressures will not be observed. Another simple device for measuring vapor pressure is the isoteniscope shown in Fig. 12.9. The isoteniscope acts as its own manometer because its operation depends upon establishing equal levels of liquid in the limbs. For this condition to be attained the vapor pressure on both sides must be equal and equal therefore to the vapor pressure of the liquid under investigation. One of the limbs is connected to a manometer so that the vapor pressure can be measured directly. The vacuum system is needed to remove all the air and dissolved gases from the system and for equalizing the liquid levels in the limbs. The detailed procedures for use of an isoteniscope are described in several physical chemistry laboratory manuals.[12,13]

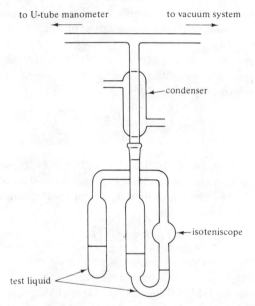

Figure 12.9. Isoteniscope for measuring the vapor pressure of a liquid.

12.1.8. Polarimetry

Optical Activity

Ordinary light may be considered to consist of an electric vector and a magnetic vector oscillating at right angles to each other and to the direction of propagation of the wave. If the direction of the electric vector remains in one plane throughout one wavelength the light is said to be plane or linearly polarized. The mechanism of interaction of light with a molecule (i.e., absorption) normally depends on a coupling of the electric vector with the molecule.

Molecules that exist as two isomeric forms that are nonsuperimposable mirror images are called enantiomers. These two different forms interact differently with polarized light, rotating the plane of polarization in opposite directions; that is, they are optically active. Those isomers which rotate the plane of polarization to the right (as one looks at the dial of the polarimeter; i.e., clockwise) are referred to as dextrorotatory (d, or +); while those which rotate it to the left are levorotatory (l, or −). Such compounds are often found among carbon compounds that contain a carbon atom substituted with four different groups (i.e., an asymmetric carbon atom), and among octahedral coordination compounds of polydentate ligands.

The optical rotation of a solution is measured with a polarimeter. Polarimetry is a common technique used to characterize optically active isomers. The magnitude of the rotation is a criterion of optical purity. If an optically active compound takes part in a chemical reaction and the activity changes as a result of the reaction, the polarimeter may often be used to follow the kinetics of the reaction; e.g., inversion of sucrose.[3] Molecular structures also can be investigated by the technique of optical rotatory dispersion (ORD)[14], the variation of optical rotation with wavelength.

The quantities used in polarimetry are the *specific rotation* $[\alpha]_\lambda^t$ and the *molar rotation* $[M]_\lambda^t$. These quantities, which are functions of wavelength and temperature, are defined as follows:

$$[\alpha]_\lambda^t = \frac{100(\alpha)}{(\ell)(c)} \qquad (12.28)$$

where α is the measured angle of rotation, l is the length of the sample tube in decimeters, c is the concentration of solute in grams per 100 ml of solution,* λ refers to the particular wavelength of light used, and t is the temperature.

The molar rotation is defined as

$$[M]_\lambda^t = \frac{M[\alpha]_\lambda^t}{100} \qquad (12.29)$$

where M is the molecular weight of the compound under investigation.

A schematic diagram of a polarimeter is given in Fig. 12.10. The light source is usually a sodium lamp (for ORD the visible and/or U.V. spectrum is scanned) and the light is polarized by passing it through a Nicol prism (two calcite crystals cemented together). The polarized light then passes through the solution in the sample cell and then through a second Nicol prism called the analyzer. The analyzer can be rotated about the axis of the instrument and it will allow passage of a maximum amount of light when its principal axis is aligned with the direction of the polarized light. Thus, when the sample cell is empty or contains an optically inactive solution, light will pass through when the axis of the analyzer is aligned with that of the first prism, but will not pass when they are "crossed." The amount of rotation of the analyzer required to establish maximum light transmission when the

* Some definitions of specific rotation express the concentration in grams per milliliter.

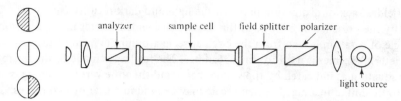

Figure 12.10. Schematic diagram of a polarimeter. Circles on the left represent the field of view: top and bottom, out of balance; middle, null position.

sample is present represents the amount of rotation caused by the sample. Because the change from aligned to crossed is not sharp, a practical polarimeter must be slightly more complex than described above to give accurate readings. Some of the key features are illustrated in Fig. 12.10. In this particular design a second Nicol prism (field splitter) covers half of the first one with its main axis slightly out of alignment with that of the first. When the analyzer is "crossed" with the polarizer, half of the field of view is bright; the other half is dark. A slight rotation of the analyzer will "cross" it with the field splitter and the bright and dark sides are interchanged. The null position exists when both sides are equally bright.

Some instruments use two field splitters and divide the field into three segments. The null point is when the central portion has the same brightness as the two outer ones. ("Darkness" might be a better descriptive term, because the null is read from the angle at which the polarizer and analyzer are crossed. A large range of angles near the alignment angle will give approximately the same brightness, so this position is not sensitive enough for practical use.)

Various kinds of polarimeter sample tubes are available. All are cylindrical, with optically flat glass end plates. They are filled through a side opening or by unscrewing the end. It is very important that no air bubbles be allowed to remain in the light path. The usual tube length is 1 dm., (i.e., 10 cm) but longer or shorter tubes are useful for special cases, e.g.; short cells when solvent absorption of light may interfere, or very long cells to increase the amount of sample in the light path with dilute solutions.

A commonly used cell employs screw-on ends. This consists of a glass cover plate, gaskets, and metal collar. The tube should be washed thoroughly, one end screwed on, and the tube filled so that the liquid surface is slightly above its end. The remaining cover plate then is slid into position so that no bubbles are trapped. The cover plate must be clean and dry on the outside, and must be handled with care. Tubes should be closed tightly to prevent leakage, but not so tightly as to break them or to produce strain in the cover glass that can interfere with accurate measurements by inducing apparent optical activity.

When the light source is in place and properly warmed up (about $\frac{1}{2}$ hr for sodium lamps) place the filled tube in the polarimeter trough. Rotation of the polarimeter head should produce a bright and a dark segment, or two bright and one dark segments in the field of view, depending on the instrument. These segments will interchange with further rotation. The eyepiece can be focussed to give a sharp view of the fields. Lack of distinction between the light and dark regions usually is caused by the presence of bubbles in the tube or suspended matter in the solution. The angle of rotation is read from the circular scale around the head of the polarimeter. A vernier is used for finer readings. The reading is made to the nearest scale graduation as indicated by the vernier zero, and then the vernier reading is added to the initial reading. The main scale may be read in either of two places, 180° apart, thus permitting a convenient check on reading errors.

A reading is taken with the pure solvent in the tube and then the tube is rinsed and filled with the solution to be studied and the rotation is measured in the same way. The polarimeter reading obtained with pure solvent is subtracted from that with the sample to give α. Several readings of the null point should be taken for each sample, and the average of these values should be used. After use, the tube should be washed thoroughly and reassembled loosely.

12.1.9. Osmometry

Osmosis refers to the fact that if a solution is separated from a reservoir of pure solvent by a semipermeable membrane (a membrane through which solvent but not solute molecules can pass), solvent will flow into the solution unless opposed by a pressure on the solution. The value of the applied pressure necessary to prevent osmotic flow, or the hydrostatic pressure head which could be generated by the process of osmosis when equilibrium is reached, is called the osmotic pressure of the solution.

12.1.9.1. Membrane Osmometry

Membrane osmometers are often used for the determination of molecular weights in polymer solutions. The osmometer consists of two compartments containing solvent and solution, separated by a membrane. A connection is made to a capillary to allow the osmotic pressure to be balanced by a hydrostatic pressure, or to some other source of external pressure.

The relationship between the osmotic pressure (π) and the molecular weight (M) is given by

$$\lim_{C \to 0} \frac{\pi}{C} = \frac{RT}{M} \qquad (12.30)$$

where C is the concentration of solute in moles/liter, T the temperature, and R the gas constant. The large effect produced compared with other colligative properties makes this particular property very useful for solutions of substances of high molecular weight, such as polymers.

Various designs of osmometers are available.[15,16,17] Common organic membrane materials are collodion, gel cellophane, polyvinyl alcohol, and polyurethanes.

12.1.9.2. Vapor Phase Osmometry

One of the most widely used methods for determining molecular weights is that of vapor phase osmometry,[15] which does not depend on measurement of osmotic pressure, but rather on the lowering of the vapor pressure of the solvent in the solution. When pure solvent and a solution of a relatively nonvolatile solute in that solvent are placed in a closed system, a transfer of solvent from the pure solvent to the solution occurs by a vaporization − condensation process, due to the differences in chemical potential. The adiabatic transfer of solvent produces a temperature differential between the pure solvent and the solution due to the heat of vaporization involved in the evaporation and condensation of the solvent. The temperature differential is proportional to the difference in vapor pressures of the pure solvent and the solution and hence is proportional to the solute concentration and the type of solvent used.

A thermistor commonly is used to detect the difference in temperature, the temperature difference being measured as a resistance difference, ΔR. Droplets of the solution and the pure solvent are placed on two thermistors suspended in the thermostatted chamber of the osmometer. The vapor pressure osmometer is calibrated with a series of solutions of known concentration in a given solvent. The molar concentration and hence molecular weight of an unknown solute in the same solvent than can be obtained from the ΔR vs. concentration plot.

12.2. SOLIDS

The two most important physical properties of solids which will be considered here are melting point and density. Some of the key methods for determining these quantities are described below. It should also be pointed out that the crystalline form of a solid compound is an important characteristic and these are often tabulated in handbooks which list physical properties of compounds.

12.2.1. Melting Point

The melting point of a substance is the characteristic temperature at which crystalline solid and liquid are in equilibrium. However, if decomposition occurs, the value will not be the true melting point. The melting point can be determined easily and repeatedly on a very small amount of sample. In extreme cases the sample, if it does not decompose on melting, may be recovered for further use. The melting point of a pure solid is depressed by impurities and therefore the temperature range over which melting occurs is increased if the sample is not pure. Use of this is made in purity evaluation and also in the determination of molecular weights in cryoscopic measurements,[18] using the relation:

$$\Delta T_f = K_f m_B = K_f \left[\frac{\text{wt of } B}{\text{mol. wt of } B} \bigg/ \frac{\text{wt of solvent}}{1000} \right] \quad (12.31)$$

where K_f is the freezing point depression constant, m_B is the molality of the solute, and ΔT_f is the measured freezing point depression.

Determination of the melting point requires a thermometer (thermocouple for solids melting higher than ~500°C) and a means of heating the sample at a steady, controlled rate. Heating can be achieved in a liquid bath or by placing the sample in an apparatus that is heated electrically in which air is the heat transfer medium. The liquid bath usually takes the form of a Thiele tube (Fig. 12.11), a specially designed apparatus containing a high-boiling, colorless oil. (Mineral oil often is used up to temperatures at which it begins to evolve fumes (~200°C); silicone oils are available for use at higher temperatures, but they are more expensive.) When the lower side portion is heated with a burner, convection produces even heating of the thermometer bulb without the need for stirring. The sample is held in a capillary tube and is attached with a rubber band to a thermometer bulb. The thermometer bulb is placed just below the upper arm of the circulating loop.

Electrically heated melting point equipment is available in several designs but the essential features are an electrically heated block or oil bath into which a

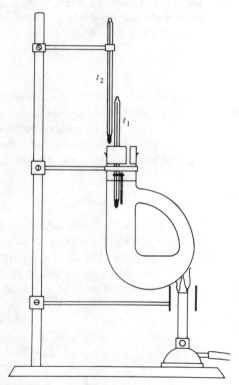

Figure 12.11. Thiele melting point tube. The two thermometers t_1 and t_2 are used to determine the stem correction (see Sec. 13.1).

thermometer and the sample in the capillary tube can be inserted. A commonly used design is the Meltemp®. The temperature and rate of heating can be controlled by varying the voltage input. A simple optical system permits the viewing of three capillary tubes simultaneously.

Other designs such as the Fisher-Johns and the Nalge-Axelrod apparatus are described in several texts.[19] In the Fisher-Johns apparatus the sample is placed in the depression of an electrically heated aluminum block, between two microscope cover glasses; the Nalge-Axelrod equipment uses a microscope for observation of the melting process in an electrically heated block. A capillary is not needed for use with either of these pieces of equipment because the sample usually is held between glass slides.

Capillary tubes, when used, are made of glass which is sealed at one end. Commercially available 2-mm capillaries are satisfactory for most purposes. Tubes can be pulled from glass tubing or a disposable pipet if necessary (see Sec. 6.4). The closed end must be completely sealed. The sample is introduced by pressing the open end of the capillary into a small pile of the crystals or powder, inverting the tube and shaking the sample to the bottom by tapping the tube on the table top, by gently stroking it with a file, or by dropping the sample tube (closed end down) through a long vertical length of glass tubing resting on the desk top. A $\frac{1}{8}$-in column of solid in the bottom of the tube is sufficient. The tube is attached to the thermometer in the Thiele apparatus with a thin rubber ring sliced from a piece of rubber tubing, with a small rubber band or even with tape. (Be sure that the rubber band does not go below the surface of the hot oil or it may break.) The end of the tube with the sample should be as close as possible to the thermometer bulb. The temperature may be raised fairly rapidly until it reaches about $10°C$ below the expected melting point. The rate of heating in the vicinity of the melting point should be $1°C/min$, because heat transfer through glass is slow, and an equilibrium temperature is to be observed. If the approximate melting point is not known, it is wise to obtain an approximate value by rapid heating and then to repeat the determination at the proper heating rate.

The observed melting point, as it is usually determined, actually is the range of temperature, from the point at which the first drops of liquid phase are observed until the crystals melt entirely. If too large a sample is used, the time required for complete melting will be longer, and the apparent melting point range will be larger. For pure crystals a range of $0.5°C$ or less should be observed.

As noted previously, a broad melting range of several degrees is usually an indication of the presence of impurities (including solvent). Impurities generally will cause broadening and lowering of the melting point. (This may be seen by examining the shape of the usual phase diagram for two solids and a solution, i.e., a two-component mixture.) A sharp melting point, unchanged by recrystallization, may be used therefore as one criterion of purity (but with caution). Conversely, a *mixed melting point* experiment may be used in the identification of a crystalline compound. The unknown compound is mixed intimately with a known compound having the same melting point. If the compounds are the same, the melting point of the mixture will not be depressed from that obtained when either is determined alone. Although this method is a useful way of confirming the identity of an unknown, it is not infallible. For the determination of a mixed melting point it is desirable to test several compositions of the mixture (e.g., 1 : 1, 1 : 2, 2 : 1) because there are sometimes certain compositions where no depression is observed and this could lead to erroneous conclusions.

The melting point depression principle (Eq. (12.21)) can be used to determine approximate molecular weights by the Rast method.[19] In this technique the melting point depression for natural d-camphor is normally used, because K_f is exceptionally large. A known weight (~50 mg) of the compound and of camphor (~5 g) are fused in a small, clean, dry test tube. Gentle heating is essential to fuse the two compounds, taking care not to sublime the camphor. The mixture is then crushed into a fine powder and the melting point is determined for it and for the pure d-camphor by any of the methods described above. The molecular weight (mol-wt), of the compound is calculated using Eq. (12.21) which, in the case of camphor, may be rewritten as:

$$\text{mol. wt} = \frac{39.7 \times W_{compound} \times 1000}{\Delta T \times W_{camphor}} \tag{12.32}$$

where ΔT is the freezing (melting) point depression and 39.7 is the molal freezing point depression constant for d-camphor.

A modification of the Rast method in which the camphor and the compound are first dissolved in methylene chloride, which is then evaporated, has been reported recently.[20] This approach has the advantage that the problem of sublimation is avoided.

A melting point determination is only as accurate as the thermometer used, and before taking an unknown melting point, the thermometer should be calibrated by checking the apparent melting points of a few compounds of known, sharp melting points. The following compounds, when pure, are convenient for checking a thermometer:

Compound	Melting point	Compound	Melting point
ice	$0°C$		
p-dichlorobenzene	$53°C$	salicyclic acid	$159°C$
acetanilide	$114°C$	p-nitrobenzoic acid	$239°C$

A calibration curve, thermometer reading vs. true melting point can then be made.

12.2.2. Density

Densities of solids can be measured by finding the volume of liquid which a known weight of solid will displace (the solid must be insoluble in the liquid). A

Weld pycnometer is filled with an inert liquid of known density, which must be less than that of the solid, and weighed. A sample of solid is weighed into the empty, dry pycnometer, and the reference liquid is again added until the pycnometer is filled. The difference between the weights of liquid present with and without the solid present can be converted to volume displaced if the density of the liquid is known. This volume corresponds to the volume of the solid. The major source of error comes from air bubbles on the solid surface and the lack of wetting in pores. Powders must be agitated, and liquids of low surface tension and good wetting tendencies are preferred.

12.3. VAPORS (GASES)

The only physical property of a vapor which will be discussed here is vapor density. Two methods are available for this determination; the Dumas method and the Victor-Meyer method. Both these methods are described in detail in several textbooks.[13]

(a) Dumas Method

This method is used in Vol. 2 in the vacuum system experiment (Chap. 41). A bulb of known volume is evacuated and weighed. It is then filled with vapor at a known temperature and pressure and reweighed. The volume of the bulb is calculated from the weight of water required to fill it.

A typical procedure is to evacuate the bulb, and to freeze and degas the substance to be measured on a vacuum line. The bulb is weighed, reattached to the line, and the substance is allowed to vaporize. When an adequate pressure is reached as indicated by a manometer, the bulb is opened, the pressure and bulb temperature recorded, and the bulb closed and reweighed.

It is not necessary to employ a vacuum system if the liquid is not too volatile. A small quantity of liquid may be placed in the bulb, and heated until completely vaporized. The bulb is then sealed by drawing out the neck or closing it with a stopcock. The pressure in the bulb will equal atmospheric pressure.

The vapor density can be used to obtain a value for the molecular weight, M of the vapor, assuming it obeys the ideal gas law, by using the relationship

$$M = \rho \frac{RT}{P} \tag{12.33}$$

where ρ is the vapor density, T is the absolute temperature, P is the pressure, and R is the gas constant. Deviations arise when the compound dissociates or associates in the vapor state; e.g.,

$$N_2O_4 \rightleftarrows 2NO_2$$

or

$$2CH_3CO_2H \;\rightleftharpoons\; CH_3-C\overset{\displaystyle O-H\cdots O}{\underset{\displaystyle O\cdots H-O}{\Big\langle}}C-CH_3$$

The vapor density may often be used to study such equilibria.

(b) Victor-Meyer Method

In this method a known weight of the liquid is vaporized in a heated tube and the vapor is allowed to pass into a gas buret or eudiometer tube, where it displaces a measured volume of water at a known temperature and pressure. This volume, adjusted for the partial pressure of water, is corrected to S.T.P. and the vapor density is obtained by dividing the weight of the liquid taken by the volume of the vapor. The whole system must be warm enough to prevent condensation of the vapor.

REFERENCES TO CHAPTER 12

1. W. Swietoslawski and J. R. Anderson in "Physical Methods of Organic Chemistry," 3d ed. A. Weissberger (ed.), Interscience, New York, 1959. Vol. I, part 1, chap. VII.

2. R. W. Hakala, *J. Chem. Educ.*, **29**, 616 (1952).

3. N. Bauer in "Physical Methods of Organic Chemistry," 2d ed., A. Weissberger (ed.), Interscience, New York, 1949. Vol. 1, part II, chap. XX.

4. J. F. Swindells, R. Ullman, and H. Mark in "Technique of Organic Chemistry," 3d ed. A. Weissberger (ed.), Interscience, New York, 1959. Vol. I, part 1, chap. XII, Physical Methods of Organic Chemistry.

5. A. W. Adamson, "Physical Chemistry of Surfaces," 2d ed., Wiley-Interscience, New York, 1967.

6. P. Debye, "Polar Molecules," Dover Publications, New York, 1929.

7. G. Hedestrand, *Z. Physik Chem*, **B2**, 428 (1929).

8. F. A. Guggenheim, *Trans. Faraday Soc.* **45**, 714 (1949).

9. E. S. Gould, "Mechanism and Structure in Organic Chemistry," Holt, Rinehart and Winston, New York, 1959, chap. 3.

10. H. Müller, *Physical Z*, **34**, 689 (1933).

11. C. P. Smyth in "Technique of Organic Chemistry," 3d ed., A. Weissberger (ed.), Interscience, New York, 1960, vol. 1, part 3, chap. 39.

12. F. A. Bettelheim, "Experimental Physical Chemistry," Saunders, Philadelphia, 1971.

13. D. P. Shoemaker and C. W. Garland, "Experiments in Physical Chemistry," 2d ed., McGraw-Hill, New York, 1967.

14. C. Djerassi, "Optical Rotatory Dispersion," McGraw-Hill, New York, 1960.

15. E. A. Collins, J. Bares, and F. W. Billmeyer, Jr., "Experiments in Polymer Science," Wiley-Interscience, New York, 1973.

16. B. H. Zimm and I. Myerson, *J. Am. Chem. Soc.*, **68**, 911 (1946).

17. J. V. Stabin and E. H. Immergut, *J. Polymer Sci.*, **14**, 209 (1954).

18. E. L. Skau, J. C. Arthur, and H. Wakeham in "Physical Methods of Organic Chemistry," 3d ed. A. Weissberger (ed.), Interscience, New York, 1959, vol. 1, part 1, chap. VII.

19. R. L. Shriner, R. C. Fuson, and D. Y. Curtin, "The Systematic Identification of Organic Compounds. A Laboratory Manual," 5th ed., Wiley, New York, 1964.

20. S. Wawzonek, *J. Chem. Educ.*, **49**, 399 (1972).

Temperature Measurement and Control

13.1. TEMPERATURE MEASUREMENT

Temperature is an important parameter in many chemical processes, and means of measuring and sometimes controlling temperature are essential. Temperature is measured by its effect on some physical property, such as volume, resistance or potential. The thermodynamic temperature scale is established with a helium gas thermometer, and is based on the ideal gas law. However, a gas thermometer is not used for general measurements because it is quite unwieldy.

Temperature measuring devices are calibrated by reference to certain "fixed points," such as the melting point of a pure substance at 1 atm pressure. A number of such systems, the temperatures of which are well established on the thermodynamic scale, are listed in Table 13.1. In fact, because gas thermometers are less accurate than many other devices, the practical International Temperature Scale is based on *defined* values of fixed points and interpolation formulae.

It often is necessary to check calibrations in the laboratory using such values. When available, temperature measuring devices calibrated by the National Bureau of Standards can be employed to calibrate an instrument which is used routinely.

For reliable measurements, good thermal contact between the measuring device and the object the temperature of which is to be measured is essential, and the device must have time to reach equilibrium with the object. Detectors which are subject to chemical attack (e.g., a thermocouple) must have a protective sleeve. In such cases, heat transfer can be improved by the presence of an inert oil in the sleeve.

Rough estimates of temperature may be made by the senses. For example, as an object is heated, it passes through stages in which it glows dull red, through

TABLE 13.1

Some Reference Points for Temperature Calibration

Process (all at 1 atm pressure)	Temperature °C
Boiling point of hydrogen	−252.753
Boiling point of nitrogen	−195.806
Boiling point of oxygen	−182.962
Freezing point of mercury	−38.862
Freezing point of air-saturated water	0.000
Triple point of water	0.01
Boiling point of water	100.000
Freezing point of indium	156.634
Freezing point of lead	327.502
Freezing point of zinc	419.5
Boiling point of sulfur	444.674
Freezing point of antimony	630.74
Freezing point of silver	961.938
Freezing point of gold	1064.43
Freezing point of platinum	1772.000

bright red, orange, yellow and finally white. These differences often are used as a rough but adequate guide for the extent of heating needed in noncritical ignition processes. Temperatures slightly above ambient may be estimated roughly by touch, with experience and caution.

A number of different physical properties are used in practical temperature measuring devices. Some are called thermometers, while others are given other names, e.g., thermocouples, or pyrometers. The most common devices are discussed briefly below.[1,2]

13.1.1. Liquid-in-glass Thermometers

The *mercury-in-glass* thermometer is the most commonly encountered temperature measuring device. This device has a range of −20°C to +500°C (+650°C with special models) although commonly a given thermometer will cover only part of this span. Readability may range from a degree or two to 0.001 degree. High sensitivity requires a large volume of mercury and a fine capillary, so that movement of the mercury in the capillary is significant for small temperature changes. Thus the length of the thermometer easily becomes unwieldy; for this reason, precision thermometers normally cover a comparatively small range of temperatures. Some thermometers, notably the Beckmann type, have an adjustable range. They are intended for precise measurement of temperature differences (e.g., in calorimetry) rather than of absolute values, and are sometimes called differential thermometers (Fig. 13.1). A Beckmann thermometer is set to the desired range by adjusting the amount of mercury in the reservoir. The thermometer is heated until excess mercury is pushed above the capillary and into the

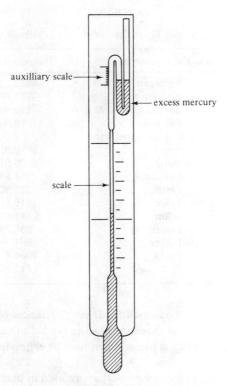

auxilliary scale

excess mercury

scale

Figure 13.1. Beckmann thermometer.

reservoir region (usually an auxillary scale indicates how much heating is required), then inverted and tapped to break the thread. The reservoir is constructed so that the excess mercury is kept away from the end of the capillary. On cooling, the thread will lie in the desired region of the scale. Because the range is usually only 5 or 6°C, some care and usually several attempts are required for a proper setting. Excess mercury can be returned to the capillary by tilting the thermometer to bring the mercury to the capillary-reservoir junction, warming so that the mercury in the capillary rises to join the mercury in the reservoir, and cooling carefully. This procedure will be necessary if too much mercury is removed, or to make use of the auxillary scale.

Most common thermometers will give different readings depending on their extent of immersion in the material the temperature of which is to be measured. Exceptions are those with insulating jackets around the capillary. Ordinarily, thermometers are calibrated for a certain extent of immersion; either total, or 76 mm are common. If used under other immersion conditions, the thermometer should be recalibrated or a *stem correction* applied. A total immersion thermometer will give a low reading if the stem is at a lower temperature than the bulb, and a high reading under the opposite condition. An approximate value for the stem correction is

$$S = 0.000154(t_e - t_s)N \qquad (13.1)$$

where t_e is the observed temperature, t_s is the temperature at the mid-point of the exposed mercury column, and N is the length in degrees of the exposed mercury column (i.e., the length above the level of immersion). The value 0.000154 is the difference between the coefficients of expansion of glass and mercury. Such corrections may amount to 4 or 5 degrees for temperatures near 200°C.

Lower temperatures than those at which a mercury thermometer can function may be measured with thermometers which contain organic liquids, e.g., alcohol (usually to −50°C), toluene (to −100°C) or hydrocarbons (to −200°C).

13.1.2. Bimetallic-strip Thermometers

Devices of this kind make use of the differences in coefficients of thermal expansion of different metals. When the temperature rises, initially identical lengths of the metals expand to different lengths. If they are welded together, differential expansion will cause the strip to bend, with the metal of higher expansivity lying on the outside of the curve as the temperature increases. A coil will tend to loosen or tighten by an amount which is proportional to temperature change. The position of a pointer attached to the coil can be calibrated in terms of temperature. Such thermometers are not very accurate, but they are rugged.

13.1.3. Resistance Thermometers

The electrical resistance of a metal varies with temperature. This phenomenon is the basis for the platinum resistance thermometer, that has very high accuracy and precision, and is used to define the International Temperature Scale over part of its range. It is useful to about 630°C. Such a device consists of a coil of platinum wire of about 2.5 or 25 Ω resistance, mounted in a suitable protective casing. Resistance changes for such coils are about 0.01 or 0.1 Ω per degree, respectively, and a special high precision bridge is required for the resistance measurements. Such thermometers are expensive, but widely used in thermochemical work. Other metals are sometimes used for the resistance elements.

13.1.4. Thermistor Thermometers

Thermistors are composed of semiconductor materials with a high negative temperature coefficient of resistance. As with a resistance thermometer, the electrical resistance value is an indication of temperature. Because resistance values are higher and vary more rapidly with temperature, the measuring apparatus is simpler and less expensive than with the platinum resistance thermometer, although the accuracy also is lower. A thermistor is more subject to drift and calibration changes than other devices. However, the detection element can be very small, so that it will reach temperature equilibrium quickly.

13.1.5. Thermocouples

If wires of two dissimilar metals are joined at both ends, and the junctions are held at different temperatures, a potential difference will be generated between them. This potential can be measured with a suitable detector inserted into one of the arms, as in Fig. 13.2(a). In this configuration, one junction, the reference junction, is kept at a fixed temperature, such as an ice-water bath. The observed potential is then proportional (but not necessarily linearly proportional) to the temperature of the other junction. Tables of potential-temperature data are available in handbooks and other sources for a variety of widely used combinations of metals.

Alternatively, the arrangement in Fig. 13.2(b) can be used. In this case, the terminals of the detector serve as the reference junction. Normally, these are at room temperature, and this must be taken into account when reading the temperature from the tables. In using a thermocouple, some care must be taken not to introduce spurious thermal potentials by having the terminals at different

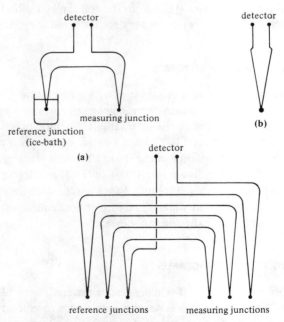

Figure 13.2. (a) Schematic representation of a thermocouple, (b) a thermocouple in which the connections to the detector are at room temperature and serve as the reference, (c) a multijunction (three-junction) thermocouple.

temperatures, or by introducing thermal potentials between the thermocouple and a lead wire.

The potentials generated in thermocouple materials are quite small (millivolts for modest temperature differences), and when higher sensitivities are required, it is common to connect two or more thermocouples in series to form a multijunction thermocouple (sometimes called a thermopile) as in Fig. 13.2(c). The potential of an n-junction thermocouple is n times the potential of a single-junction thermocouple of the same kind.

The potential of a thermocouple may be determined with a millivoltmeter, and such devices calibrated directly in temperature are called pyrometers. If current is drawn, however, the output potential will change, and precise readings require high input resistance detectors. Potentiometers often are used, as are electronic recorders, for continuous monitoring of temperature. For high accuracy, thermocouples must be calibrated rather than relying on potential-temperature tables.

Different thermocouple materials are used in different temperature ranges. The choice is based partly on sensitivity and partly on stability of the metals under the conditions of use. Some common combinations are given in Table 13.2.

Thermocouple wires must be insulated electrically from each other except at the measuring junctions, and the junctions of multijunction couples must not touch. Cloth or enamel is adequate at low temperatures; fiberglass or asbestos sleeves or porcelain insulator tubes are used at higher temperatures. Protection of the wires from corrosion by air at higher temperatures by closed glass or quartz sleeves or metal jackets frequently is necessary. When measuring temperatures of

TABLE 13.2

Some Common Thermocouple Types

Metals	Temperature Interval for normal use (°C)	Comments
Copper-Constantan[a]	−190 to +300	Cu is easily oxidized in air at higher temperatures
Iron-Constantan	−190 to +760	Iron is easily oxidized in air at higher temperatures
Chromel[b]-alumel[c]	−190 to +1,100	
Platinum-10% rhodium-platinum	0 to +1,450	Lower temperature coefficient than others; can be used up to 1,700°C briefly
Tungsten-3% rhenium-tungsten-25% rhenium	0 to +2,600	

[a] an alloy of 55% Cu, 45% Ni.
[b] an alloy of 90% Ni, 10% Cr.
[c] an alloy of 96% Ni, 2% Al and 2% Mn.

liquids, the junctions always should be protected from direct contact with the liquid to reduce the possibility of chemical attack; glass or other inert tubes can be used, preferrably containing a little oil to assist in heat transfer.

The junctions are made usually by welding the metals together. Copper-constantan couples are most easily made, because they may be welded by placing the ends, well twisted together, in the tip of the blue reducing flame of a glass-blowing torch until the metals melt and form a bead. Care must be used not to oxidize the metal. For low temperature use, a simple solder joint is satisfactory. In any case, care must be taken that the wires are fully cleaned of any insulation and enamel which may be used to coat them.

13.1.6. Optical Pyrometer

A black body radiator, (or a reasonable approximation) emits radiation of which the intensity and wave-length distribution is characteristic of the temperature. At high temperatures, much of this radiation lies in the visible region of the spectrum and objects at different temperatures will appear different in color and brightness (e.g., red-hot is cooler than white-hot). An optical pyrometer is a device for measuring high temperature by matching the emitted intensity of the source to the intensity of a standard (usually a heated filament). Current through the filament is adjusted to make its intensity identical to that of the object as seen in the pyrometer optical system. The current is calibrated in terms of temperature. This type of device is useful for very high temperatures, and permits measurement with no direct contact to the system.

13.2. TEMPERATURE CONTROL

Many experiments require some means of controlling temperature. The simplest method is to use an appropriate "bath." From the phase rule, at a given pressure a single component system in which two phases are in equilibrium will exist only at a single temperature. Various fixed-temperature baths depend on this principle, for example, liquid-vapor (liquid N_2); solid-vapor (dry ice – inert liquid); solid-liquid (ice-water). Eutectic mixtures in two-component systems are used also. It must be remembered that for truly constant temperatures, all phases in equilibrium must be present in a well-mixed bath. For example, an ice-water bath used as a zero degree reference for a thermocouple should be a mixture of both ice and water, not merely a vessel filled with ice, which may take considerable time to reach $0°C$.

13.2.1. Multi-phase Baths

In many cases, baths such as the ones described below are used simply to maintain a low temperature, not necessarily a particular constant value. Purity, mixing, and other conditions must be carefully considered in the latter instance.

The examples below are among the most common, but they are certainly not the only ones used.

1. *Liquid Nitrogen*: This is a common laboratory coolant, used particularly to cool traps on high vacuum lines. The normal boiling point is $-196°C$, and it can be handled conveniently in vacuum-jacketed vessels (Dewar flasks). It will condense many volatile materials, including oxygen from the air (b.p. $-183°C$). **Liquid oxygen is very dangerous, especially if it condenses in the pure state (e.g., in a tube immersed in the liquid N_2 bath), because it can react explosively with organic materials. Condensation of oxygen into the liquid nitrogen bath itself can be reduced by loosely plugging the mouth of the vessel with glass or cotton wool. Liquid nitrogen is cold enough to damage body tissues, and it must be handled carefully.**

2. *Dry ice-acetone*: The temperature of this bath is $-78°C$, which is the sublimation temperature of carbon dioxide. The dry ice and acetone must be mixed carefully or foaming will occur and the liquid will spill from the container. There are two methods of preparing these baths, and both require that some of the dry ice be pulverized. One method is to fill the Dewar partially with powdered dry ice and very slowly pour in the acetone. If a grinder is not available, chunks of dry ice can be wrapped in a towel and pounded with a hammer. A second method is to fill a Dewar flask about half full with acetone and then slowly add very small amounts of the powdered dry ice with a spatula. The initial additions will result in considerable foaming that must be permitted to subside before adding more dry ice. Soon, however, the evaporation of CO_2 will become much slower and some dry ice will accumulate on the bottom of the flask, after which larger pieces of dry ice may be added. The dry ice which evaporates must be replaced periodically to keep the flask fairly well filled with dry ice.

   Because acetone acts only as a heat transfer medium, any liquid with a low freezing point could be used. In fact, the flammable nature of acetone has prompted many laboratories to use other liquids such as trichloroethylene. The vapors of this, or of any chlorinated hydrocarbon, are very toxic, however. The method of preparation of the bath is the same as with acetone. **Dry ice is cold enough to damage flesh; it should be handled with gloves.**

3. *Slush Baths*: Because a solid in equilibrium with its liquid phase can only exist at a constant temperature (at a constant pressure), cold baths of various temperatures can be prepared by partially freezing suitable liquids. To prepare a low-temperature slush bath, partially fill a Dewar flask with the liquid and add liquid nitrogen while stirring vigorously (*Caution!*) until a thick slush (applesauce consistency) is achieved. Care must be taken not to permit a crust to form, which might break the flask. A large number of slush baths are listed in Ref. 3.

4. *Ice-water baths*: This is a very common bath, very often used to maintain a temperature of $0°C$. This bath will be accurate only if the ice and water are

pure, because impurities lower the freezing point. This bath should be made of crushed ice and water.

5. *Ice-salt baths*: A mixture of ice, water, and NaCl will attain a temperature near $-22°C$, the eutectic temperature for this two-component system. Such a bath must be well mixed, with solid NaCl, ice, and water saturated with NaCl all present to obtain this temperature. The bath may be prepared by mixing about 25% salt, 75% crushed ice in alternate layers. Other salts, e.g., $CaCl_2$, can be used to give even lower temperatures.

6. *Boiling liquids*: A bath of boiling water will provide a temperature of $100°C$ (at 1 atm), but the temperature control is unlikely to be good unless super-heating is prevented. Water, and many other liquids, may be used to maintain an appropriate vessel at a constant temperature if the vapors from the boiling liquid are allowed to condense on its surface. A reflux system is necessary.

13.2.2. Thermostatic Baths

Baths in which the temperature of a liquid is fixed by controlling the heating or cooling permit one to select almost any desired temperature. Such controlled temperature baths may use water as the liquid from slightly below room temperature to about $60°C$. Above this temperature, evaporation becomes a problem. An inert oil, often a silicone oil, also can be used over this range and up to $250°C$ or more. Molten salts, (e.g., KNO_3-NaNO_3 eutectic, $218-500°C$), or metals (Woods metal, $61-500°C$; solder $220-650°C$) also find use as liquids in high temperature baths. **All high temperature baths must be used with caution.** Air baths also can be used, but heat transfer is much less efficient.

A controlled temperature bath must contain the following components. First, some means of effective stirring is essential, so the temperature will be uniform throughout the bath. Various types of stirring propellers or circulating pumps are used. Second, heating elements are required. Numerous styles of immersion heaters are available. A light bulb can be used in oil or water baths near room temperature. For precise temperature control, the heater chosen should not be too powerful, because the temperature will overshoot the desired setting. At high temperatures it is common to use one heater which is permanently on to maintain the bath somewhat below the desired temperature, and to use a second heater for control. Commercial constant temperature baths often have several power settings, with the highest used for high temperatures or to raise the temperature more rapidly. Third, the most vital component is the regulator. Normally, power is controlled by a device that closes or opens the electrical circuit to the heaters as required to maintain the desired temperature, usually through an intermediate relay.

Mercury-filled regulators are common and may be very precise. Basically these devices are built on the principles of a thermometer, with a large mercury reservoir and a narrow tube along which the mercury expands as the temperature

electrical leads to relay

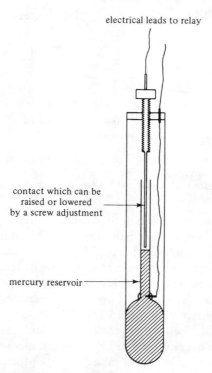

contact which can be
raised or lowered
by a screw adjustment

mercury reservoir

Figure 13.3. Schematic drawing of a mercury temperature regulator.

rises (Fig. 13.3). An electrical contact is sealed into the regulator below
the capillary, while a fine wire on a screw adjustment serves as the second contact.
When the circuit is closed, the heater current is turned off. The height of the
contact can be adjusted so that contact is made at the desired temperature. In
some designs, a scale is present that indicates the set-point, although usually a
separate thermometer must be used for exact temperature readings, and final
adjustment to a desired temperature must be done by trial. For models without a
temperature scale, an approximate calibration can be made by allowing the bath to
come to temperature, making one turn of the adjustment screw, and noting the new
equilibrium temperature. In this way, the number of turns necessary to change the
temperature a desired number of degrees can be found. In many commercial
regulators, the system is completely sealed to eliminate the oxidation which
otherwise occurs as the contact is made and broken in air, and the screw adjustment
is operated magnetically. Some of these devices are constructed somewhat like a
Beckmann thermometer in that the amount of mercury in the bulb can be changed
to give a compact length but high sensitivity over a short, variable range.
Home-made regulators are encountered commonly, especially when the highest
sensitivity is desired.

Bimetallic thermoregulators (bimetallic thermometers with adjustable

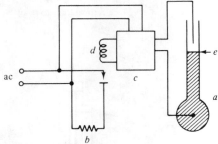

Figure 13.4. Schematic diagram of constant temperature bath control circuit: (*a*) regulator; (*b*) heater; (*c*) amplifier; (*d*) relay. (*c* and *d* often are part of one unit.)

As the contact at *e* is made and broken, the relay *d* opens and closes the circuit to the heater.

contacts activated by the expansion of the bimetallic strip) are also used. They are less precise, but more rugged than a typical mercury regulator.

The current which can be carried by a regulator is small; consequently it is used to control a relay which handles the heater current. Thus, a fourth component of a constant temperature bath usually is an electronic relay (Fig. 13.4).

The uniformity and constancy of the temperatures attainable with a liquid bath are highly variable, but are of the order of ±0.2°C with typical commercial baths and regulators, to 0.001°C with a very sensitive regulator and efficient stirring and insulation. Baths to be operated near or below normal room temperature also need a cooling coil − this is a coil of copper tubing through which cold water can be passed slowly. When tap-water is used, a device for controlling the water pressure to ensure a uniform flow rate is helpful.

Other means of controlling temperature are available that employ the electrical output from thermocouples or thermistors to operate electronic control circuitry that activates the heating elements. Simple on-off controls, or more elaborate devices that gradually reduce the power as the desired temperature is approached (proportioning controllers) also are found. These units reduce surges in temperature. Still more elaborate equipment with such detectors permit temperatures to be changed automatically in specified ways (temperature programmers).

Finally, low temperature baths can be constructed in which thermoelectric cooling elements are used in analogy to heating elements above ambient temperature. Circulation of pre-cooled liquids, (brine, ethylene glycol) also may be used for this purpose.

REFERENCES TO CHAPTER 13

1. W. F. Forsythe, "Temperature: Its Measurement and Control in Science and Industry," Reinhold, New York, 1962.

2. J. M. Sturtevant in "Technique of Organic Chemistry," 3d ed., A. Weissberger (ed.), vol. 1, Physical Methods of Organic Chemistry, part 1, chap. 6. Interscience, New York, 1959.

3. R. E. Rondeau, *J. Chem. Engng. Data,* **11,** 124 (1966).

Vacuum Systems

INTRODUCTION

Laboratory applications of reduced pressures ("vacua") are frequent, ranging from suction filtrations and simple vacuum distillations to elaborate high vacuum systems operated at very low pressures. Vacuum systems are used extensively for physical chemical measurements and syntheses when dealing with volatile and/or air-sensitive materials. Vacuum applications require some means of attaining and measuring the low pressure, as well as the vacuum apparatus itself.

14.1. SOURCES OF VACUUM

14.1.1. Water Aspirator

A high velocity jet of water entering a narrow tube will entrain air and remove it from the system. A fast stream of water is necessary to obtain the lowest pressure. A schematic diagram of a water aspirator is shown in Fig. 14.1; metal, glass, and plastic models are available. This device is very inexpensive and quite effective for modest vacuum requirements — e.g., suction filtrations, or routine vacuum distillations. A properly designed pump of this kind can produce a pressure as low as 12–15 torr under good conditions. The pressure attainable is limited by the vapor pressure of the water, if all other variables are optimized. The use of a rubber hose as a splash suppressor for the fast water stream is permissible if it does not interfere with the full flow of water from the exit tube.

All water aspirators are subject to "suck-back"; i.e., water is sucked into the evacuated system if the water is turned off, or if the water pressure drops

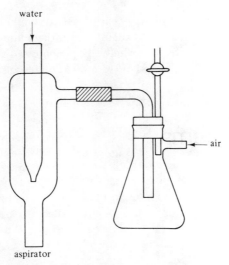

Figure 14.1. A water aspirator and trap.

significantly. Check valves in some designs help to overcome suck-back problems. However, use of an empty flask as a trap between the aspirator and the experimental system as shown in Fig. 14.1 is strongly recommended. In any event, air should be admitted to the system before the flow of water is turned off.

14.1.2. Mechanical Pumps[1,2]

For the production of pressures lower than those attainable with an aspirator, a mechanical vacuum pump is used. Mechanical pumps usually consist of an eccentric rotor moving in a cylinder, as illustrated in Fig. 14.2. As the inner rotor spins the spring-loaded moveable vanes are in contact continuously with the inside wall of the larger cylinder. Thus, gas is sucked first into the cylinder, then compressed and forced out through the exit. The rotor and cylinder are immersed in an oil of low vapor pressure which acts both as a lubricant and as a seal to prevent leakage. This oil should be changed whenever it becomes dirty or contaminated to ensure good performance. Most pumps have an oil drain on the front or side. Fresh high-vacuum oil is added through the air exhaust. Flushing oils

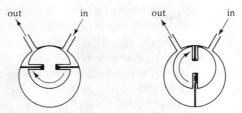

Figure 14.2. Schematic illustration of a mechanical vacuum pump.

may be used to clean very dirty pumps, which should not be flushed with organic solvents. Excess oil will spray from the exhaust vent if the pump is over-filled. The proper oil level usually is indicated in a window in the pump. Air should be admitted to the pump when it is turned off to prevent oil being sucked into the apparatus.

Gases or vapors from the vacuum systems should not be allowed to pass through the pumps, because they may cause corrosion of the mechanical parts of the pump or contaminate the oil. Volatile materials dissolved in the oil will prevent reaching the lowest attainable vacuum. For this reason mechanical pumps must be protected with a trap which is cooled with dry ice — acetone or liquid nitrogen to condense volatile materials. Such cooling baths are discussed in Sec. 13.2.

Pumps are specified in terms of the pressure that they can produce, and also by their pumping rates. A good mechanical pump working on a leak-free system which is free of volatile materials, can achieve a vacuum on the order of 10^{-3} torr. The pumping rate refers to the volume of gas which the pump can exhaust in a given time at a given inlet pressure, and therefore relates to how quickly the desired vacuum can be achieved.

Most vacuum pumps are driven by a belt from an electric motor. **An exposed belt represents a significant hazard, because clothing and hands are caught in it easily. Such pumps should be used only with a proper belt guard installed. Frayed and worn belts should be replaced.** It is good practice to place the pump in a metal tray to collect any oil which may leak from the pump.

14.1.3. Diffusion Pumps[1,2,3]

In a diffusion pump (Fig. 14.3), an oil of low vapor pressure or mercury is boiled and the vapor is passed through a nozzle near the inlet. Gas molecules diffuse

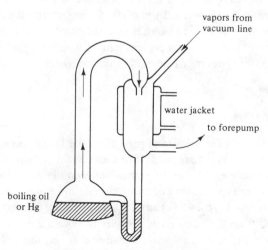

Figure 14.3. A diffusion pump.

from the system into this region, where they collide with molecules of the oil or mercury and are carried to the discharge section of the pump. The vapors of the pump oil or of Hg are condensed, and the molecules of the other gases are removed at the outlet with a mechanical pump before they can diffuse back into the system. Mechanical pumps often are referred to as forepumps, or backing pumps because of this application. Diffusion pumps will not operate effectively unless the overall pressure is kept low with a forepump, to allow free passage of the vapor of the pumping fluid.

Diffusion pumps can attain much lower pressures and pump at much faster rates at low pressures than can mechanical pumps. In a good system a diffusion pump can bring the pressure to 10^{-6} torr. For pressures lower than this value special pumps (e.g., ion pumps) must be employed.

The need to protect a diffusion pump from reactive vapors is even greater than for a mechanical pump. A liquid nitrogen trap is essential. This trap also helps to reduce diffusion of oil or mercury vapors from the pump into the system. (Care must be taken to ensure that O_2 is not condensed from the air into the liquid N_2 trap; see Sec. 13.2.) In addition, large amounts of air passing through the pumps may cause oxidation of the pumping fluid. Thus, the diffusion pump should be bypassed until most of the air in the line has been removed by the forepump. The diffusion pump should be evacuated by the forepump before the pumping fluid is heated. If the diffusion pump is water cooled, be sure water is flowing through the condenser.

14.2. PRESSURE MEASUREMENT

The most common unit used to express the pressure in a vacuum line is the torr, previously designated as mm of Hg. For high vacuum work a more convenient unit is the micrometer (μm), where 1 μm = 10^{-3} torr (micron is used for micrometer in the older literature). A brief description of a few of the devices commonly used to measure pressure is given below. A more complete discussion of these and several other types may be found in the references.

14.2.1. Manometers[3]

The simplest device for measuring pressure is the U-tube mercury manometer shown in Fig. 14.4. Two types are used. If one end of the U tube is closed, as in Fig. 14.4(a) the pressure equals the difference between the mercury levels in the two arms. In the other type, instead of being closed, one end of the manometer is open to the atmosphere (Fig. 14.4(b) and (c)). The reading (corrected as below) must be subtracted from the barometric pressure to obtain the absolute pressure in the system. Another design of a closed-end manometer was shown in Fig. 10.18(d).

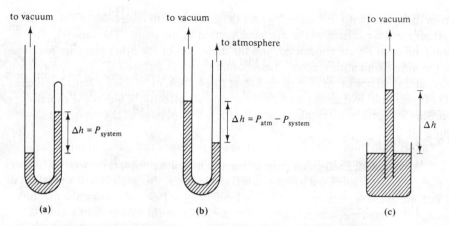

Figure 14.4. Various forms of manometers: (a) a closed-end manometer of classical U shape, (b) An open-end manometer of classical U shape, (c) Another form of an open-end manometer.

For accurate work the observed pressure, P, should be corrected for temperature to yield the pressure in torr at $0°C$, P_0, by use of the formula:

$$P_0 = P - P\left[\frac{\alpha t - \beta(t - t_s)}{1 + \alpha t}\right] \quad (14.1)$$

where α is the volume coefficient of expansion for mercury (1.82×10^{-4} deg^{-1}), β is the coefficient of linear expansion of the scale, t is the temperature ($°C$) of the manometer, and t_s is the temperature at which the scale was graduated. For paper or wooden scales, β may be neglected, and for low pressures (less than about 100 torr) the first correction begins to approach the reading error of most simple manometers.

A manometer generally is useful in the pressure range of 1–1000 torr. For precise work the position of the menisci are read with a cathetometer and large diameter (at least 20 mm o.d.) tubing is used to minimize distortion of the menisci. The useful range of a manometer may be extended to about 0.03 torr by using a suitable oil rather than mercury, because

$$P_p = \frac{\text{density of oil}}{13.59} h \quad (14.2)$$

where h is the difference between the oil levels in millimeters. Unfortunately oil-filled manometers are troublesome to use; wetting the walls of the manometer causes slow response, and the size becomes unwieldy.

Some manometers require a knowledge of the atmospheric pressure before an absolute pressure value can be obtained. Accurate atmospheric pressure measurements are normally made with a mercury barometer. This device is a special form of a closed-end manometer. The mercury in the lower reservoir, which is open to the atmosphere, can be adjusted to a fixed level by an adjusting screw; the level is indicated by a pointer that should just touch the mercury surface. The mercury level in the barometer tube is then read with a sliding vernier scale. Most barometers are graduated to give correct readings at $0°C$, and a temperature correction to allow for expansion at other temperatures should be used (Eq. (14.1)).

14.2.2. McLeod Gauges[1,2,3]

A typical McLeod gauge may be used to measure pressures between 10^{-6} and 2 torr. A diagram of such a gauge is given in Fig. 14.5. The principle of operation of a McLeod gauge is governed by Boyle's Law, $P_1 V_1 = P_2 V_2$. Allowing the mercury level to rise traps gas at an unknown pressure, P, from the system into a known large volume, V (the bulb and capillary), and compresses it into a known small volume (in the capillary), where the pressure can be measured.

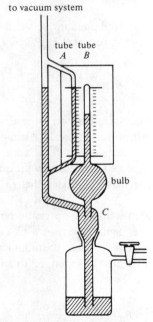

to vacuum system

tube tube
A   B

bulb

C

Figure 14.5. A McLeod gauge. The pressure in the system is given by:

$$P = \frac{\text{Vol. of gas in capillary tube B}}{\text{Vol. of bulb}} \times \begin{array}{l}\text{difference in height} \\ \text{of mercury in tubes} \\ A \text{ and } B.\end{array}$$

The scale is calibrated to read pressure directly when the mercury in A is level with the top of B.

For measurement of low pressures (e.g., below 0.1 torr) the mercury level is allowed to rise until the level in the reference capillary (A) reaches the upper reference point, which is even with the top of the closed capillary (B). If the difference between the mercury levels in the reference and closed capillary is h and the cross section of the closed capillary is a, then the volume of gas in the closed capillary is ha. The pressure in the closed capillary is $h + P$, but because P is very much smaller than h it may be neglected. Thus, the pressure in the system may be determined from Boyle's Law as $P = ah^2$. This scale is often referred to as the quadratic scale of the McLeod gauge, because the pressure is proportional to h^2, rather than to h.

For measurement of higher pressures the mercury level is allowed to rise until the level in the closed capillary reaches the lower reference mark. In this case the volume in the closed capillary is always a constant, v, so that if the difference in the mercury levels is h', the pressure in the system may again be found from Boyle's law as $P = (v/V)h'$. This is called the linear scale of the McLeod gauge.

The operation of a McLeod gauge of this type is controlled by carefully admitting air or vacuum to the lower mercury reservoir through a 3-way stopcock. The McLeod gauge first is evacuated to the pressure in the vacuum system. All of the mercury must be below level C, and the lower bulb is evacuated as necessary with an auxilliary pump. To take a pressure reading, air is *slowly* admitted by the 3-way stopcock to force the mercury up to the zero line; the pressure then can be read from the level in the other tube. The mercury is lowered by re-evacuating the lower bulb. It is essential that all such operations be carried out slowly, because a rapidly moving column of mercury possesses considerable momentum and can break the glass. When not in use the McLeod gauge should be closed from the system with a stopcock.

Tilting McLeod gauges (see Sec. 10.3) are smaller gauges used when less accurate measurements are adequate. These gauges are attached to the vacuum system with ground glass joints, about which they can be rotated. The gauge is evacuated in a horizontal position. When rotated to the vertical position, the mercury traps the residual gas and compresses it into a capillary as in the normal design.

The chief advantages of a McLeod gauge are the wide pressure range over which it may be used and the fact that the readings are absolute; that is they depend only on the dimensions of the gauge and not on the properties of the gases present (except that they must approximate ideal gas behavior). There are some serious drawbacks to the use of such gauges, however. Because the vapor pressure of mercury is about 10^{-3} torr at room temperature, the gauge must be isolated from the system with a trap to prevent mercury vapor from entering the system. (Fortunately, the rate of vaporization is quite slow.) Also, a McLeod gauge cannot be used to measure the pressure of a gas which would condense when compressed into the capillary, that is when h (or h') exceeds the room-temperature vapor pressure of the substance. Finally, because of its slow response it is very difficult to use a McLeod gauge to monitor pressure as a function of relatively short times, for example in leak detection.

14.2.3. Other Gauges[1,4]

There are many relative pressure measuring devices that are commonly used for measuring low pressures. Their main advantage is rapid, simple measurements. If the absolute pressure is needed, the gauges require calibration, both to correct for instrumental drift, and because they depend on the nature of the gases in the system. Three of the more commonly used gauges are described below.

1. *Thermocouple gauge.* This gauge depends on the variation of thermal conductivity of a gas with pressure. It is related to the thermal conductivity detector for gas chromatographs which depends on the variation of thermal conductivity with composition (see Sec. 10.6). A constant current is passed through a resistance wire which has a thermocouple attached to the center. Because the thermal conductivity of the gas changes with pressure, the temperature of the wire also changes with the pressure. The pressure is read as a function of the thermocouple output. The scales on most commercial meters have been calibrated to read the pressure of air directly. If the absolute pressure of another gas is desired a calibration chart or multiplication factor must be used. Thermocouple gauges are rugged and inexpensive. Their useful range of pressure measurement is generally from about 0.1 to 10^{-3} torr.
2. *Pirani gauge.* This device is similar to a thermocouple gauge but the resistance of the wire is used to measure its temperature. A second, completely sealed resistance wire is used as a reference. A Pirani gauge is more subject to drift than a thermocouple gauge, but its range is larger, typically $0.3-10^{-4}$ torr.
3. *Ionization gauge.* This type of gauge measures pressure in terms of a current flowing between electrodes. The operation is based on the fact that electrons emitted by a heated filament and attracted to another electrode will ionize the gas through which they pass. The current that flows depends on the amount of ionization produced, that in turn depends on the pressure and the composition of the gas. These gauges are useful for very low pressures. A typical gauge has a pressure range of $10^{-3}-10^{-8}$ torr, but the range can be extended to pressures as low as 10^{-11} torr.

14.3. VACUUM APPARATUS

Some applications of vacuum require only simple apparatus. An aspirator for filtration or distillation may be connected through an empty trap to standard laboratory glassware with rubber tubing. However, thin walled tubing is likely to collapse; heavier walled vacuum tubing should be used. **Glassware under vacuum can implode with serious consequences, and therefore, care must be taken in selection of such glassware to eliminate apparatus with even small cracks. Standard glassware with flat surfaces (e.g., Erlenmeyer flasks) must *never* be evacuated, because they are not safe for such use.** Heavy-walled suction flasks are obvious exceptions to this rule.

More sophisticated vacuum lines are employed for manipulation of volatile and air-sensitive compounds, and are useful for synthetic and purification procedures, and for physical measurements.[4] Sections which are demountable by means of ground glass joints, and stopcocks to permit isolation of particular parts of the system, usually are incorporated, as are traps, gas storage bulbs, and pressure measuring devices. Stopcocks used in vacuum work are often of the hollow-plug type illustrated in Fig. 14.6, that are designed to minimize leakage. Generally, large-bore stopcocks are preferred, because they permit more rapid evacuation of the system. The plug of a vacuum stopcock should always be used with the same barrel; care should be taken not to interchange plugs when they are removed for cleaning. Ground glass surfaces to be used in high vacuum systems must be treated with care to maintain leak-free connections. They must be kept free of grit, dust, and corrosive chemicals such as alkali. All ground glass connections in a vacuum system must be greased, both to prevent their "freezing", and to prevent leaks. For low vacua, inexpensive stopcock greases are adequate, but for high vacuum work higher quality, low vapor pressure greases specified for use in high vacuum are required. The properties of some common kinds of high vacuum grease are given in Table 14.1. Joints or stopcocks may be greased by applying several strips of grease to the ground glass surface, and rotating the parts together so that a transparent, streak-free film is formed. Excess grease should be avoided. Prior to greasing, joints should be cleaned of old grease with an appropriate solvent (toluene, hexane). Stopcocks should be turned with a slow, steady motion, and all manipulations and dismantling should be done with care so that excessive strain is not exerted on the glass. Warming with a "heat-gun" is helpful when the grease has stiffened.

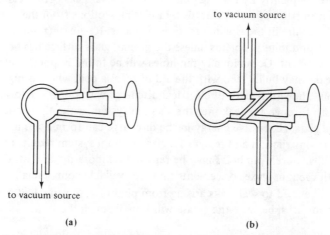

to vacuum source

to vacuum source

(a)

(b)

Figure 14.6. Two types of vacuum stopcocks: (a) a hollow-plug type shown in the open position, (b) a type with an oblique bore, shown in the position in which the bulb of the stopcock is being evacuated while the stopcock is in the closed position.

TABLE 14.1

Some Widely Used Greases for Vacuum Work

Grease	Vapor pressure at 20°C (torr)	Recommended use	Temperature range for general applications (°C)
Apiezon H	10^{-9}	General purpose high vacuum; wide temperature range	−15−250
Apiezon L	10^{-11}	General purpose high vacuum	10−30
Apiezon M	10^{-8}	Moderate vacuum	10−30
Apiezon N	10^{-9}	High vacuum, especially for stopcocks	10−30
Apiezon T	10^{-8}	General purpose high vacuum, good at high temperatures	0−120
Dow-corning high vacuum grease (silicone)	$<10^{-6}$	General purpose; more resistant to organic solvents than hydrocarbon greases above	−40−200

Poor performance of a vacuum system may be caused by contaminated pump oils, but more frequently by leaking joints or stopcocks. Leaks often can be isolated to a particular portion of the apparatus by isolating sections from the pumps and seeing if the pressure in the isolated section increases. Re-greasing normally will solve stopcock and joint problems unless the ground glass surface has been damaged by careless treatment. Occassionally pin-holes will be found in poorly made glass seals; such leaks may be located with the aid of a Tesla coil, which is high frequency discharge coil. The visible path of electrical discharge will follow the lower resistance path through the leak into the vacuum system as the tip of the coil is held near the glass. (Excessive energy in the discharge can form holes in thin areas of the glass.) Construction and repair of a glass vacuum system requires some glass blowing skill. However, pin holes may be repaired temporarily, and other seals can be made with vacuum waxes or cements that are available commercially. Stopcock grease cannot be used to seal leaks arising from pin holes, or scratched or warped ground glass surfaces, because the grease will flow through the flaw.

REFERENCES TO CHAPTER 14

1. D. P. Shoemaker, C. W. Garland, and J. I. Steinfeld, "Experiments in Physical Chemistry," 3d ed., McGraw-Hill, New York, 1974.

2. F. Daniels, R. A. Alberty, J. W. Williams, C. D. Cornwell, P. Bender, and J. E. Harriman, "Experimental Physical Chemistry," 7th ed., McGraw-Hill, New York, 1970.

3. R. E. Dodd and P. L. Robinson, "Experimental Inorganic Chemistry," Elsevier, Amsterdam, 1975.

4. W. L. Jolly, "The Synthesis and Characterization of Inorganic Compounds," Prentice-Hall, Englewood Cliffs, New Jersey, 1970.

Spectroscopic Techniques

15.1. INTRODUCTION

Spectroscopy is the study of the absorption, emission or scattering of electromagnetic radiation by matter. The information obtained from spectroscopic measurements varies with the type of spectroscopy and the equipment used, but some of the most common applications include:

- determination of the quantum mechanical structure (e.g., energy levels) of atoms and molecules
- determination of the geometry or the exact structure of matter, including the distances and angles between atoms in molecules and crystals
- qualitative identification of the molecules and/or atoms present in a sample, including the identification of functional groups
- quantitative determination of the molecules and/or atoms present in a sample
- investigation of inter- and intra-molecular interactions such as hydrogen bonding
- investigation of the kinetics of chemical reactions.

Although many of the above applications can be achieved by traditional chemical methods, some cannot. Furthermore, spectroscopic methods will generally require less sample, which usually can be recovered and provide the information more rapidly and more precisely than traditional methods. There are, however, situations for which the reverse is true, and spectroscopic methods should be thought of as being complementary to, rather than a replacement for, the more traditional chemical methods. It is therefore imperative that the student become familiar with spectroscopic techniques and some of their applications and limitations as soon as possible. A complete treatment of the various types of spectroscopy would require several volumes and is clearly not possible here. In this chapter some of the general features of spectroscopy will be described, followed by a brief discussion of the methods and applications of the spectroscopic techniques most frequently encountered in the chemical laboratory. For a more complete treatment or description of special applications, the student is urged to consult one of the numerous books available on the various types of spectroscopy or their applications.[1-3]

References to 15.1

1. A. Weissberger (ed.), "Technique of Organic Chemistry," Interscience, New York, 1956, vol. IX; see also 3d ed., 1959, vol. 1, part 3.
2. J. C. P. Schwartz (ed.), "Physical Methods in Organic Chemistry," Holden Day, San Francisco, 1964.
3. H. H. Willard, L. L. Merritt, Jr., and J. A. Dean, "Instrumental Methods of Analysis," 5th ed., Van Nostrand, New York, 1974.

15.2. ELECTROMAGNETIC RADIATION AND SPECTROSCOPY

The theory of the interaction of electromagnetic radiation with matter is complicated and beyond the scope of this book. Fortunately, most applications of spectroscopy are empirical in nature, so that a detailed understanding of these interactions is unnecessary at this point. Before proceeding, however, it is necessary to review briefly some fundamental properties of electromagnetic radiation. In general, electromagnetic radiation can be characterized by its wavelength, λ, or frequency, ν, and these are related by the well known formula:

$$\nu = c/\lambda \tag{15.1}$$

where c is the velocity of light, which in a vacuum is 3.00×10^{10} cm/s (3.00×10^8 m/s). The frequency generally is expressed in Hertz, Hz (1 Hz = 1 cycle/s). The wavelength may be expressed in several units depending upon the spectral region, but the most commonly used units are the ångstrom, Å, the nanometer, nm (formerly called a millimicron, mμ), the micrometer, μm (formerly called the micron), and less commonly, the centimeter. These are simply related to each other, 1 μm = 1000 nm = 10,000 Å = 1×10^{-4} cm.

In addition to wavelength and frequency, electromagnetic radiation often is characterized by its wavenumber, $\bar{\nu}$, which is the number of waves per centimeter. Because the wavenumber is merely the reciprocal of the wavelength expressed in centimeters, the units are cm^{-1}, or reciprocal centimeters, although these are sometimes referred to as Kaysers, K, or, less accurately, simply wavenumbers. The wavenumber is a measure of frequency, and according to Eq. (15.1) multiplying the frequency expressed in cm^{-1} by the speed of light (in cm/s) yields the frequency in Hz. Finally, it should be recalled that the energy of the radiation is independent of its intensity and is proportional to its frequency according to the relationship given by Planck:

$$E = h\nu \tag{15.2}$$

where Planck's constant, h, has the value of 6.626×10^{-27} erg · s (6.626×10^{-34} J · s). Because the energy is linearly proportional to the frequency and inversely proportional to the wavelength, the wavenumber often is used also as a measure of the energy. It should be noted that the energy calculated from Eq. (15.2) is the energy per molecule (or atom); to obtain the energy per mole, as, for example, is usually encountered in thermodynamic calculations, the energy obtained from Eq. (15.2) must be multiplied by Avogadro's number. As an example of the above relationships, light which has a wavelength of 500 nm (or 5000 Å) has a wavenumber of $1/5000 \times 10^{-8}$ cm = 20,000 cm^{-1}, a frequency of 6.00×10^{14} Hz, and an energy of 3.98×10^{-12} erg (3.98×10^{-19} J). This corresponds to 2.40×10^5 J/mol (57.3 kcal/mol).

Although in theory there are no limits on the wavelengths possible for electromagnetic radiation, only relatively small regions of the total spectrum are used in spectroscopy. Figure 15.1 illustrates the major divisions of the electromagnetic spectrum, together with the types of spectroscopy encountered in these regions. Even in these regions the absorption or emission of radiation is a highly selective process. When a molecule or atom absorbs or emits radiation, its energy changes by the amount given by Eq. (15.2). Because atoms and molecules may exist only in certain discrete energy levels (i.e., the levels are quantized), only certain wavelengths can be absorbed or emitted. This high degree of selectivity results in a unique absorption or emission spectrum for each atom or molecule and is one of the properties which makes spectroscopic methods so useful. Some, but certainly not all, of the applications of the more common types of spectroscopy are presented in Table 15.1. Of these, those most frequently encountered in the chemical laboratory are nuclear magnetic resonance (NMR) and those involving the

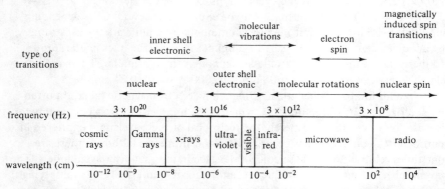

Figure 15.1. The electromagnetic spectrum. Notice that the scale is logarithmic. The boundaries between the types of transitions are not sharp.

TABLE 15.1

A Brief Summary of Some Common Applications of Spectroscopic Methods

Method	Principle	Usual State of Sample	Information
Gamma ray emission spectroscopy	Emission from nucleus in an excited nuclear state	Solid or solution	Qualitative and/or quantitative identification of radioisotopes.
Mössbauer spectroscopy	Resonance absorption of gamma rays by nuclei incorporated in a crystal lattice	Solid	Qualitative and/or quantitative identification of isotopes. Oxidation state of isotopes. Nature of chemical bonds or crystal lattice.
X-Ray diffraction	Scattering of X-rays	Solid: powder or single crystal	Determination of complete (e.g., bond lengths, angles, etc.) molecular or crystal structure. The interpretation is often very difficult. Powder spectra may be used for identification of compounds but for crystal structure their use is limited to cubic crystals.
X-Ray fluorescence spectroscopy	Emission of X-rays by atoms due to de-excitation of inner shell electrons	Solid or solution	Qualitative and/or quantitative identification of elements.

Continued

269

TABLE 15.1–Continued

Method	Principle	Usual State of Sample	Information
Ultraviolet and visible absorption spectroscopy	Absorption of light leading to excitation of electronic states	Dilute solutions; solids (usually done by reflection techniques)	*Atomic* absorption: normally used for quantitative determination of an element. *Molecular* absorption: can yield information about the electronic structure of the molecule (e.g., energy levels and bonding scheme), or about metal-ligand interactions in inorganic complexes. Some information about the types of functional groups and degree of conjugation present may also be determined, but generally it is not as useful as IR and NMR. High sensitivity is possible making it very useful for quantitative studies including analyses, equilibria or kinetics.
Ultraviolet and visible emission spectroscopy	Emission of light caused by the de-excitation of electronic states	Solid or solution (glassy solution in special cases, particularly for phosphorescence)	*Atomic* emission: very sensitive and used for the quantitative and/or qualitative determination of elements. *Molecular* emission (usually fluorescence, but occasionally phosphorescence); most often used for quantitative studies. The sensitivity can be high but the molecule must fluoresce (or phosphoresce) in an accessible region, which limits the method primarily to conjugated systems.
Infrared spectroscopy	Absorption of infrared radiation leading to the excitation of vibrational states of molecules	Solid; liquid; solution; vapor	Molecular structure. Usually used to show the presence and position of functional groups as well as the molecular skeleton. For small molecules detailed structural information is possible (e.g., bond lengths and angles). Has

Technique	Description	Sample	Information / Applications
Raman spectroscopy	Similar to IR, but the transitions occur as the result of inelastic scattering of visible or ultraviolet radiation	Solution; solid (less commonly)	some quantitative applications but generally they are more difficult than ultraviolet-visible techniques. Similar to IR, above. Transitions which will not occur (or are very weak) in IR frequently may be observed in the Raman spectrum, and *vice versa*.
Microwave spectroscopy	Absorption of microwave radiation resulting in transitions between rotational states of molecules	Vapor	Structural information such as conformation, bond lengths and angles, dipole moment determination.
Electron spin resonance	Absorption of microwave radiation resulting in transitions between the magnetic spin levels of unpaired electrons in a magnetic field	Vapor; solutions (usually dilute); solid (usually glassy single crystals) solutions (in special cases)	Existence and number of unpaired electrons. Structural information, especially as applied to atomic and molecular orbital calculations of electron density. Metal-ligand interactions. Kinetics of radical reactions.
Nuclear magnetic resonance	Absorption of radio-frequency radiation leading to transitions between the magnetic spin levels of certain nuclei in a magnetic field	Concentrated solution; solid (in special cases)	Usually applied to protons but other nuclei, especially those with a nuclear spin of 1/2 such as ^{13}C and ^{19}F are becoming increasingly important. Overall structure (e.g., types and positions of functional groups) of organic molecules. Can be used for quantitative studies (e.g., equilibrium or kinetics), but much less sensitive than UV-visible techniques.

absorption of radiation in the ultraviolet, visible, and infrared regions. The rest of this chapter will be primarily concerned with these techniques. X-ray diffraction, which involves the scattering, rather than the absorption or emission of radiation, is discussed in Chap. 16.

15.3. QUANTITATIVE MEASUREMENTS

15.3.1. Absorbance and the Beer-Lambert Law

To perform quantitative measurements a relationship between the amount of radiation absorbed and the concentration of the absorbing species is required. Such a relationship is the Beer-Lambert Law, usually referred to as simply Beer's Law. If I_0 and I represent, respectively, the intensity of the radiation at a particular wavelength before and after it passes through a length, b, of the solution, and if the concentration of the substance responsible for the absorption is c, then Beer's Law may be written:

$$-\log(I/I_0) = abc \tag{15.3}$$

The proportionality constant, a, is known as the *absorptivity*, and it is a property of the substance that varies with the wavelength of the radiation. When the concentration is expressed in moles/liter and b is in centimeters, a is known as the molar absorptivity and is often written as ϵ. The absorptivity formerly was known as the extinction coefficient, and this term is still frequently employed. The quantity I/I_0 is known as the *transmittance*, T, and it is often multiplied by 100 and referred to as percent transmittance. The Beer-Lambert Law states that for a given solute the negative log of the transmittance at a particular wavelength is proportional to the thickness and concentration of the solution.

The *absorbance* of the solution, A, is defined as the negative log of the transmittance, and Beer's Law generally is written in the more convenient form:

$$A = abc \tag{15.4}$$

One of the more important properties of the absorbance is that it is usually additive. If a solution contains more than one absorbing substance, and if there is no interaction between these substances, the total absorbance of the solution at any given wavelength will be the sum of the individual absorbances at that wavelength. Thus, for a two component system the total absorbance at a particular wavelength will be

$$A = a_1 b c_1 + a_2 b c_2 \tag{15.5}$$

Beer's Law is a limiting law and as such it is applicable only to dilute solutions, usually less than 0.01 M. In more concentrated solutions interactions between solute molecules may become important and cause deviations from the linear relationship between absorbance and concentration. In concentrated solutions variations in the refractive index also become important. The index of refraction determines the amount of radiation that will pass through the solution and how much will be lost at the interfaces between the sample and container; thus variations in the refractive index also will cause deviations from Beer's Law. Even in dilute solutions, however, deviations sometimes occur. These deviations can be either chemical or instrumental in nature. Chemical deviations occur when the solute either interacts with the solvent or undergoes a chemical reaction. Instrumental deviations generally are a result of using radiation that is not sufficiently monochromatic, i.e., the range of frequencies present is too large. Instrumental deviations of this type can be minimized by taking the measurements in a region in which the absorptivity is not changing rapidly, such as at the top of a peak in the absorption curve. This approach is illustrated in Fig. 15.2. In infrared spectroscopy it is usually necessary to use a rather large range of frequencies (i.e., bandwidth), and because the absorption peaks are generally quite sharp, Beer's Law is often not obeyed. Similarly, atomic absorption lines are so narrow that it is necessary to use an emission line from the element being investigated to obtain a source of radiation that is sufficiently monochromatic for quantitative work. Another instrumental problem is scattered light, and this is also more severe in infrared spectrometers than those used in the ultraviolet or visible region.

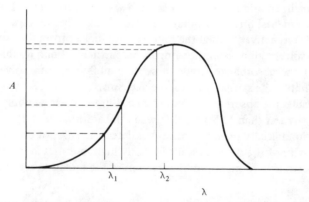

Figure 15.2. The effect of bandwidth on a quantitative measurement. Notice the large uncertainty in the absorbance when monitoring at wavelength λ_1, as opposed to the maximum at λ_2.

15.3.2. Procedures for Quantitative Absorption Studies

Below is a list of some of the more important procedures that should be followed in quantitative work.

1. Prepare a calibration curve, that is, a plot of A vs. concentration for several known concentrations. This task should be done even for systems which have been reported to obey Beer's Law, because deviations from linearity can arise

from impurities or from the instrument being used. Although it is desirable for the calibration curve to be linear, a non-linear curve can be used, if necessary, but first be certain that the cause of the curvature is unavoidable. Sufficient data points should be taken to define the calibration curve, and the range should be large enough so that the unknown concentrations are included within its limits. To minimize environmental effects the compositions of the calibration solutions should be as close as possible to those of the unknown solutions; that is, all species should be present in approximately the same amounts in both standards and unknowns.

2. Choose a wavelength which corresponds to a maximum, i.e., a peak, in the absorption spectrum. It is a good idea to examine the absorbance at a few different wavelengths to verify the position of the peak, because a slight error in calibration of the instrument can result in a discrepancy between the observed and literature values of a peak position. The bandwidth of the radiation also should be as narrow as is practical.

3. Make certain that the cells and equipment are absolutely clean and that no scratches are present on the cells. Any residual solvent used for cleaning must be thoroughly rinsed from the cell with the solution to be used before making the measurements, and the cells must not be handled in places that will be in the path of the radiation.

4. The instrument should be properly balanced *at each wavelength*. For single beam instruments, this means that the transmittance should be set to read 0% and 100%, respectively, when the radiation is blocked from the detector and when the solvent alone is in the sample compartment. For a double beam instrument the absorbance should be set to read 0 when pure solvent is present in both the sample and reference beams.

5. For best results the absorbance should be kept in the range of about 0.2–0.7, and certainly less than 1. Notice that when the absorbance is 2, only 1% of the radiation initially present reaches the detector. This not only means that the sensitivity of the detector must be fairly high, but that any scattered light present becomes a more important fraction of the radiation reaching the detector.

15.3.3. Emission Methods

When a molecule or atom is excited from the lowest energy level to a higher energy level by means of radiation, a flame, or an electric arc, some or all of the excitation energy may be re-emitted as radiation. Because the number of excited species, and hence the amount of emission, depends on the concentration of the substance, the intensity of this emission can be used in principle to measure this concentration. The emission methods most commonly used for quantitative work are flame photometry for atoms and fluorescence spectroscopy for molecules with significant conjugation. In recent years the technique of x-ray fluorescence has become increasingly important for the quantitative analysis of elements.

The procedure for emission work is similar to that used in absorption spectroscopy and will not be given in detail here. The usual procedure is to construct a calibration curve and then to make measurements on the unknowns.

15.4. SPECTROMETERS

Although the exact nature of the spectrometer components vary with the spectral region and application, all spectrometers have several features in common. The usual arrangement of these components for an instrument used for absorption studies is shown in Fig. 15.3. For emission work the position of the wavelength selector is changed so that it is between the sample and the detector. Some of the more general features of spectrometers are discussed briefly below.

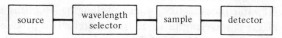

Figure 15.3. A schematic diagram for a typical absorption spectrometer. For emission studies the positions of the sample and wavelength selector often are reversed.

15.4.1. Sources

The purpose of the source is to provide a reasonably powerful beam of radiation in the spectral region of interest. The requirements of the source depend on the actual application. If it is desired to obtain the absorption spectrum, that is, the variation in the absorption with wavelength, then the source should be continuous and provide radiation of all wavelengths in the region of interest. On the other hand, if it is desired merely to measure the absorbance at a particular wavelength, then the source need not be continuous. In atomic absorption, for example, the source is normally an emission line of the element being determined. In emission work generally there are few requirements for the source other than its ability to elevate the molecules or atoms to the appropriate excited state. In the case of atomic emission studies the exciting source is often an electric arc or flame.

15.4.2. Wavelength Selectors

In most spectroscopic experiments it is necessary to isolate a narrow band of radiation from the rest of the radiation emitted by the source. The width of this band often is given in terms of the *effective bandwidth,* that is, the range bounded by the wavelengths for which the intensity of the radiation has fallen to one-half of its maximum value. Clearly the bandwidth is related to the spectrometer's resolution, that is, its ability to separate closely spaced peaks in the spectrum. In general the bandwidth should be narrow compared to the width of the spectral peaks of interest, but notice that a decrease in bandwidth is usually accompanied by a decrease in the intensity of the radiation, making detection more difficult. Wavelength selection may be accomplished with filters or monochromators, or a combination of the two.

The simplest and most common filters are absorption filters, which merely absorb the undesired wavelengths and transmit the rest. Such filters may be either "cutoff" that absorb all radiation above or below a certain wavelength while transmitting the rest, or they may transmit a band of radiation. However, the bandwidth for absorption filters is quite large. Interference filters, which are both delicate and more expensive, have relatively narrow bandwidths (as low as 10 nm) as well as a higher transmittance at the desired wavelengths. Filters generally are used when wavelength selection is not critical and will not be changed during the experiment, as, for example, in fluorescence analysis.

Monochromators are devices that disperse polychromatic radiation by either a prism or a diffraction grating and then permit the desired wavelengths to pass through an exit slit while blocking the other wavelengths. Rotating the prism or grating permits different wavelengths to be transmitted. This process is illustrated for a prism monochromator in Fig. 15.4. The bandwidth of monochromators is much smaller generally than that of even interference filters and depends on the degree of dispersion and the slit width. Because the desired wavelength and the bandwidth can be controlled easily with monochromators, they are particularly useful for spectrometers.

Because the effective bandwidth of a monochromator is controlled by the slit width, it is appropriate to discuss this adjustment briefly. There are two slits on a monochromator, one at the entrance and the other at the exit. The entrance slit is required to render the entering light reasonably parallel, while the exit slit actually limits the spectral region being transmitted by the monochromator. Narrowing the

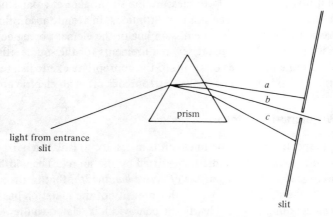

Figure 15.4. A schematic of a prism monochromator. The prism dispenses the polychromatic radiation according to wavelength. Only radiation around the desired wavelength, i.e., ray *b*, is able to pass through the slit. The wavelength of ray *a* is longer (i.e., redder) and that of ray *c* is shorter (bluer) than the desired wavelength.

slits will produce radiation that is more nearly monochromatic, but at the expense of intensity. Usually the entrance and exit slits should be the same width for best results. On many recording spectrometers the slit widths are "programmed" to vary automatically as the spectrum is scanned in such a way that the signal at the detector is held constant.

15.4.3. Detectors

There are several devices that might be used to detect radiation. For visible light the simplest device is the eye, and the instrument is then referred to as a *spectroscope*. Another possibility, which is clearly not limited to the visible region, is a photographic plate, in which case the instrument is referred to as a *spectrograph*. The term *spectrometer* is reserved for the cases in which the radiation is collected and converted to an electrical signal. The nature of these collection devices obviously depends on the spectral region of interest and will not be discussed here. The main point is that the electrical signal can then be displayed on a meter, oscilloscope, or recorder. One final note on nomenclature should be made. An instrument that permits quantitative measurement of the intensity of the radiation often is referred to as a *photometer*. Because most spectrometers have this capability, the terms spectrometer and spectrophotometer often are used interchangeably.

15.4.4. Single and Double-Beam Spectrometers

In absorption spectroscopy only the absorption of radiation by the species under investigation is of interest. Changes in the intensity of the radiation caused by other factors such as scattering or absorption of the radiation by other species (e.g., the solvent or impurities) must be subtracted from the change produced by the entire sample. Because absorbances are additive, this correction may be accomplished simply by subtracting the absorbance measured for the sample cell and solvent from that of the cell and solution to be studied. A simpler method is to adjust the controls initially to read zero and 100% transmittance, respectively, when the beam of radiation is blocked from the detector and when only the sample cell and solvent are present in the beam. Because this procedure must be repeated at each wavelength, it is rather tedious. Furthermore, the success of both methods depends upon the stability of the electronics of the photometer over the length of time required to perform the steps. Modern spectrometers — especially recording units — eliminate these problems by splitting the radiation from the source into two beams, one of which passes through the entire sample (e.g., the solution), while the other passes through a reference substance (e.g., the solvent). The intensity of the two beams may then be compared directly to yield the desired absorption. In this way fluctuations in the source intensity or absorption common to both samples will not be detected. The primary disadvantage of this method is that the operator may not be aware that the solvent (or sample cell) is strongly absorbing the radiation in a particular region, although some instruments indicate when the reference beam energy is too low. When this occurs, very little radiation can reach the detector and erroneous results may be obtained. For the same reason, the common practice in infrared spectroscopy of "increasing" the transmittance by partially blocking (attenuating) the reference beam with an object must be done with care. Another problem with double-beam instruments is the possibility of the reference cell absorbing or scattering the radiation to a different degree than the cell in the sample beam. This problem may be corrected by using "matched" cells and obtaining a "baseline" by recording the spectrum of the cells when both are filled with pure solvent.

15.5. ULTRAVIOLET-VISIBLE SPECTROSCOPY

15.5.1. Introduction

The absorption or emission of radiation in the visible or ultraviolet region of the electromagnetic spectrum results in changing the energy level of one or more electrons in the molecule or atom. In the case of molecules this change in electronic energy may also be accompanied by a change in the rotational and/or vibrational energy as shown in Fig. 15.5. As the size of the molecule increases, the number of ways in which the vibrational energy can be distributed also increases, and it

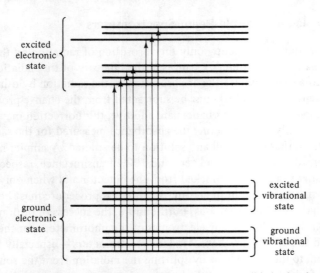

Figure 15.5. An illustration of some of the transitions possible in the electronic spectrum of a molecule. For simplicity only one excited vibrational state is shown for each electronic state, and the rotational states are indented. The relative spacing between the rotational and vibrational levels is greatly exaggerated, as is that between the vibrational and electronic levels. Finally, only the transitions from the lowest rotational level are shown, although at most temperatures transitions from other levels also are common.

becomes increasingly difficult to resolve the individual transitions. Hence, in most molecules the ultraviolet-visible spectrum consists of relatively broad bands. In the easily accessible spectral regions, that is, wavelengths between about 200 nm and 850 nm, the transitions normally occur between non-bonding or pi-bonding orbitals and pi-antibonding orbitals for organic molecules, and between d orbitals in transition metals. These last transitions often occur at visible wavelengths, yielding distinctively colored solutions. On the other hand, most transitions in organic molecules correspond to energies in the ultraviolet region, and as a result their solutions usually are colorless. The values of the molar absorptivities are measures of the probability of the electronic transitions occurring, and to some degree are characteristic of the type of transition involved. They are discussed in 15.5.4.

It is possible to obtain information about the electronic structure (e.g., the energies of the molecular or atomic orbitals) of a molecule or atom from its ultraviolet-visible spectrum. This type of knowledge is particularly useful in the case of transition metal complexes, because it can yield information about the coordination geometry, metal-ligand interactions and bonding. Generally, however, the applications of electronic spectroscopy are more practical in nature and are designed to tell what and/or how much of a particular substance is present. Some general features of electronic spectroscopy are discussed below.

15.5.2. Radiation Sources

There are several types of radiation sources available for use in the ultraviolet and visible regions of the electromagnetic spectrum, and it is worthwhile briefly to mention those that are most commonly encountered. As noted previously, the selection of a source depends in part upon the application. Thus, in the case of quantitative atomic absorption studies, the absorption lines are so narrow that it is necessary to use emission from the element being examined as the source. Such emission usually is obtained from a hollow cathode tube, although for low melting elements, such as the alkali metals, a vapor discharge tube occasionally is used. For molecular systems it is usually sufficient to use a continuous source in conjunction with a monochromator.

In the visible and near infrared region (i.e., between about 320 nm and 3000 nm) of the electromagnetic spectrum, by far the most common source is a tungsten (or tungsten-iodine) incandescent bulb. These lamps provide a continuous source of radiation characteristic of a blackbody radiator at a temperature on the order of $3000°C$. In the ultraviolet region the source is normally a hydrogen or deuterium (which produces a higher intensity) low pressure discharge tube. These sources also provide continuous radiation, although at wavelengths greater than 360 nm emission lines are superimposed on the continuum of the hydrogen discharge.

Other sources which may be used include low pressure mercury lamps and high pressure mercury or xenon lamps. Low pressure mercury lamps provide sharp emission lines with little continuous radiation in both the ultraviolet and visible regions, and are often used as wavelength calibration sources. On the other hand, high pressure lamps generally provide a fairly continuous source of radiation and the highest intensities are obtained with these lamps.

15.5.3. Solvents and Cells

Most routine work in the ultraviolet-visible region is done with solutions. If quantitative work is to be done, or if it is desired to obtain molar absorptivities, the concentrations of the absorbing species must be known. Otherwise, it is only necessary to have a concentration which will keep the peaks of interest on scale (in no case should the absorbance exceed 2). Care must be taken to select a solvent which is transparent in the region of interest, and that no absorption or emission is caused by impurities. Many commonly used solvents will be unsatisfactory in the ultraviolet region because of their own absorption bands. Table 15.2 lists several common solvents and the minimum wavelength at which they should be used in a 1 cm cell. The problem of impurities can be reduced by using a commercially available spectroscopic grade solvent, although these also should be examined carefully before use.

Similarly, the cells (cuvettes) must not be made of a material which absorbs in the region of interest. Glass cells may be used in the visible region, but for wavelengths below about 320 nm it is necessary to use cells made of fused silica or fused quartz. In recent years plastic cells also have been used for non-critical work in the visible region. The cells may be either rectangular or cylindrical, and for most work those with a path length of 1 cm are used, although other sizes are available. Longer path length cells are used for very dilute solutions (or gases) or when the transitions are especially weak. Shorter path lengths can be used when examining regions in which the solvent absorbs. When the quantity of solution is small, special micro cells, which still have a path length of 1 cm, can be used. Alternatively, spacers, which decrease the volume but not the path length of rectangular cells, are available. The cells are expensive, particularly those made of quartz, and must be handled with extreme care. They should be washed with soap and water or solvent and they should not be dried in an oven, which might change the path length. Strong acids should be avoided, and strong bases, which would attack the cell, must never be used. In addition, the cells must be protected from becoming scratched, and they should be handled only by the sides (often frosted) which are not in the light path. Fingerprints should be removed with solvent immediately.

15.5.4. Transitions

Many transitions appear in atomic spectra, but the nature of them is rarely of general interest and they will not be considered here. The types of transitions which appear in molecular spectra have more significance for most purposes. Transitions involving the excitation of electrons from π bonding to π antibonding orbitals ($\pi \rightarrow \pi^*$) are intense, having molar absorptivities of $10^3 - 10^5$. They usually are in the range 200–170 nm for localized π bonds, but lie at lower energies (longer wavelengths) for conjugated systems. Transitions involving the excitation of non-bonding electrons to π^* orbitals (n $\rightarrow$ π^* transitions) are generally weaker, with molar absorptivities on the order of $10-10^3$. The energies tend to be lower than those of $\pi \rightarrow \pi^*$ transitions, but still lie in the ultraviolet region of the spectrum. Complexes of transition metals exhibit transitions which are often described as arising from the excitation of electrons from one d orbital to another as a result of splitting of the energies of the d orbitals by the field of the ligands (d-d transitions); they can also be regarded as n $\rightarrow$ σ^* transitions and are often referred to as ligand field bands. These transitions typically have molar absorptivities on the order of $1-10^2$, and often lie in the visible region of the spectrum. The energies of $\sigma \rightarrow \sigma^*$ transitions are too high to be observed with a typical spectrometer and will not be discussed further here. In most molecules other than transition metal complexes, n $\rightarrow$ σ^* transitions also are at very high energies. Typically, they occur at the short wavelength end of the ultraviolet spectrum, and are often referred to as *end absorption* because the intensity continuously increases as the wavelength is decreased. Many of the more common solvents contain oxygen or halogen atoms, and the resulting end absorptions from the n $\rightarrow$ σ^* transitions limits the useful range of these solvents, as shown in Table 15.2.

Charge transfer transitions refer to those in which the distribution of the electron is greatly changed on excitation. Such a transfer may be between two different molecules (in which case we say a charge transfer complex has been formed) or it may simply involve transfer between two different groups in a single molecule, such as a metal ion and a ligand. In any case, the excitation results in the effective transfer of the electron from a donor to an acceptor site.

TABLE 15.2

Ultraviolet Cutoff Wavelength for Selected Solvents

Solvent	Cutoff (nm)
acetone	330
acetonitrile	210
aliphatic hydrocarbons (including cyclic)	210
benzene	280
carbon tetrachloride	265
chloroform	245
p-dioxane	220
ether	220
ethanol	210
methanol	210
pyridine	305
toluene	285
water	180

The cutoff wavelength above is the wavelength for which the absorbance is approaching unity when the solvent is in a 1 cm cell.

15.5.5. Qualitative and Structural Applications

Atomic emission spectroscopy can be used to analyze qualitatively (and in some cases quantitatively) a sample for most elements. The technique is simple and very sensitive, often being used to detect elements in the parts per million range. The presence of an element is best demonstrated by comparing the emission spectrum obtained from the sample with a known spectrum of the element in question, although tabulated spectral data such as that found in the Handbook of Chemistry and Physics or in the text by Brode[1] may be used.

With the exception of atomic emission mentioned above, ultraviolet-visible spectroscopy is generally more concerned with quantitative analysis than it is with qualitative analysis or structural determinations. Although the electronic spectrum of a molecule can be an aid in indicating the presence of certain functional groups* in that molecule, more often it is useful in indicating when such functional groups are absent. This is because electronic transitions usually involve several bonds so that it is difficult to associate the absorption (or emission) with particular functional groups, as is done for example in infrared spectroscopy. Furthermore, the bands usually are fairly broad and unresolved so that several compounds may yield quite similar spectra. Thus, although exact agreement (including molar absorptivities) in the spectra for a known and unknown compound is good evidence as to the identity of the compound, electronic spectra normally are used only in conjunction with other spectral or chemical data to eliminate or verify the possibility of certain structures.

Some structural information also may be obtained from the types of transitions which occur (or do not occur) in the electronic absorption spectrum. The molar absorptivities may be used to distinguish $\pi \to \pi^*$ from $n \to \pi^*$ transitions, although this must be done with caution; symmetry selection rules, for example, may result in some $\pi \to \pi^*$ transitions having relatively low absorptivities, as occurs in benzene. A better method is to change the solvent. As the polarity of the solvent is increased, $n \to \pi^*$ transitions generally will be shifted to shorter wavelengths, while $\pi \to \pi^*$ transitions will not be shifted, or they will be shifted to slightly longer wavelengths.[2]

The absence of any absorption at wavelengths above 200 nm indicates that no conjugated systems are present and strongly implies that there are also no non-bonding electrons in the system. A molecule containing two conjugated double bonds will generally have strong $\pi \to \pi^*$ absorption in the 220–250 nm region. As the degree of conjugation increases, the absorption will shift to longer wavelengths. Generally each additional conjugated double bond increases the wavelength of the absorption band about 30 nm, so that strong absorption at about 260, 300, or 330 nm is often caused by respectively, three, four, or five unit conjugated systems.

* The part of the molecule which is responsible for the absorption or emission of radiation is often referred to as the *chromophore*. Although this term is used in other types of spectroscopy, it is most often associated with electronic spectroscopy.

A colored organic compound is usually highly conjugated with at least four units of conjugation present, although occasionally the color may be caused by the presence of certain functional groups. The lack of any absorption above 200 nm also implies the absence of any lone-pair electrons, because such molecules would be expected to have $n \rightarrow \pi^*$ absorption in this region. Finally, the presence of certain functional groups will shift the absorption bands of the parent molecule in a reasonably predictable manner.[3]

In contrast to organic compounds, inorganic complexes of transition metals often are highly colored, and the absorption spectrum can provide useful information about the structure as well as the strength of the metal-ligand interactions in these compounds. The transitions in the visible region arise because the five d orbitals on the metal do not interact with the ligands to the same degree. For example, an orbital which points directly toward a ligand will be affected more by the presence of that ligand than will an orbital which is pointing between the ligands. Hence, although in the free ion the five d orbitals all have the same energy (i.e., they are degenerate), they will have different energies when the ligands are present. The actual pattern of energies obtained will depend on the geometry of the compound. Figure 15.6 illustrates the splitting pattern obtained for the commonly encountered octahedral and tetrahedral complexes. Other geometries will yield different patterns. The crystal field splitting parameter, $10\,Dq$ (or Δ), is the energy difference between the two sets of orbitals, and its magnitude depends on the strength of the interaction between the ligands and the metal ion. The lowest energy band observed in the visible spectrum will correspond approximately to $10\,Dq$. A more complete discussion of these interactions and their relationship to the visible spectrum of the compound may be found in textbooks of inorganic chemistry.[4]

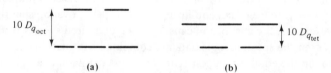

$10\,D_{q_{\text{oct}}}$ (a) $10\,D_{q_{\text{tet}}}$ (b)

Figure 15.6. The splitting pattern observed for the d orbitals of a transition metal complex in (*a*) an octahedral and a (*b*) tetrahedral ligand field. Notice that the splitting is smaller in the tetrahedral case and that the levels are inverted.

15.5.6. Quantitative Applications

Electronic spectroscopy is an extremely useful method for quantitatively determining the amount of a substance present in a sample. The sensitivity is normally very high because in absorption studies the absorptivities are usually large, while in emission measurements very low levels of light may be detected. The methods employed have been discussed in a previous section and will not be reproduced here. Instead some of the more common applications will be described briefly. As indicated previously, emission methods generally are limited to atoms or to molecules having a fairly high degree of conjugation. Because absorption methods are more widely used, particularly in undergraduate laboratories, the discussion shall be restricted to this area.

In principle, absorption methods could be used in almost any investigation in which it was necessary to know the concentration (or ratio of concentrations) of the species present. Two of the more commonly encountered cases are kinetic and equilibrium studies.

15.5.6.1. Kinetics

If one of the reactants or products absorbs energy at a wavelength at which the other species present do not absorb energy, the rate of change in the concentration of the absorbing substance can be followed by monitoring the absorbance at the corresponding wavelength. Although it is possible often to monitor the reaction directly in the sample cell, the more common procedure is to take aliquots of the reaction mixture at various times, dilute them quantitatively, and determine the absorbance of the resulting solutions. With a conventional recording spectrophotometer the reaction times that can be examined vary from seconds to days or more.

15.5.6.2. Equilibria

If the absorption peaks of the species involved in an equilibrium are sufficiently separated, electronic spectroscopy may be used to determine their concentrations and hence the equilibrium constant. Furthermore, under certain conditions electronic spectroscopy may be used to obtain the stoichiometry of, and even test for the existence of, an equilibrium. If two peaks overlap slightly, it often is possible to subtract the absorption caused by the undesired species from the total absorption. Often the spectroscopic data are combined with data from other sources to obtain the equilibrium constant. The most common example of this approach is in the study of acid-base equilibria, where a pH measurement is used to determine the concentration of H^+ while the absorbance gives another equilibrium species. It should be noted that if n species are in equilibrium, then generally it is only necessary to determine the concentration of $n - 1$ of the species, because the remaining concentration may be obtained from the law of conservation of mass. For example, if the amount of the acid initially present corresponds to a concentration of $[HA]_0$ and the concentration of the anion is found spectroscopically to be $[A^-]$ at a pH corresponding to $[H^+]$, then the concentration of HA present at this pH is $[HA] = [HA]_0 - [A^-]$. The equilibrium constant is thus given by: $K = [A^-][H^+]/[HA]_0 - [A^-]$.

The existence of an equilibrium of the type $B + C \rightleftarrows D$ can often be demonstrated by examining the behavior of an *isosbestic point*. Such a point occurs at a wavelength for which the molar absorptivities (ϵ) for two of the components in equilibrium, e.g., D and B, are equal, and absorption from the other component(s) is negligible. When these conditions are fulfilled the absorbance at this wavelength will depend only on the sum of [B] and [D] but will be independent of their ratio, and hence independent of the amount of C which might have been added to shift the equilibrium. This explanation may be verified easily by examining Beer's Law for this case. The absorbance at any wavelength at which absorption by C is negligible is given by:

$$A = \epsilon_B b [B] + \epsilon_D b [D] \tag{15.6}$$

If $\epsilon_D = \epsilon_B = \epsilon$, this equation may be rewritten as

$$A = \epsilon b ([B] + [D]) \tag{15.7}$$

Therefore, the absorbance at this wavelength will remain constant if the sum of [B] and [D] is constant. If the absorption peaks of B and D are widely separated, then the isosbestic point will occur at a wavelength for which the absorbance is at or near zero, and hence will not be useful. Similarly, the relationship will not hold if a third species (e.g., C) is also absorbing in this region, except through an accidental combination of circumstances. Thus, the existence of an isosbestic point is good evidence that two species are in equilibrium, but it is not absolutely conclusive. An example of an isosbestic point is given in Fig. 15.7.

Quantitative studies often require a knowledge of the absorptivity of the various species being investigated. Often these absorptivities may be obtained by adjusting the initial conditions of the solutions, for example, in the case of acid-base equilibria only the protonated form will be present in very acidic solutions; i.e., when the pH of the solution is at least two units below the pK, while

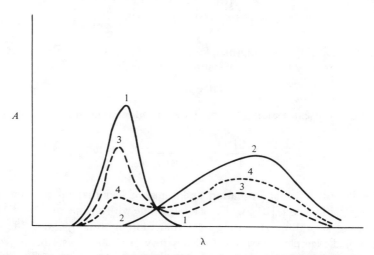

Figure 15.7. Illustration of an isosbestic point for two species in equilibrium. In all solutions the sum of the concentrations of B and D is constant. (1) Pure B, (2) pure D, (3), and (4) mixtures of B and D.

for strongly basic solutions, i.e., when the pH is at least two units greater than the pK, only the unprotonated form will be present. Thus, the absorption spectra and absorptivities for the protonated and unprotonated forms may be obtained by adjusting the pH of the solution. Often, however, it is not convenient or possible to determine such absorptivities directly. Several indirect methods are available, and one of the most commonly used is that developed by Benesi and Hildebrand.[5] Although this method was applied originally to the study of charge transfer complexes formed between aromatic molecules and halogens, it may be applied to any equilibrium such as that previously described, i.e., $B + C \rightleftarrows D$. The equilibrium constant for this reaction may be written as

$$K = [D]/[B][C] \tag{15.8}$$

This may be expressed in terms of the initial concentrations $[B]_0$ and $[C]_0$ as

$$K = \frac{[D]}{([B]_0 - [D])([C]_0 - D)} \tag{15.9}$$

If the initial conditions are adjusted so that, e.g., $[B]_0$ is much greater than $[C]_0$ then $[B]_0$ will necessarily also be much greater than $[D]$. Thus, in this case, Eq. (15.9) may be approximated as

$$K = \frac{[D]}{[B]_0([C]_0 - [D])} \tag{15.10}$$

or

$$\frac{1}{K[B]_0} = \frac{[C]_0 - [D]}{[D]} = \frac{[C]_0}{[D]} - 1 \tag{15.10a}$$

The concentration $[D]$ may, in principle, be obtained by applying Beer's Law at a wavelength where only D absorbs, because at this wavelength

$$[D] = A/\epsilon_D b \tag{15.11}$$

Substitution of this expression into Eq. (15.10a) and rearrangement yields

$$\frac{b[C]_0}{A} = \frac{1}{\epsilon_D K[B]_0} + \frac{1}{\epsilon_D} \tag{15.12}$$

This is the Benesi-Hildebrand equation, and it may be seen that a plot of the left side of the equation vs. $1/[B]_0$ will be a straight line with a slope of $1/(\epsilon_D K)$ and an intercept of $1/\epsilon_D$, thus permitting a determination of both ϵ_D and K.

As a final example of how spectroscopy may be applied to studies of equilibria, one of the more commonly used methods for determining the stoichiometry of a complex will be described briefly. The above discussion assumed a simple 1 : 1 complex, but often more complicated stoichiometries are observed such as $Z + nY \rightleftarrows ZY_n$. One of the simplest methods to obtain the stoichiometry is the method of continuous variations, often called Job's method. In this method the absorbance of several solutions is measured at a wavelength where the complex absorbs strongly but the other species do not. The solutions are prepared so that the *relative* amounts of the complexing species are changed, but the total number of moles of these species is not; that is, the total number of moles of Z and Y initially present is constant, but their ratio is varied. A plot of the absorbance vs. the mole fraction of Y initially added (that is, the ratio $[Y]_0/([Y]_0 + [Z]_0)$) will yield a curve such as that shown in Fig. 15.8. The sides of the triangle may be extrapolated until they intersect, and the mole fraction at which this occurs (0.5 in Fig. 15.8) yields the empirical formula of the complex. This method will yield erroneous results if more than one species is present such as ZY_n and ZY_m. This possibility may be investigated by repeating the experiment at a different wavelength and/or a different value for the sum $[Z]_0 + [Y]_0$. If more than one complex is present these additional experiments generally will yield different results.

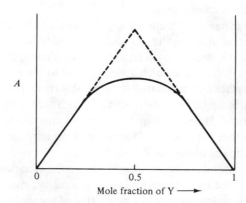

Figure 15.8. Job's plot for the reaction $Z + Y \rightleftarrows ZY$. The sum of the concentrations of Z and Y has been held constant. The extrapolated sides of the triangle intersect at a mole fraction of 0.5, corresponding to a stoichiometry of $Z : Y = 1 : 1$.

References to 15.5

1. W. R. Brode, "Chemical Spectroscopy," Wiley, New York, 1939.

2. H. H. Jaffe and M. Orchin, "Theory and Applications of Ultraviolet Spectroscopy," Wiley, New York, 1962, pp. 186–189.

3. J. C. P. Schwartz (ed.), "Physical Methods in Organic Chemistry," Holden Day, San Francisco, 1964.

4. See for example, F. A. Cotton and G. Wilkinson, "Advanced Inorganic Chemistry: A Comprehensive Text," Interscience, New York, 1972, chap. 20.

5. H. A. Benesi and J. H. Hildebrand, *J. Am. Chem. Soc.*, **71**, 2703 (1949).

15.6. INFRARED SPECTROSCOPY

15.6.1. Introduction and General Principles

Infrared spectroscopy is a powerful technique for characterizing chemical bonds in both organic and inorganic compounds. Among its applications are identification of organic compounds, study of chemical bonding in organic and inorganic compounds, and measurement of bond lengths and angles.

The infrared spectrum includes radiation with frequencies between 12,500 cm^{-1} (0.8 μm) and 25 cm^{-1} (400 μm), but only radiation with frequencies between 4,000 cm^{-1} (2.5 μm) and 625 cm^{-1} (16 μm) is used at present for routine work. This frequency range is often referred to as the "sodium chloride region" or the "rock salt region" of the infrared spectrum, because sodium chloride can be used as an optical material at these frequencies. The region below 625 cm^{-1} often is referred to as the far infrared region. Frequencies arising from bonds of heavy elements occur here. Far infrared spectra are useful in structural studies of some inorganic compounds but the spectra are not of interest in the routine identification of organic compounds.

Absorption of infrared radiation by a molecule raises the molecule to an excited vibrational state in which the molecule is vibrating with a larger amplitude than in the ground state. Because each energy state of a molecule corresponds to a definite energy, only frequencies which are equivalent to the energy difference between the ground state and an excited state are absorbed. The resulting absorption spectrum is a series of bands, each of which represents a transition to an excited vibrational state. Molecular vibrations are classified as stretching or bending, depending upon whether the motion of the atoms is primarily along the bond axis or primarily perpendicular to the bond axis.

It is sometimes useful further to classify molecular vibrations with respect to whether or not the molecule undergoes a loss of symmetry, especially for stretching vibrations. Vibrations which preserve the symmetry of the molecule are referred to as symmetric vibrations and those which lower the symmetry are known as asymmetric vibrations. The symmetric and asymmetric stretching vibrations of the H_2O molecule are illustrated in Fig. 15.9, the direction of motion of the atoms being denoted by the arrows. The frequency of the asymmetric stretching vibration is higher than that of the corresponding symmetric vibration, and if the two

symmetric stretching   asymmetric stretching

Figure 15.9. Symmetric and asymmetric stretching vibrations of the H_2O molecule.

vibrations are distinguishable in an infrared spectrum, it is customary to specify the symmetry of each.

Many of the molecular vibrations which are important in infrared spectroscopy can be approximated as involving the distortion of only one bond or two or three bonds of the same type, and this fact is the basis for naming infrared absorption bands. In addition to the nature of the chemical bonds, it is customary to identify the vibration as a bending or a stretching and if it is useful, to state the symmetry of the vibration. As an example, the C–H stretching vibrations of toluene are usually subdivided into the two general categories of aliphatic C–H stretching and aromatic C–H stretching. The aliphatic and aromatic C–H stretching vibrations are not resolved clearly into asymmetric and symmetric bands by the spectrometers used for routine infrared spectra, so that the symmetry of these bands is not usually given. On the other hand, the N–O stretching vibrations of the nitro group give rise to two well resolved absorption bands that are differentiated by the symmetry of the vibration, i.e., as symmetric and asymmetric N–O stretching vibrations.

Because a stretching vibration can usually be approximated as the expansion and contraction of a single type of bond, naming the two atoms is usually sufficient to identify the vibration (e.g., C–H stretching). A bending vibration involves primarily variation of a bond angle, and therefore it is necessary, at least in principle, to specify three atoms to identify a bending vibration (e.g., C–O–H bending). Often it is assumed that one terminal atom is carbon, and only the two remaining atoms are specified. Thus, the abbreviated term O–H bending is sometimes used in place of the more exact term C–O–H bending to indicate bending of the C–O–H bond. The principal disadvantage of the abbreviated terminology is the danger of confusion arising from a literal interpretation. Of course the O–H bond itself does not bend.

The vibrational frequency of a diatomic molecule, which corresponds to the frequency of the vibrational absorption band, is defined by Eq. (15.13)

$$\bar{\nu} = \frac{1}{2\pi c} \sqrt{\frac{k}{\mu}} \tag{15.13}$$

where

$\bar{\nu}$ is the frequency in cm^{-1}
k is the force constant of the bond in dyn/cm
μ is the reduced mass of the atoms in g
c is the velocity of light in cm/s

A knowledge of the effects of bond order and atomic masses on vibrational absorption frequencies as defined by Eq. (15.13) is helpful in understanding the relative frequencies of various vibrational absorption bands. The force constant k is approximately proportional to the bond order; that is, the force constant of a carbon-carbon double bond stretching vibration is approximately twice that of a carbon-carbon single bond stretching vibration. Also, the force constant of a bending vibration is approximately one fourth the value of the force constant of the corresponding stretching vibration. The aliphatic C–H stretching frequency ($2{,}950 \text{ cm}^{-1}$) is much higher than the aliphatic C–C stretching frequency ($1{,}100 \text{ cm}^{-1}$) primarily because of the effects of atomic masses, while the C=C stretching frequency ($1{,}650 \text{ cm}^{-1}$) is significantly higher than the C–C stretching frequency ($1{,}100 \text{ cm}^{-1}$) primarily because of the effects of the bond order. Similarly, the fact that the aliphatic C–H stretching frequency ($2{,}950 \text{ cm}^{-1}$) is approximately twice the aliphatic bending frequency ($\sim 1{,}400 \text{ cm}^{-1}$) results from the greater force constant of the stretching vibration.

Interpretation of infrared spectra is vastly simplified by the fact that *to a good approximation* the absorption frequencies of a given structural unit essentially are independent of the structure of the remainder of the molecule. Detection of a specific functional group by infrared spectroscopy thus becomes a matter of observing its characteristic group frequencies in the infrared spectrum. It must be recognized that in many instances one part of the molecule will influence the absorption frequency of a second part through steric or electronic effects. Such second order effects provide valuable structural information and must be borne in mind when using infrared spectroscopy. For example, the aliphatic carbonyl stretching frequency is shifted from its "normal" value of $1{,}715 \text{ cm}^{-1}$ to a value of $1{,}745 \text{ cm}^{-1}$ in cyclopentanone. Failure to consider this possibility could lead one to misidentify the cyclopentanone as an ester on the basis of the carbonyl stretching frequency. In this case, examination of the remainder of the spectrum would indicate quickly that the compound was in fact not an ester. The need to evaluate the entire spectrum to test the assignments made on the basis of characteristic group frequencies cannot be overemphasized. The absolute necessity of obtaining independent data, such as chemical properties, also must be understood clearly. Infrared spectroscopy greatly simplifies identification of organic compounds, but additional information often is vital to obtaining the correct result.

15.6.2. Obtaining Infrared Spectra

The methods of obtaining infrared spectra are relatively simple, but care in selecting appropriate sample preparation techniques is necessary to obtain high quality spectral information. Details of sample preparation techniques for liquid and solid samples are described below, and information on techniques for handling gas samples may be found in texts such as that by Colthup *et al.*[1]

Sealed cells for liquid samples are available in a range of path lengths ranging from 0.01 mm to 1 mm. The path length is maintained precisely by a plastic or metal separator which also acts as a gasket to prevent leakage or volatilization. Path lengths of the order of 0.01–0.015 mm are suitable for pure liquids and path lengths of the order of 0.1 mm are commonly used for solution spectra. Quantitative determination of a solute is possible by the use of a calibration curve, provided the same cell and spectrometer are used for the samples and the standards. Simple cells for gas samples are usually constructed with a glass body fitted with vacuum stopcocks and standard taper joints suitable for connection to a vacuum line. Windows of appropriate material cemented on the ends of the cell body are held in place by retainer rings. Path lengths ranging from 1 cm to as much as 20 m are available, although long path lengths involve an internal reflecting system and are very elaborate in construction.

Optical surfaces for infrared spectroscopy are usually fabricated from sodium chloride or potassium bromide, and special care is necessary to prevent damage to the cell surfaces. Cell windows of water-insoluble materials such as silver chloride or IRtran® are available for handling aqueous solutions, but expense and poor mechanical properties of these materials prohibit their use in routine spectral work. Sodium chloride plates and cell windows must not be allowed to come in contact with water and contact with the atmosphere must be minimized. The optical surfaces must never be touched with the fingers. Cells must be cleaned after use with a volatile organic solvent such as acetone, dried, and stored in a desiccator as quickly as possible.

Often it is necessary to compensate for moderate light scattering or absorption by the sample by using an attenuator to reduce the intensity of the reference beam. An attenuator cannot be used to correct intense scattering or absorption, because it would be necessary to reduce the intensity of the reference beam to nearly zero, and the spectrophotometer will not function reliably when this is done. Attenuation should be used conservatively. In using an attenuator, always place the sample in the instrument before the attenuator, and remove the attenuator before the sample is removed. Reversing the order will cause the pen to be driven above the 100% transmittance level and may damage the instrument.

15.6.2.1. Instrument Operation

Table-top IR spectrometers of the type used for qualitative organic analysis are simple to operate and routinely yield spectra of high quality. If the spectrum is characterized by excessive noise or if the pen responds so slowly that bands are distorted, consult the instructor. It is recommended that the performance of the spectrometer be checked occasionally by obtaining the spectrum of a known compound and comparing it to a reference spectrum, such as that of a polystyrene film.

The functions of the front-panel controls are generally self-explanatory, but two precautions need emphasis. First, never unplug the line cord. Many bench-top spectrometers are heated electrically to prevent condensation of moisture, which damages the optical surfaces. Second, before attempting to adjust the wavelength drive manually, *be certain* that the scan control is in the neutral or reset position. The wavelength drive is easily damaged and force must never be used. Specific instructions will be provided for the instrument used, and should be followed carefully.

Calibration of the frequency scale is necessary to obtain accurate values of band frequencies, and a polystyrene film, approximately 0.05 mm thick, mounted in cardboard, is an excellent standard for this purpose. The frequencies of many bands of the polystyrene spectrum are known with high precision, and a summary of the most useful absorption bands is usually provided by the instrument manufacturer. For reference, the spectrum of a polystyrene film is given in Fig. 15.10, and the frequencies of a number of bands are indicated on the spectrum. In general, calibration bands should be selected to fall as close as possible to the frequency region of interest, and the reference band should be recorded on the chart prior to obtaining the sample spectrum. For general survey spectra, the band at $1,601$ cm^{-1} (6.24 μm) is commonly used. Some adjustment of the chart may be necessary to obtain the proper calibration, but the minor effort required greatly simplifies the measurement of characteristic group frequencies.

After calibrating the wavelength scale, place the pen in the retracted position, remove the wavelength calibration standard, place the wavelength scan control in the neutral or reset position, and manually set the instrument to the starting wavelength. Adjust the 100% control to bring the transmittance to the 90–100% range and place the sample in the spectrometer. With the pen in the retracted position, slowly scan the spectrum manually and observe the approximate intensities of the absorption peaks. Ideally, the baseline should be in the 80–100% transmittance range and the most intense peaks should fall in the 5–40% transmittance range. The peaks should not be so intense that the transmittance decreases to zero for an appreciable interval, because this prevents accurate measurement of the frequencies of the absorption peaks. If the preliminary scan indicates that the absorption bands are either too weak or too intense, adjust the sample concentration or the path length accordingly.

Once the absorption bands are of the proper intensity, set the wavelength to the starting value, place the pen on the chart paper, and set the wavelength scan control to the scan position. If more than one scan speed is available, the fastest scan speed will give a spectrum of adequate quality for most purposes and the highest scan speed should be used.

Many students attempt to correct weak absorption or excessive background scattering in infrared spectra by placing an attenuator in the reference beam, when the proper solution is to prepare the sample more carefully. The limitations of using an attenuator are discussed in Sec. 15.6.2, and you should recognize that using an attenuator will not compensate for a poorly prepared sample.

15.6.2.2. Liquid Samples

Spectra of liquids boiling above approximately 90°C usually can be obtained using a capillary film between two sodium chloride plates. One or two drops of sample are placed on a plate, the second plate is pressed firmly in place, and the two plates, cushioned with rubber gaskets, are clamped in a metal holder. Care must be taken to prevent cracking the plates through excessive or uneven application of

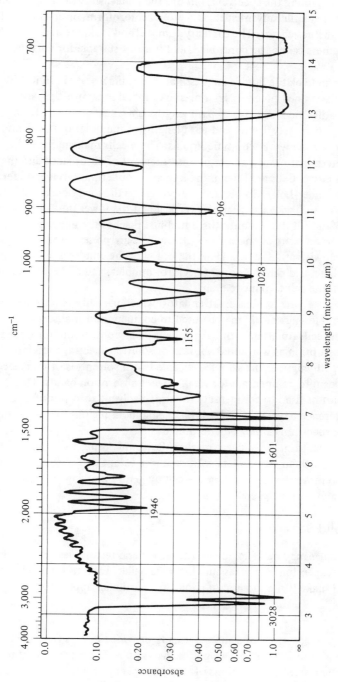

Figure 15.10. Infrared spectrum of polystyrene film showing frequencies of reference bands.

293

pressure. After obtaining the spectrum, remove the plates, slide them apart, and wash them with a volatile solvent such as acetone, chloroform, or hexane, and return them to the desiccator. The capillary film method is simple and rapid, but the sample thickness cannot be controlled, and it is not suitable for volatile compounds.

Solutions and volatile liquids are best handled using a sealed cell in which the path length is maintained precisely by a plastic or metal separator that also prevents leakage and volatilization. A range of path lengths is available, ranging from 0.01 mm to approximately 1 mm, and path lengths of the order of 0.015 mm are appropriate for routine work with pure liquids. These cells make possible quantitative determination of a solute by measurement of absorbance and the use of a calibration curve. Cells are filled using a small hypodermic syringe, preferably one with a metal Luer-Lok® tip. Attach the syringe to the lower port, hold the cell in an upright position, with the windows vertical and the lower port directly below the upper port, and fill the cell until the liquid just reaches the upper port. Inspect the cell to be certain there are no bubbles. If bubbles are present, gently withdraw the liquid and slowly refill the cell. Hold the cell with the windows horizontal, remove the syringe, and cap the ports. Check for bubbles, and if they are present, empty and refill the cell to remove them.

For spectra of pure liquids, an attenuator, which partially blocks the radiation in the reference beam, may be used to compensate for absorption and scattering by the cell. An attenuator cannot be used to compensate for strong scattering or absorption as was noted above. For solution spectra, a reference cell with the same pathlength as the sample cell, filled with solvent, is often placed in the reference beam to compensate for weak solvent absorption bands. This technique cannot be used to compensate for strong solvent absorption bands because the reference beam effectively is blocked and, in regions of strong solvent absorption, the spectrum *cannot* be interpreted.

After the spectrum is obtained, remove the sample using a syringe, rinse the cell several times with an appropriate solvent, and several times with acetone or chloroform to remove the solvent. Dry the cell by gently drawing air through it with a syringe and return the cell to the desiccator.

15.6.2.3. Solid Samples

Solid samples may be handled using three general techniques; as a solution, as a mull, or as a suspension in a potassium bromide pellet. Each technique has certain advantages that make it the method of choice for particular applications.

(a) Solution method

The solution spectrum of a solid is usually obtained using a 0.1-mm cell and a solution containing at least 5% by weight of the sample. The solvent must not obscure important wavelength regions of the infrared spectrum. Fig. 15.11 denotes the frequency ranges over which common solvents are sufficiently transparent as

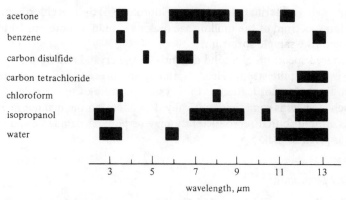

Figure 15.11. Infrared characteristics of some common solvents. Regions of strong infrared absorbance in 0.1-mm cells (except water, 0.01 mm) are shown as the shaded areas. Longer cell paths will broaden the regions of absorption, and in some cases introduce new regions where absorption is significant.

white bars, and those ranges over which the solvents absorb too strongly to be useful as black bars. All solvents obscure significant portions of the infrared spectrum, and if it is necessary to obtain the entire spectrum of the sample, more than one solvent must be used. As an example, securing the spectrum of a sample first in carbon tetrachloride and then in bromoform permits examination of nearly the entire sodium chloride region of the infrared spectrum.

The study of weak absorption bands or of samples of low solubility requires appropriate adjustment of concentrations and cell path lengths. If the solubility of the sample is sufficient, usable spectra can be obtained simply by using solutions of higher concentration. For compounds of low solubility, it is possible to use cells of longer pathlength (e.g., to 1 mm) and to reduce the sample concentration accordingly if the bands of interest are not too close to the solvent bands.

(b) Mull techniques

The mull technique, in which the sample is dispersed in mineral oil, is useful for qualitative spectra, and it requires only simple equipment. The sample is prepared by grinding a few milligrams thoroughly with a drop or two of mineral oil (Nujol) using a small agate mortar and pestle. Thorough grinding is necessary to obtain useful spectra, and some practice is necessary, particularly in choosing satisfactory proportions of the sample and the mulling agent. The mull is scanned between sodium chloride plates as a capillary film (see liquid samples), and an attenuator may be needed in the reference beam. The most common fault is to use a film of mull which is too thin. Normally, the peaks of the mulling agent should be off-scale so that the peaks of interest will have a good intensity. Not all compounds give satisfactory mulls, and the strong C–H bands of the mineral oil may obscure sample absorption bands. The latter problem can be eliminated by using another mulling agent such as fluorinated kerosene (Fluorolube®) or hexachlorobutadiene. The absorption spectrum of the mulling agent alone should be determined for comparison with the sample spectrum.

The infrared spectrum of a solid depends on the crystal structure and the dependence is often quite strong. Hence, grinding and pressing the sample can alter the spectrum through modification of the crystal structure of the sample. The extent of such changes is sometimes critically dependent on the manner in which the sample is ground so that reproducibility may be poor. A detailed discussion of such effects is given by Baker.[2]

(c) KBr pellet technique

Probably the most widely used method for routine examination of solid samples by infrared spectroscopy is the KBr pellet technique in which the sample is suspended in a transparent (to IR radiation) disc of KBr. Approximately 2 mg of the compound and 200 mg of KBr are ground together using a mechanical grinder. The proper amount of KBr can be measured quickly and with sufficient accuracy using a small test tube calibrated to deliver the necessary quantity. To obtain a satisfactory spectrum, it is necessary to reduce the particle size below that which will scatter incident radiation excessively (less than 2 μm). A small ball mill mounted in a dental amalgamator (Wig-L-Bug®) is used routinely, although manual grinding is possible. The necessary quantities of material are placed in the vial with the mixing ball; the vial is capped and clamped in the amalgamator. Usually grinding for 30 sec is sufficient. Much longer grinding with the plastic vials can lead to contamination. It is necessary to use infrared grade KBr because reagent grade KBr contains nitrate ion which contributes to the infrared spectrum. The particle size of reagent grade KBr is also too large to permit adequate particle size reduction in a short time. Reagent grade KBr can be used if it is pre-ground in a mortar and dried in an oven at 150°C, if the impurity peaks do not interfere. A spectrum of the KBr used should always be obtained for comparison with the sample spectrum. The KBr must be kept in a desiccator or in an oven to prevent absorption of excessive moisture, which prevents formation of a usable pellet. Small amounts of moisture in the KBr result in broad but weak absorption near 3,300 cm^{-1} and can usually be tolerated.

The pellet is formed in a die under high pressure. Two types of die are in use; a simple, self-contained mechanical unit, in which the force is generated by tightening a bolt, and a more complex unit in which the force is generated with a hydraulic press.

The first type of die consists of two bolts with polished ends threading into a metal block, that serves as both die and sample holder. Thread one bolt into the block five or six turns, so that one or two threads remain showing, pour the mixed and ground KBr-sample material into the open end, and distribute it evenly by tapping the die gently. Some practice is required in obtaining an even distribution and a satisfactory amount of material. Screw the second bolt into the block and tighten the bolts using box wrenches of the proper size. Excessive force is not necessary. Allow the die to stand for at least 2 min to permit the KBr to flow and form a glass-like pellet. Remove the bolts, leaving the pellet in the block. Place the die on the special holder in the spectrophotometer, and, if necessary, place an attenuator in the reference beam. If the pellet is very cloudy, either the compound was not ground well or the bolts were not tightened sufficiently, and a poor spectrum will result because of light scattering. In such cases, regrind the material and press it into another pellet. On completion of the measurements, clean the die by rinsing it with water to remove the last traces of KBr and sample, rinse it with methanol or acetone, and dry the parts in an oven before replacing them in a storage desiccator. *Be careful* not to scratch the polished ends of the bolts.

The more elaborate die permits evacuation of the sample during pressing to minimize absorption of moisture, and is designed to be used in a hydraulic press. A typical die of this type is shown in Fig. 15.12, which also indicates the manner in which the parts of the die are assembled. Other dies will vary somewhat from this design, but the basic features of all dies of this type are similar. The anvil is placed in the bottom of the barrel of the die with the polished side upward and the barrel is placed in the base. The previously ground KBr sample mixture is poured into the barrel. The powder is distributed evenly by tapping the die, and the plunger is inserted in the barrel and rotated to spread the powder evenly. The upper anvil is inserted (polished side down) in the barrel, and the plunger re-inserted. If the plunger or the anvil does not fit smoothly in the barrel, consult the instructor. Application of excessive force may damage the die beyond repair. Place the die in the press, evacuate to a pressure of 20 torr, and apply hydraulic pressure, being certain not to exceed the maximum force allowable. The maximum allowable force is indicated on some dies. If it is not, consult the instructor before applying pressure. In general pressing for 60 sec is sufficient to form the disc.

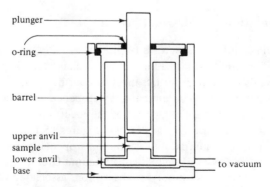

Figure 15.12. Schematic drawing of a typical infrared pellet die, showing the arrangement of the major components.

When the proper time has elapsed, remove the die from the press, separate the base from the barrel and place the barrel and plunger upside down (plunger down, barrel up) on the press. Place a metal C-ring on the barrel so that it surrounds the anvil, and use the press to force the plunger up through the barrel. When the anvil and plunger emerge from the base, separate the anvil and remove the KBr pellet with a spatula or tweezers. Mount the pellet in the sample holder and, if necessary, place an attenuator in the reference beam to compensate for moderate scattering by the pellet.

A successful pellet will be glassy in appearance, but not necessarily clear. The most common cause of failure is the use of KBr containing excessive moisture that usually prevents formation of a cohesive pellet. If the sample absorption bands are too weak, the sample may not be well ground or the sample may be too dilute. Examine the pellet for irregular white spots that indicate a poorly ground sample. The pellet should be reground, and more sample added if necessary. If the sample absorption bands are too strong, break the pellet into several portions, return part of the pellet to the grinding device, add an appropriate quantity of fresh KBr, regrind the material, and press a new pellet.

It is vital to clean pellet dies carefully to remove all traces of salt and to store the dies in a desiccator to prevent corrosion. Failure to clean the die thoroughly will result in seizure of the close fitting parts and will destroy the die. Water followed by a rinse with acetone and drying in an oven is the best procedure for removing traces of KBr and sample from dies.

The extensive grinding and pressing of the sample which is necessary in the preparation of potassium bromide pellets can alter the crystal structure of the sample and thereby can change the spectrum significantly. Spectra obtained using potassium bromide pellets should be compared only with reference spectra obtained in the same manner and under similar conditions. Baker[2] has discussed this problem in detail and has presented numerous examples of such spectral artifacts.

15.6.3. Interpretation of Infrared Spectra

Interpretation of infrared spectra is facilitated by an organized approach to the examination of spectra based on a knowledge of characteristic group frequencies. Peaks arising from impurities must be considered in this examination. For example, reaction products contaminated by stopcock grease or vacuum grease exhibit infrared spectra that are a composite of the spectra of the product and the contaminant, and that often are very misleading. Figures 15.13 through 15.15 illustrate infrared spectra of Lubriseal®, a common stopcock grease, Apiezon H®, a high vacuum grease, and Dow Corning Silicone Vacuum Grease®, three commonly used greases, to demonstrate the variety of absorptions associated with such greases.

The approach below will assist in identification of most common functional groups and structural elements in a systematic fashion. This approach divides the spectrum into four general frequency ranges, each typified by certain types of

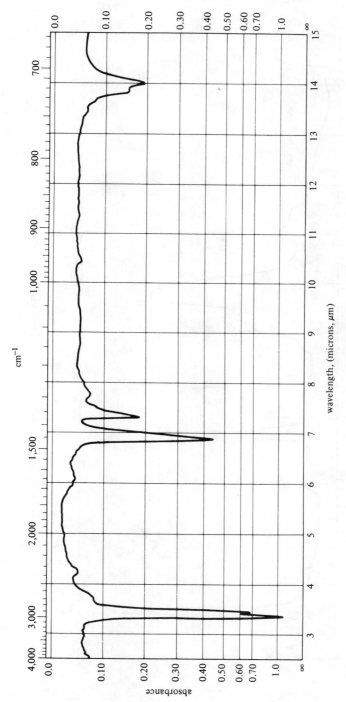

Figure 15.13. Infrared spectrum of Lubriseal®, a common stopcock grease.

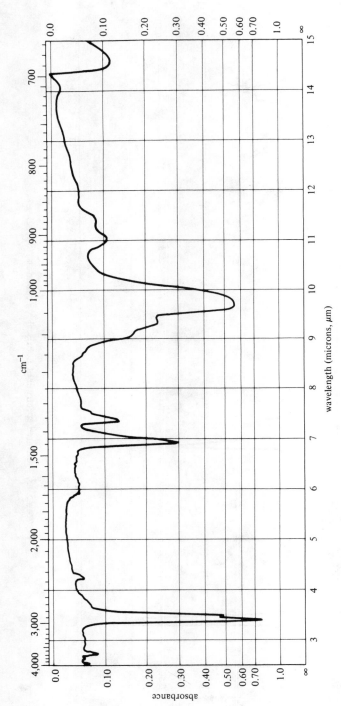

Figure 15.14. Infrared spectrum of Apiezon H®, a high vacuum grease.

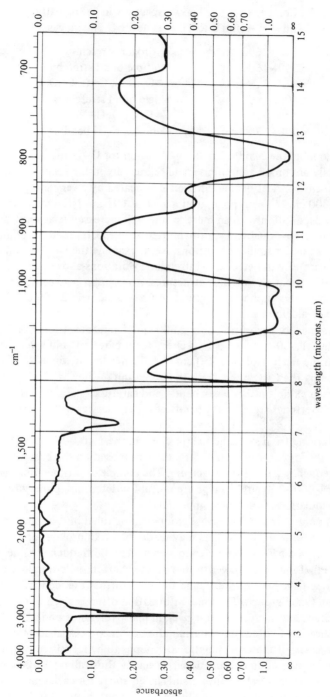

Figure 15.15. Infrared spectrum of Dow Corning Silicone Vacuum Grease®.

absorption bands. Comparison with known spectra will help greatly and a familiarity with band shapes and intensities is needed. The ranges are:

3500–2600 cm^{-1}	Hydrogen stretching; X–H
2300–1600 cm^{-1}	Multiple bond stretching;
	C≡C, C≡N, C=C, C=O
1400– 900 cm^{-1}	Single bond stretching;
	C–C, C–O, C–N
1000– 700 cm^{-1}	Complex C–H bending

First examine the 3,500–2,600 cm^{-1} region for O–H and N–H absorptions, which generally are broad and intense, and fall usually in the range 3,500–3,300 cm^{-1}. Acetylenic hydrogen is indicated by a very sharp but weak band near 3,300 cm^{-1}. With polar groups such as OH and NH which form hydrogen bonds, dilution with an inert solvent will increase the absorption frequency and narrow the band width and reference spectra used for comparison must be secured under similar conditions. Next, examine the C–H region around 3,000 cm^{-1} carefully to detect C–H stretching vibrations above 3,000 cm^{-1} that are evidence of aromatic or olefinic structures. Absorption between 2,950 and 2,850 cm^{-1} indicates aliphatic structures. A weak band near 2,720 cm^{-1} is characteristic of aldehydes.

The presence of absorption near 2,300 cm^{-1} indicates triple bonds, absorption near 1,600–1,650 cm^{-1} is characteristic of C=C and a very strong absorption in the range 1,650–1,850 cm^{-1} indicates the presence of a functional group containing the C=O bond. With the exception of the C=O group, these structures often exhibit rather weak bands, and careful examination is necessary. Occasionally overtones from strong bands of lower energies will be found in this region.

Bands arising from single bond stretching in the 1,400–900 cm^{-1} range are useful primarily for confirmation of structures proposed on the basis of bands in the higher frequency part of the spectrum. The complex C–H bending vibrations found in the 1,000–700 cm^{-1} range are useful for determining the substitution patterns of aromatic and olefinic compounds.

At this point, certain functional groups can be ruled out, others are almost certain to be present, and still others may be doubtful. Consolidation of the evidence is achieved best by reference to a characteristic frequency table, such as the one given in Table 15.3, to verify that all the characteristic frequencies of a suspected group have been checked, and to refine further the identity of the suspected functional groups. The same information often is given in the form of a correlation chart, and such charts are given in many reference books.[1,3–8] For example, it should be noted that determination of the precise absorption frequency of C=O is necessary if the C=O functional group is to be identified properly. One should also take note of any absorptions, especially those above 1,500 cm^{-1}, that do not fit the brief list of absorption bands listed above for each wavelength range. Many functional groups in the correlation table are not included in the brief listing

TABLE 15.3

Infrared Frequencies of Common Functional Groups

Vibration	Frequency, cm^{-1}		Intensity
1. *C–H stretching*			
Alkyne		3300	*s*
Alkene		3010–3090	*m*
Aromatic		3030	*v*
Alkane		2850–2960	*m–s*
Aldehyde		2720	*w–m*
2. *C–H bending*			
CH$_3$		1450	*m*
	and	1375	*m*
CH$_2$		1465	*m*
Alkenes			
Monosubstituted		1410	*s*
	and	990 and 910	*s, s*
Disubstituted, *cis*	*ca.*	690	*s*
Disubstituted, *trans*		1300	*m*
	and	965	*s*
Disubstituted, gem		1415	*s*
	and	890	*s*
Alkynes	*ca.*	630	*s*
Substituted aromatics			
Monosubstituted		750	*s*
	and	700	*s*
Disubstituted			
Ortho		750	*s*
	and	700	*s*
Meta		780	*s*
	and	700	*m*
Para		820	*s*
3. *Multiple bonded C stretching*			
Alkynes			
Monosubstituted		2120	*m*
Disubstituted		2220	*w–m*
Nitriles			
Unconjugated		2250	*m*
Conjugated		2200	*m*
Alkenes			
Monosubstituted		1645	*m*
Disubstituted, *cis*		1660	*m*
Disubstituted, *trans*		1675	*m*
Disubstituted, gem		1655	*m*
Diene		1650	*w*
	and	1600	*w*
Allene		1960	*m*
	and	1050	*m*

Continued

TABLE 15.3 (*cont.*)

Infrared Frequencies of Common Functional Groups

Vibration	Frequency, cm^{-1}	Intensity
Aromatic	1600	*m*
and	1580	*m*
and	1500	*m*
and	1450	*m*
4. *Carbonyl stretching*		
Aldehydes		
Aliphatic	1730	*s*
α,β unsaturated	1690	*s*
Aromatic	1705	*s*
Ketones		
Aliphatic	1715	*s*
α,β unsaturated	1675	*s*
Aryl	1690	*s*
Diaryl	1665	*s*
Esters		
Aliphatic	1745	*s*
α,β unsaturated	1725	*s*
Aryl	1725	*s*
Carboxylic acids		
Aliphatic	1710	*s*
Aromatic	1690	*s*
Acid anhydrides		
Acylic, saturated	1820	*s*
and	1770	*s*
Acyclic, aryl	1800	*s*
and	1740	*s*
Primary and secondary amides	1650–1690	*s*
Tertiary amides	1650	*s*
Acyl chlorides	1800	*s*
Acyl bromides	1815	*s*
5. *N–H and O–H stretching*		
Phenols, alcohols, H bonded	3200–3600	*s, b*
Phenols, alcohols, free	3600	*s*
Carboxylic acids	3500–2500	*s, b*
Amines, amides, primary	3500	*s*
and	3400	*s*
Amines, amides, secondary	3400	*s*
6. *N–H and O–H bending*		
Alcohols and phenols	1250–1350	*s*
Amines	1550–1650	*w*
7. *C–N and C–O stretching*		
Primary alcohols	1050	*s*
Secondary alcohols	1100	*s*

Continued

304

TABLE 15.3 *(cont.)*

Infrared Frequencies of Common Functional Groups

Vibration	Frequency, cm^{-1}	Intensity
Tertiary alcohols	1150	*s*
Phenols	1200	*s*
Aliphatic amines	1100	*m*
Aromatic amines	1300	*m*
8. *N—O and N—N stretching*		
Azo compounds	1600	*v*
Aromatic nitro compounds	1530	*s*
and 1350		*s*
Aliphatic nitro compounds	1560	*s*
and 1380		*s*
Nitrates	1625	*s*
and 1280		*s*
Nitroso compounds	1550	*s*
9. *Carbon-halogen stretching*		
CCl and CCl$_2$	800–500	*s*
CF	950–1300	*s*
CF$_2$ and CF$_3$	1280	*s*
and 1150		*s*
10. *Sulfur stretching*		
Mercaptans (H—S)	2550	*w*
Thiocarbonyl	1120	*s*
Sulfur-oxygen stretching		
Sulfonic acids	1050	*s*
and 650		*s*
Sulfonamides	1300	*s*
and 1180		*s*
Sulfonyl chlorides	1350	*s*
and 1170		*s*
Sulfoxides	1050	*s*

Intensities of Bands

s	strong	*w*	weak	*b*	broad
m	medium	*v*	variable		

of bond types in the three spectral regions selected for the preliminary examinations of the spectrum. Also, it must be borne in mind that the infrared spectrum of each compound is unique, and that the spectrum of an unknown sample must correspond in all respects to that of an authentic sample of the same material, obtained under comparable conditions.

15.6.4. Infrared Characteristics of Common Functional Groups

The infrared characteristics of a number of common functional groups are illustrated in the spectra given in Figs. 15.16—15.38. These spectra should be useful for reference in identification of unknowns and for gaining experience in recognizing typical absorption bands. The spectra were secured using sodium chloride prism instruments. They will be virtually identical to spectra obtained on commercial sodium chloride prism spectrometers, but will differ significantly with regard to band shapes in comparison with spectra obtained using grating spectrometers.

The sample spectra are from the Sadtler Collection of Standard Prism Infrared Spectra, which is a valuable reference source for comparison of infrared spectra. A comparable collection of grating spectra is also available from Sadtler and should be consulted for comparison of spectra obtained using grating spectrometers. In comparing spectra, be sure that the spectra to be compared were obtained under comparable conditions. Frequencies and intensities of the bands of polar and hydrogen bonded groups in a polar solvent or in the solid state can differ substantially from those in a non-polar solvent.

Only a limited number of functional groups are illustrated in the sample spectra, and for discussions of less common functional groups and interactions between functional groups, one should consult standard texts such as those by Colthup *et al.*[1] and by Conley.[3] Discussions of the application of infrared spectroscopy to identification of organic compounds are given by a number of authors including Dyer,[4] Pasto and Johnson,[5] Williams and Fleming,[6] Rao,[7] Phillips,[8] and Szymanski.[9,10]

The salient features of the sample spectra in Fig. 15.16–15.38 are summarized below to draw attention to the absorption frequencies that are associated with each structural feature. In general, only spectral features that have not been discussed in connection with an earlier spectrum will be described for each spectrum.

Figure 15.16; *2-Methylheptane.*   The strong band at 2,960 cm^{-1} results from aliphatic C–H stretching, and the medium band at 1,470 cm^{-1} arises from aliphatic C–H bending. The closely spaced doublet at 1,370 cm^{-1} and 1,390 cm^{-1} is characteristic of an isopropyl group, and in the spectrum of the unbranched analog, this doublet is replaced by a single band at 1,380 cm^{-1}. The broad, weak band at 725 cm^{-1} is characteristic of compounds having unbranched carbon chains at least four carbon atoms long.

Figure 15.17; *1-Dodecene.*   The weak band at 3,120 cm^{-1} is caused by stretching of the olefinic C–H bond, the medium band at 1,647 cm^{-1} represents olefinic C–C stretching and the strong bands at 990 cm^{-1} and 907 cm^{-1} are olefinic C–H bending vibrations. Olefins exhibit one or more bands in the 1,000–650 cm^{-1} range, and the number, frequencies and intensities of these bands are determined by the substitution pattern of the olefin as indicated by Table 15.3. A standard text should be consulted for detailed analysis.

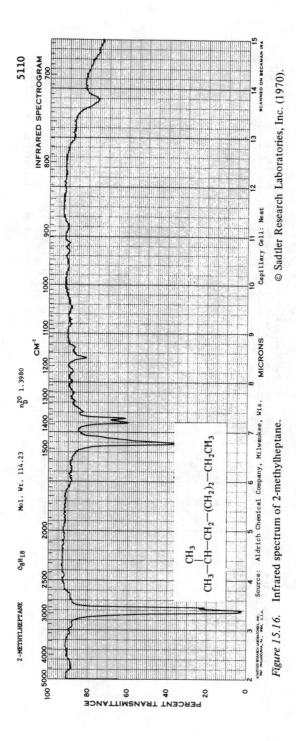

Figure 15.16. Infrared spectrum of 2-methylheptane.

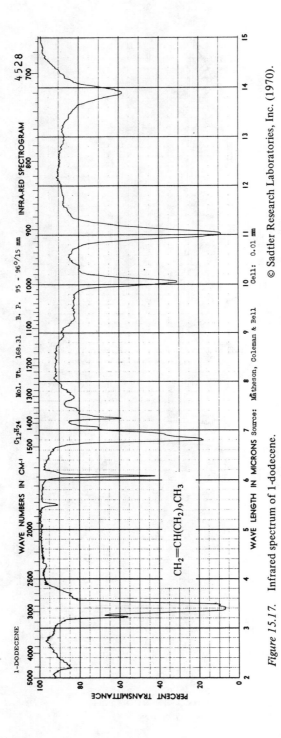

Figure 15.17. Infrared spectrum of 1-dodecene.

© Sadtler Research Laboratories, Inc. (1970).

Figure 15.18; Ethylbenzene.   The moderate band at $3,010 \text{ cm}^{-1}$, which is not completely resolved from the stronger aliphatic C–H stretching band, is caused by aromatic C–H stretching and the sharp band at $1,600 \text{ cm}^{-1}$ represents aromatic C–C stretching. The medium band at 745 cm^{-1} and the strong band at 695 cm^{-1} are complex C–H bending vibrations that are characteristic of monosubstituted aromatics. The series of weak bands between $2,000$ and $1,700 \text{ cm}^{-1}$ is also characteristic of the substitution pattern of aromatic compounds, and it can sometimes be used to determine the substitution pattern. Note the differences in this region for *o, m* and *p*-xylene below.

Figure 15.19; ortho-Xylene.   The aromatic C–H stretching band appears as the unresolved shoulder at $3,030 \text{ cm}^{-1}$. The intense C–H bending band at 735 cm^{-1} is typical of ortho-disubstituted aromatic molecules.

Figure 15.20; meta-Xylene.   The aromatic C–H stretching band at $3,050 \text{ cm}^{-1}$ is not resolved from the aliphatic C–H stretching band. The strong C–H bending band at 770 cm^{-1} and the moderate band at 695 cm^{-1} are found in meta-disubstituted aromatic molecules.

Figure 15.21; para-Xylene.   The aromatic C–H stretching band at $3,010 \text{ cm}^{-1}$ is not resolved from the aliphatic C–H stretching band. The intense C–H bending band at 795 cm^{-1} is representative of para-disubstituted atomic molecules.

Figure 15.22; 4-Ethylpyridine.   The infrared characteristics of heterocyclic molecules qualitatively resemble those of aromatic molecules. The heterocyclic C–H stretching band is not resolved from the aliphatic C–H stretching band in this case. The strong band at $1,620 \text{ cm}^{-1}$ arises from stretching of the C–C bonds of the heterocyclic ring and the intense C–H bending band at 820 cm^{-1} is characteristic of 4-substituted pyridines. The positions and intensities of C–H bending bands for substituted heterocycles are given in standard references.

Figure 15.23; 1-Hexyne.   The sharp, intense band at $3,250 \text{ cm}^{-1}$ is caused by acetylenic C–H stretching and the moderate band at $2,120 \text{ cm}^{-1}$ represents C–C triple bond stretching. The broad band at $1,250 \text{ cm}^{-1}$ and the continuously increasing absorption below 700 cm^{-1} are caused by C–H bending vibrations associated with the acetylenic group.

Figure 15.24; Di-n-Butyl Ether.   The strong, broad absorption at $1,120 \text{ cm}^{-1}$ represents C–O stretching. Except for this feature, the spectrum resembles that of an aliphatic hydrocarbon.

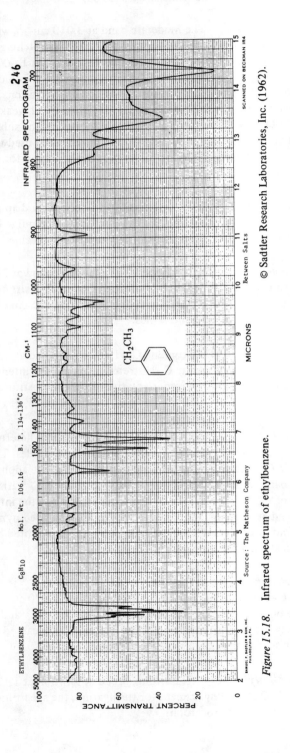

Figure 15.18. Infrared spectrum of ethylbenzene.

© Sadtler Research Laboratories, Inc. (1962).

310

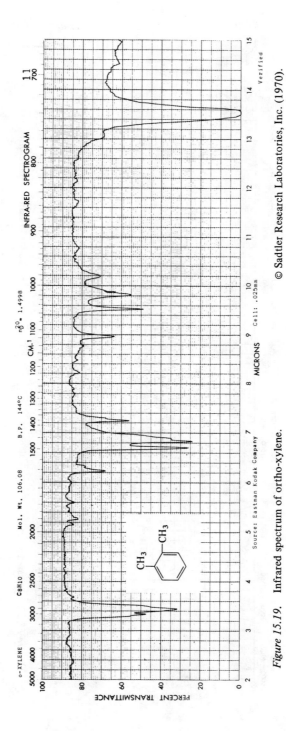

Figure 15.19. Infrared spectrum of ortho-xylene.

311

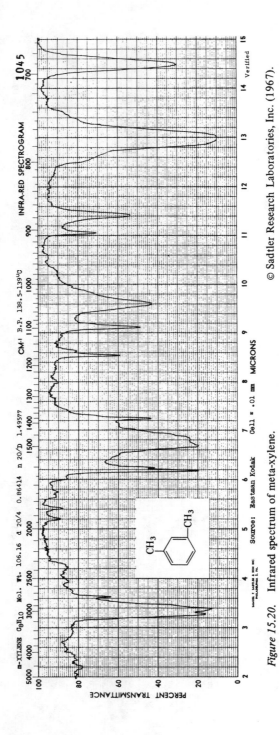

Figure 15.20. Infrared spectrum of meta-xylene.

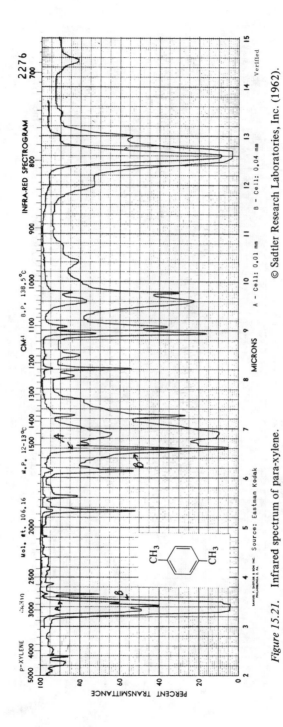

Figure 15.21. Infrared spectrum of para-xylene.

© Sadtler Research Laboratories, Inc. (1962).

313

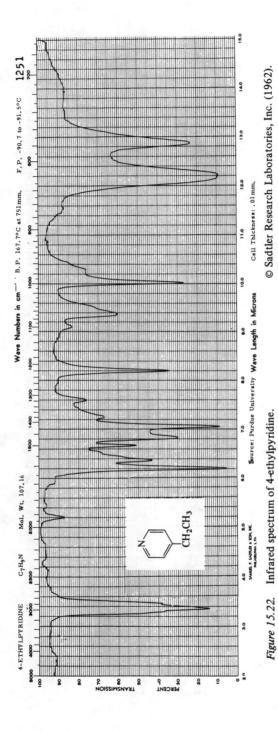

Figure 15.22. Infrared spectrum of 4-ethylpyridine.

© Sadtler Research Laboratories, Inc. (1962).

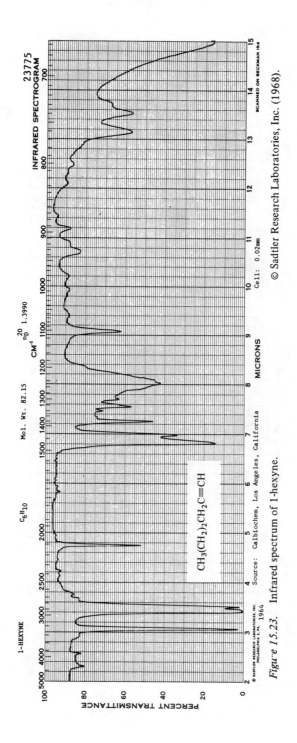

Figure 15.23. Infrared spectrum of 1-hexyne.

© Sadtler Research Laboratories, Inc. (1968).

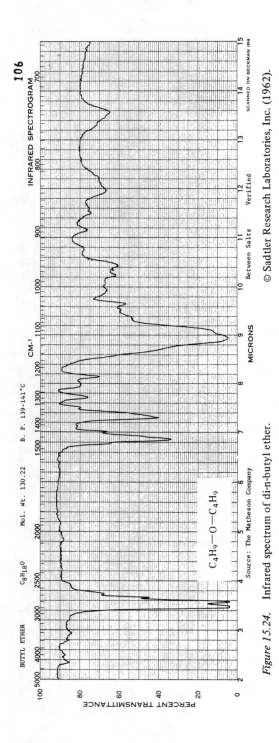

Figure 15.24. Infrared spectrum of di-n-butyl ether.

Figure 15.25; 2-Hexanol. The broad, intense absorption centered at $3,350 \text{ cm}^{-1}$ is caused by O—H stretching and the broad moderate band at $1,320 \text{ cm}^{-1}$ arises from O—H bending. The moderate, broad band at $1,100-1,200 \text{ cm}^{-1}$ represents C—O stretching, and the frequency of this band differs if the alcohol is primary, secondary, or tertiary.

Figure 15.26; meta-Cresol. The broad, intense band centered at $1,260 \text{ cm}^{-1}$ arises from C—O stretching. The aromatic and hydroxyl bands are located as noted previously. Note the increase of the C—O stretching frequency with respect to the alcohol.

Figure 15.27; n-Hexylamine. The moderate doublet band at $3,200-3,300 \text{ cm}^{-1}$ arises from symmetric and asymmetric N—H stretching and the broad band at $1,600 \text{ cm}^{-1}$ is caused by N—H bending. With primary amines the N—H stretching band often appears as a doublet, while with secondary amines, it is a singlet. The intensity and breadth of the N—H bending and stretching bands vary greatly from one compound to another. The very broad band centered at 800 cm^{-1} is a complex N—H bending vibration characteristic of primary amines, and the broad band centered at $1,060 \text{ cm}^{-1}$ arises from C—N stretching.

Figure 15.28; para-Toluidine. The incompletely resolved doublet at $3,350-3,450 \text{ cm}^{-1}$ is caused by the symmetric and asymmetric N—H stretching. Both the N—H stretching bands and the C—N stretching band at $1,260 \text{ cm}^{-1}$ are significantly higher in frequency than for aliphatic amines.

Figure 15.29; 3-Hexanone. The intense carbonyl stretching band at $1,705 \text{ cm}^{-1}$ is characteristic of aliphatic acyclic ketones. Carbonyl stretching frequencies are strongly influenced by ring size in cyclic compounds and by conjugation; reference spectra should be consulted. The very weak band at $3,410 \text{ cm}^{-1}$ is an overtone of the fundamental carbonyl stretching vibration.

Figure 15.30; n-Heptanal. The carbonyl band at $1,715 \text{ cm}^{-1}$ of an aliphatic aldehyde is virtually indistinguishable from that of an aliphatic ketone. The aldehyde can be distinguished by the moderate band at $2,710 \text{ cm}^{-1}$ which is caused by the stretching of the aldehydic C—H bond.

Figure 15.31; Propyl Propanoate. The carbonyl stretching frequency at $1,735 \text{ cm}^{-1}$ distinguishes an aliphatic ester from a ketone. The broad intense band at $1,170 \text{ cm}^{-1}$ and the moderately intense band at $1,080 \text{ cm}^{-1}$ arise from carbon-oxygen stretching and are characteristic of esters.

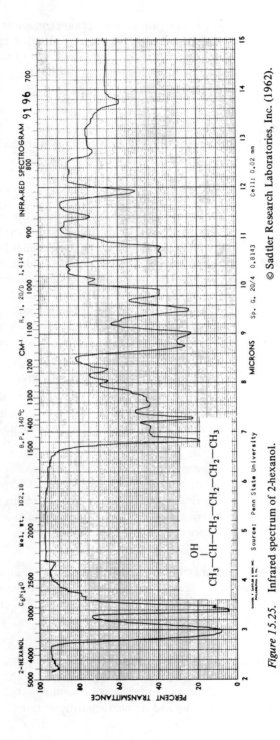

Figure 15.25. Infrared spectrum of 2-hexanol.

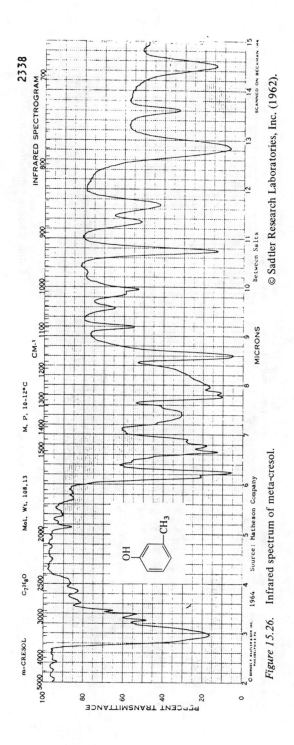

Figure 15.26. Infrared spectrum of meta-cresol.

© Sadtler Research Laboratories, Inc. (1962).

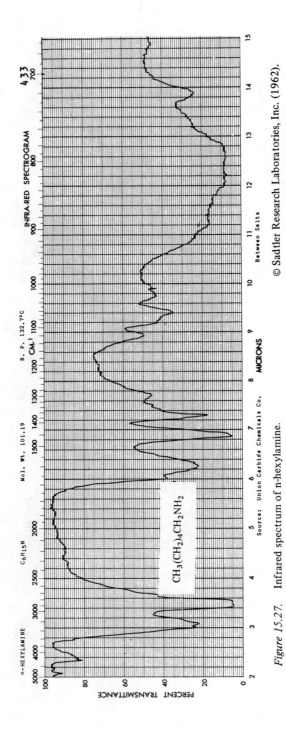

Figure 15.27. Infrared spectrum of n-hexylamine.

© Sadtler Research Laboratories, Inc. (1962).

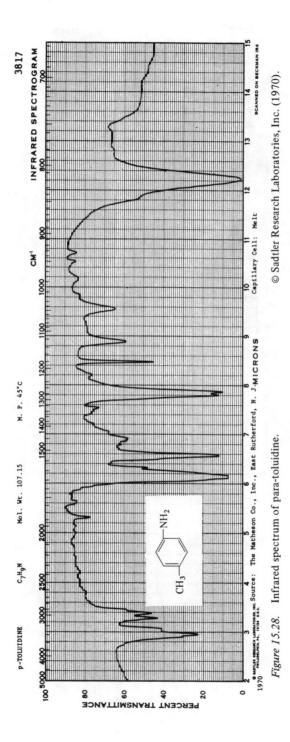

Figure 15.28. Infrared spectrum of para-toluidine.

321

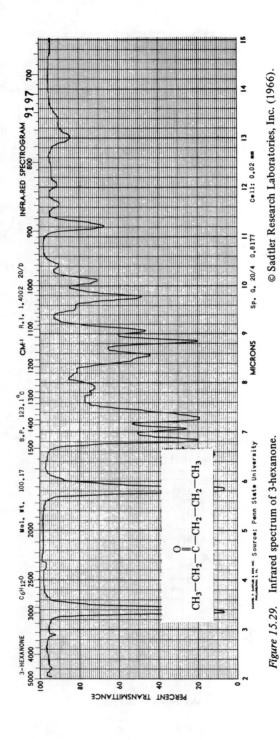

Figure 15.29. Infrared spectrum of 3-hexanone.

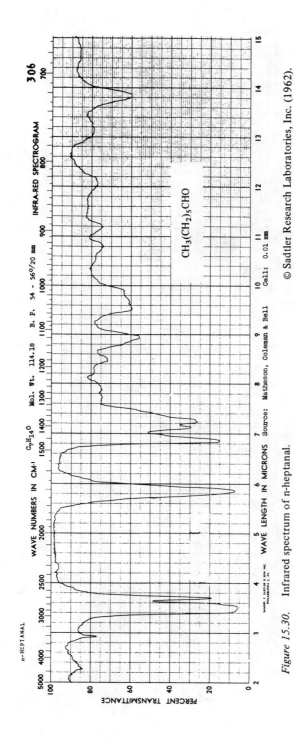

Figure 15.30. Infrared spectrum of n-heptanal.

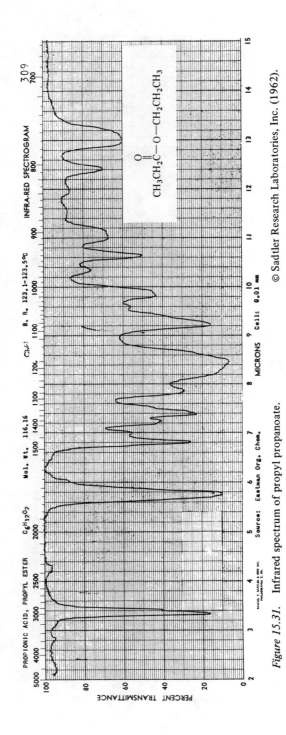

Figure 15.31. Infrared spectrum of propyl propanoate.

© Sadtler Research Laboratories, Inc. (1962).

324

Figure 15.32; n-*Hexanoic Acid.*   The carbonyl stretching frequency of 1,700 cm^{-1} is similar to that of ketones, but the intense, broad O–H stretching between 3,600 cm^{-1} and 2,500 cm^{-1} is distinctive for carboxylic acids. Moderate bands near 1,400 cm^{-1}, 1,250 cm^{-1} and 900 cm^{-1} arise from complex O–H bending and C–O stretching vibrations. The carbonyl and O–H frequencies are very sensitive to H-bonding.

Figure 15.33; *Propionic Anhydride.*   The intense doublet at 1,750 cm^{-1} and 1,800 cm^{-1} arises from the symmetric stretching of the two carbonyl groups and it differentiates anhydrides from other carbonyl compounds. The frequencies of these bands are sensitive to conjugation and to ring size in cyclic anhydrides. The intense band at 1,040 cm^{-1} caused by C–O stretching is characteristic of aliphatic anhydrides.

Figure 15.34; *Benzamide.*   The sharp, intense bands at 3,350 cm^{-1} and 3,160 cm^{-1} are caused by asymmetric and symmetric N–H stretching. The sharp, intense band at 1,680 cm^{-1}, known as the amide I band, is attributed to carbonyl stretching, while the slightly less intense band at 1,620 cm^{-1}, known as the amide II band, arises from N–H bending. The frequencies of the N–H stretching band and the amide I and II bands are very sensitive to hydrogen bonding and hence to the physical state of the sample, and this observation must be borne in mind when comparisons are made.

Figure 15.35; *Valeronitrile.*   The sharp, moderate absorption at 2,250 cm^{-1} represents stretching of the C≡N triple bond, and the remainder of the spectrum is comparable to that of an aliphatic hydrocarbon.

Figure 15.36; *meta-Nitrotoluene.*   The intense, broad bands at 1,340 cm^{-1} and 1,530 cm^{-1} are caused by the symmetric and asymmetric N–O stretching vibrations of the nitro group.

Figure 15.37; *para-Nitrobenzenesulfonic Acid.*   The strong doublet centered at 1,220 cm^{-1} and the moderate singlet at 1,060 cm^{-1} are characteristic of S=O stretching vibrations. Sulfonic acids are extremely hygroscopic and the presence of water causes strong absorption at 3,300 cm^{-1} and a broad diffuse band at 1,700 cm^{-1} and broadens the bands in the 1,260–1,020 cm^{-1} range.

Figure 15.38; *para-Toluenesulfonamide.*   The intense bands at 1,330 cm^{-1} and 1,150 cm^{-1} are characteristic of S=O stretching in sulfonamides. The N–H stretching and bending vibrations are comparable to those of primary amines.

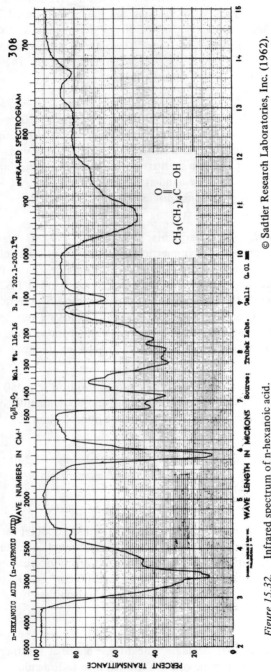

Figure 15.32. Infrared spectrum of n-hexanoic acid.

326

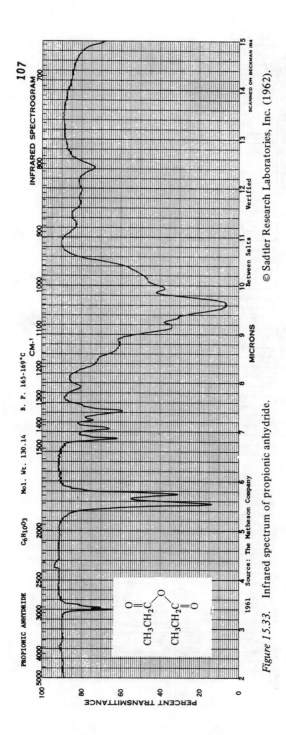

Figure 15.33. Infrared spectrum of propionic anhydride.

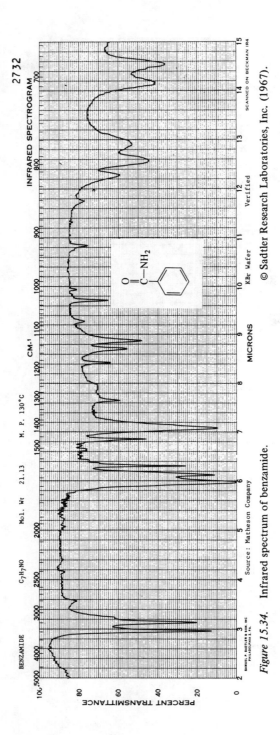

Figure 15.34. Infrared spectrum of benzamide.

328

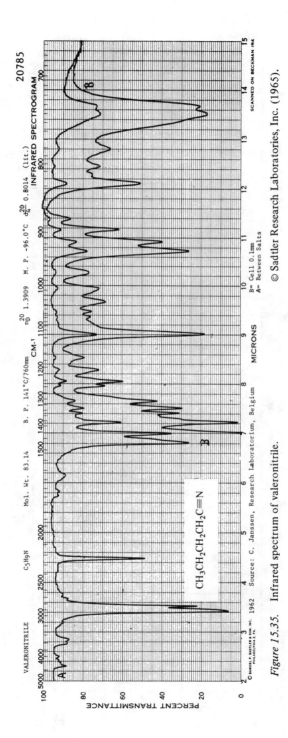

Figure 15.35. Infrared spectrum of valeronitrile.

329

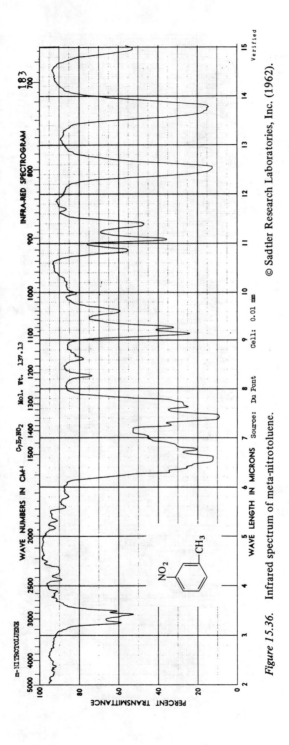

Figure 15.36. Infrared spectrum of meta-nitrotoluene.

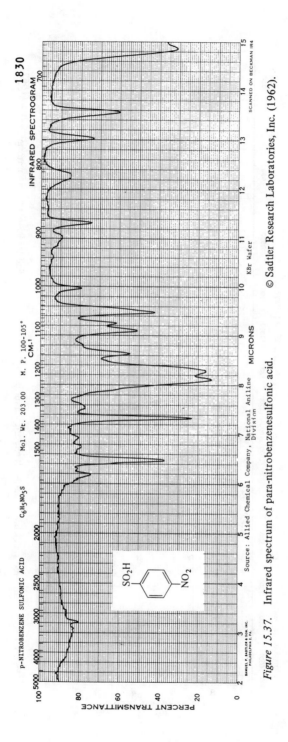

Figure 15.37. Infrared spectrum of para-nitrobenzenesulfonic acid.

331

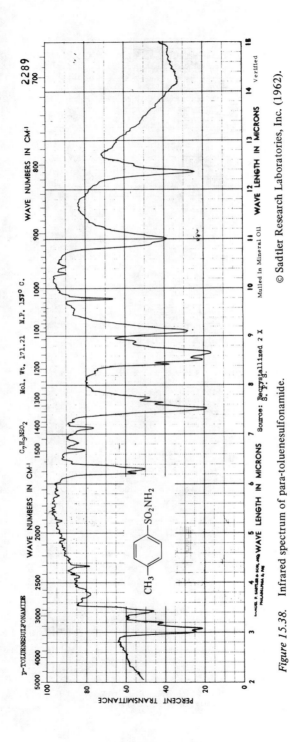

Figure 15.38. Infrared spectrum of para-toluenesulfonamide.

© Sadtler Research Laboratories, Inc. (1962).

References to 15.6

1. N. B. Colthup, L. H. Daley, and S. E. Wiberley, "Infrared and Raman Spectroscopy," Academic Press, New York, 1964.
2. A. M. Baker, *J. Phys. Chem.*, **61**, 450 (1957).
3. R. T. Conley, "Infrared Spectroscopy," Allyn and Bacon, Boston, 1966.
4. J. R. Dyer, "Applications of Absorption Spectroscopy of Organic Compounds," Prentice-Hall, Englewood Cliffs, New Jersey, 1965.
5. D. J. Pasto and C. R. Johnson, "Organic Structure Determination," Prentice-Hall, Englewood Cliffs, New Jersey, 1969.
6. D. H. Williams and I. Fleming, "Spectroscopic Methods in Organic Chemistry," 2d ed., McGraw-Hill, New York, 1973.
7. C. N. R. Rao, "Chemical Applications of Infrared Spectroscopy," Academic Press, New York, 1963.
8. J. P. Phillips, "Spectra-Structure Correlations," Academic Press, New York, 1964.
9. H. A. Szymanski and R. E. Erickson, "Infrared Band Handbook," 2d ed., Plenum, New York, 1970.
10. H. A. Szymanski, "Interpreted Infrared Spectra," Plenum, New York, 1964, vols. 1−3.

15.7. PROTON NUCLEAR MAGNETIC RESONANCE SPECTROSCOPY

15.7.1. Introduction

A nucleus with a non-zero nuclear spin quantum number, I, possesses a magnetic moment, and when such a nucleus is placed in a magnetic field, the magnetic moment must assume one of a limited number of orientations with respect to the external field. There are $2I + 1$ such orientations or spin states, and each is characterized by a magnetic quantum number, m_I, that may take the values $I, I - 1, I - 2, \ldots, -I$. The simplest case, which applies to the proton, is that in which I equals $\frac{1}{2}$ and there are two nuclear spin states corresponding to m_I values of $+\frac{1}{2}$ and $-\frac{1}{2}$. The energy difference between the two spin states, ΔE, is proportional to the external magnetic field, and when the nuclei are exposed to radiation with an energy of ΔE, energy is absorbed as shown in Fig. 15.39.

Nuclear magnetic resonance spectroscopy, usually referred to simply as NMR, is the study of such transitions, and it has been used widely to study chemical bonds, the kinetics of certain reactions, and as a means of identifying organic compounds. The present discussion is limited to proton NMR and its application to the identification of organic compounds. Examples of other applications of NMR including NMR of other nuclei are given by Jackman and Sternhell,[1] Mathieson,[2] Pople et al.,[3] and Lynden-Bell and Harris.[4]

Proton NMR can provide detailed information concerning molecular structure which is not obtainable by other techniques. The utility of NMR as a structural tool

333

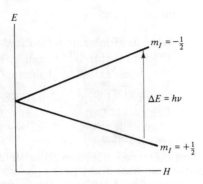

Figure 15.39. Nuclear spin energy level diagram showing the effect of the magnetic field strength, H, on ΔE, the transition energy for a nucleus with I equal to $\frac{1}{2}$.

arises because the energy difference between the two nuclear spin states of a proton in a given external field is influenced to a small but easily measurable extent by the chemical environment of that proton. Thus, one can distinguish protons in different chemical environments within a molecule and determine the relative numbers of each type of proton. The structural relationship of the various subunits of a molecule may be established by analysis of the spectral fine structure caused by interactions between protons on neighboring atoms. The NMR spectrum of a compound is characteristic of that compound, just as the physical properties and the infrared spectrum are characteristic. Information obtained from the NMR spectrum usually complements that derived from the physical properties and other spectral studies, and it should be used in conjunction with such data.

15.7.2. Proton NMR Spectroscopy

Proton NMR spectroscopic studies usually are characterized by the frequency of the radiation which is employed. Most routine studies are performed at present using 60 MHz radiation, although both higher and lower frequencies are possible. Higher frequencies result in simpler spectra in general, but with a marked increase in the complexity and cost of the instrumentation. Considerable savings in instrumentation costs are possible by reducing the frequency to 30 MHz, and several 30 MHz instruments have been designed for instructional purposes. For technical simplicity, the radiation frequency is fixed, and the magnetic field is varied over the desired range. Liquid samples of low viscosity can be studied "neat," i.e., without solvent, whereas viscous liquids or solids must be dissolved in a suitable solvent so that sufficient resolution is obtained to permit detection of the spectral fine structure.

15.7.3. Chemical Shift Values

Protons in different chemical environments are isolated or "shielded" from the external magnetic field to a degree that is determined by the functional group to which they are bonded and to a lesser extent by the nature and steric relationships of neighboring functional groups to the functional group in question. The nature of nuclear shielding is discussed in Refs. 1–4. In general, each chemically distinguishable type of proton in a molecule absorbs energy at a slightly different value of the external magnetic field, because it is shielded to a slightly different degree. Thus as the magnetic field is varied over a small range (on the order of 100 to 200 mG), each type of proton in turn will absorb radiation at a characteristic value of the applied field. The resulting NMR spectrum consists of a series of peaks or resonance absorption bands, each of which corresponds to a particular type of proton, and appears at a characteristic value of the external magnetic field.

In the simplest case, which is usually observed when the protons of each type are separated from protons of other types by four or more single bonds (including the C–H bonds), the resonance peaks are singlets, as illustrated by the spectrum of methoxyacetonitrile, CH_3-O-CH_2-CN, in Fig. 15.40. The peak labelled A corresponds to the CH_2 group of methoxyacetonitrile, the peak labelled B corresponds to the CH_3 group of methoxyacetonitrile, and peak C corresponds to

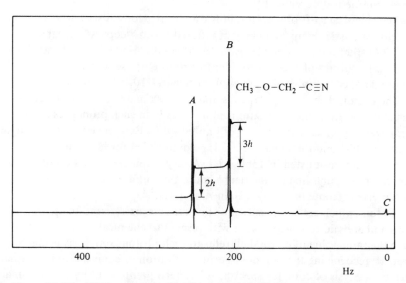

Figure 15.40. NMR spectrum of methoxyacetonitrile, illustrating singlet resonance peaks for each proton type. Peaks A and B represent the CH_2 protons and the CH_3 protons of methoxyacetonitrile respectively. Peak C represents TMS. The upper trace represents the integral of the spectrum, which corresponds to the areas under the peaks.

tetramethylsilane, $(CH_3)_4Si$, a reference compound added to permit standardization of the magnetic field. Tetramethylsilane, abbreviated TMS, is useful because all 12 protons exhibit the same chemical shift and the NMR spectrum is therefore a strong singlet. The relative number of protons of each type in the sample molecule is determined conveniently by electronic integration of the areas under the absorption peaks. The integral of the spectrum is shown as the upper trace in Fig. 15.40, and the height of each step is directly proportional to the number of protons of each type.

A proton is characterized in NMR spectroscopy by its chemical shift value, δ, a dimensionless ratio which indicates the displacement of its resonance peak from the resonance peak of a reference substance, as defined by Eq. (15.14). Values of δ are expressed in parts per million (ppm);

$$\delta = \frac{H_{reference} - H_{sample}}{H_{reference}} \times 10^6 \qquad (15.14)$$

and it is understood, unless otherwise stated, that the reference substance is TMS. Resonance peaks that occur at a higher magnetic field than TMS will have negative chemical shifts and those which occur at a lower magnetic field than TMS have positive chemical shifts. Protons that have small values of δ (1—3 ppm) are said to be "shielded," and they often are referred to as being "upfield." Protons which have large values of δ (>5 ppm) are said to be "deshielded," and they often are referred to as being "downfield."

The chemical shift of a proton in NMR spectroscopy is similar in application to the characteristic group frequency in infrared spectroscopy as a means of identifying structural features of organic compounds. A brief list of δ values of representative groups of protons for identification purposes is given in Table 15.4, and more detailed compilations are given in Refs. 1—10.

The chemical shift value depends in a complex manner on the structure of the molecule in question, and correlations of δ values with such properties as the electronegativity of substituents are of limited value. Reference to more complete correlation tables and reference spectra is recommended for identification of functional groups not listed in Table 15.4. A very extensive collection of reference spectra has been published by Sadtler[11] and a more limited, but very useful collection has been published by Varian Associates.[12]

Protons which are chemically indistinguishable are referred to as a *set* of protons and are said to be *equivalent* with respect to chemical shift, because they all have the same value of δ. We shall refer to such protons simply as being equivalent, recognizing that a more rigorous definition of equivalence is necessary for precise analysis of complex spectra, as has been pointed out by Silverstein and Silberman.[13] Thus, a methylene group is a set of two equivalent protons, and the two methyl groups of an isopropyl group form a set of six equivalent protons.

Although the δ values of most protons are relatively independent of the solute concentration and the nature of the solvent, the δ values of certain protons are not. The chemical shift values of protons that are subject to

TABLE 15.4

Chemical Shift Values of Representative Protons Relative to TMS

Proton Type	δ, ppm
CH_3-	0.9
$-CH_2-$	1.3
$-CH-$	1.5
$CH_3-C=C$	1.6
CH_3-CO	2.2
CH_3-Ar	2.3
CH_3-O-R	3.3
$CH_3-N<$	2.3
CH_3-C-Cl	1.8
CH_3-C-Br	1.4
$>C=C<^H_H$	5.3
$-C\equiv C-H$	3.1
ArH	7.0–9.0
R–CHO	9–10
ArCHO	10
RCOOH	11–12

concentration-dependent hydrogen bonding, such as those of amines, alcohols, and phenols, are influenced to a significant degree by the concentration of the solute and the nature of the solvent. Hence, the δ values of hydrogen-bonded protons must be interpreted with caution to avoid erroneous conclusions.

Chemical shifts also are strongly affected by the presence of paramagnetic materials in solution — that is, by materials which have upaired electrons such as free radicals and some transition metal ions. This effect can be used to evaluate the magnetic moment and number of unpaired electrons of the paramagnetic substance.

15.7.4. Spin-Spin Splitting

15.7.4.1. First Order Spin-Spin Splitting

Most NMR spectra are considerably more complex than the spectrum of methoxyacetonitrile shown in Fig. 15.40. Usually the resonance peaks are split into multiplets of two or more closely spaced lines by the interaction of the spins of neighboring protons. The resultant splitting of the resonance peaks, which is known as spin-spin splitting or spin-spin coupling, provides valuable information concerning the number and types of protons which are coupled. Spin-spin splitting does not change the order of the resonance peaks in the NMR spectrum, nor does it affect the total relative areas under the peaks.

The coupling of two sets of protons is characterized by the coupling constant, J, which is equal numerically to the separation of adjacent lines in each multiplet, expressed in units of Hz. The value of J is determined by the structural relationship of the two sets of protons, and it is independent of the value of the magnetic field.

If the value of J in Hz is less than approximately one sixth of the difference between the chemical shift values of the two coupled sets expressed in Hz, the sets are said to be *weakly coupled* and the resultant splitting is said to be *first order*. The relative intensities of the peaks within each multiplet correspond closely to the relative probabilities of the various possible spin orientations of the neighboring protons, and hence first order splitting patterns are analyzed readily.

Because the difference between the values of the two sets of coupled protons is large relative to the value of J, first order splitting patterns are referred to as $A_n X_m$ patterns, where the subscripts indicate the number of protons in the A set (downfield) and the X set (upfield). The general term "AX pattern" is often applied loosely to such spectra to indicate that the splitting is first order, without defining the number of protons in each set. The NMR spectrum of 1,1,2-trichloroethane, shown in Fig. 15.41, provides a simple example of first order splitting. The AX_2 pattern is composed of a doublet at 3.95 ppm representing the methylene protons

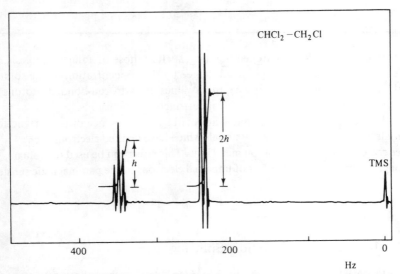

Figure 15.41. The NMR spectrum of 1,1,2-trichloroethane showing the splitting of the CH peak and the CH_2 peak by spin-spin coupling. The areas under each peak are unaffected by spin-spin coupling.

and a triplet at 5.77 ppm representing the methine proton. The ratio of the area of the methylene peak to the area of the methine peak, as determined from the integral trace is 2 : 1.

The resonance peak of a set of protons is split into a multiplet by the magnetic moments (nuclear spins) of the protons of neighboring sets that influence the magnetic field at the site of the first set. At any instant, the probability that the magnetic moment of any given proton is aligned with the external field (the low energy state) is approximately 0.5. Thus, in a sample of 1,1,2-trichloroethane, the methine protons of one half of the molecules reinforce the external magnetic field at the site of the methylene protons, and the methylene absorption peak of these molecules is shifted slightly downfield. The methine protons of the other half of the sample molecules oppose the external magnetic field at the site of the methylene protons, and the methylene absorption peak of these molecules is shifted slightly upfield.

The influence of the methylene protons on the magnetic field at the methine proton can be determined by a simple extension of this approach. The nuclear spins of the two methylene protons can take on four different configurations, each of which has a probability of $(0.5)^2$ or 0.25. The four configurations are shown in Fig. 15.42, where the arrows pointing upward indicate spins aligned with the magnetic field. The effect of the two methylene protons on the magnetic field at the methine proton is simply the sum of the contributions of the individual protons. If the methylene protons are in configurations 2 or 3, the two magnetic moments cancel. Hence, the methine resonance peak of one half of the molecules appears at the same external field strength as predicted by the chemical shift value of the methine group. If the methylene protons are in configuration 1, the two magnetic moments weaken the magnetic field at the methine proton slightly, so that the methine peak for one fourth of the molecules is shifted slightly upfield. Similarly, if the methylene protons are in configuration 4, the two magnetic moments reinforce the magnetic field at the methine proton slightly, and the methine peak of one fourth of the molecules is shifted downfield an equal amount. The result is that the methine peak appears as a closely spaced triplet, centered at

Configuration	Nuclear Spin Orientation
1	↑↑
2	↑↓
3	↓↑
4	↓↓

Figure 15.42. The four configurations of the nuclear spins of a set of two protons in a magnetic field. Arrows pointed upward indicate spins aligned with the external field. Each configuration has a probability of 0.25.

the value of the external magnetic field corresponding to the chemical shift value of the methine proton. The three lines are separated by J Hz and have relative intensities of $1 : 2 : 1$.

This analysis can be extended to the splitting patterns encountered when larger sets of protons are coupled. For example, a methyl group splits the peak of neighboring set of protons into a quartet, and the relative intensities of the lines are $1 : 3 : 3 : 1$. The first-order splitting of a resonance peak by a neighboring set of n equivalent protons can be described by the two general rules below.

1.   The peak is split into a multiplet of $n + 1$ lines.
2.   The relative intensities of the $n + 1$ lines are equal to the coefficients of the successive terms obtained by expanding the binomial $(a + b)^n$.

It is important to recognize that if two sets of protons are coupled to each other, the value of J estimated from the splitting of one peak must equal the value of J estimated from the splitting of the other peak. This fact often simplifies the analysis of a spectrum by facilitating the determination of which sets of protons are coupled.

A more complex splitting pattern occurs when a single proton is weakly coupled to two different protons, and the coupling constants are unequal. The resulting first order pattern, which consists of three quartets in which the peaks are of equal intensity, as illustrated in Fig. 15.43 which shows the NMR spectrum of furan-2-aldehyde. Such patterns are labeled AMX to indicate that three different protons are involved and that they are coupled weakly. Because the spin-spin coupling of the protons in an AMX group is weak, AMX patterns can be interpreted by direct extension of the principles described previously for simple first order spectra. The splitting of each peak into a quartet reflects coupling of the proton with *each* of the other two protons. Thus, for example, the A peak is split into a doublet by coupling with the M proton, and each of the peaks of the doublet is split into a doublet by coupling with the X proton. The four lines are equally intense because the M proton and the X proton act independently, and the probability that the spin of a single proton will reinforce the external field is equal to the probability that it will oppose the external field.

15.7.4.2. Higher-Order Spin-Spin Splitting

As the value of J in Hz is made progressively greater than approximately one sixth of the difference between the chemical shift values of two sets of coupled protons, the relative peak intensities within the multiplet deviate increasingly from the values predicted using the first order rules. Such patterns are referred to in general as A_nB_m patterns with appropriate subscripts to denote the number of protons in the A set (downfield) and in the B set (upfield). The protons in the two sets are then said to be strongly coupled, and the splitting patterns can be quite complex if either set consists of more than one proton. Reference to a textbook on

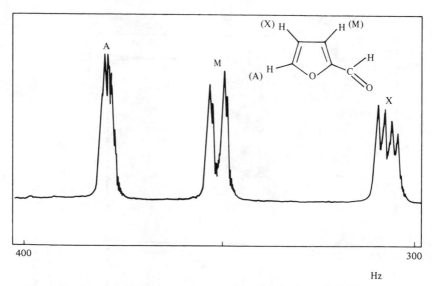

Figure 15.43.   NMR spectrum of furan-2-aldehyde, showing an AMX pattern resulting from the first order coupling of three heterocyclic protons. The letters adjacent to the formula indicate the multiplets in the spectrum. The aldehyde proton resonance is not shown and the spectrum has been expanded to cover the full width of the chart to show the splitting more clearly.

NMR spectroscopy is recommended for information on the appearance and analysis of such spectra.[1-10] The *AB* pattern, which arises from strong coupling of a single proton of each type, is distinctive in appearance and readily analyzed, however. A typical example of the *AB* pattern is found in the peaks of the aromatic protons of *p*-chloroacetophenone, which is shown in Fig. 15.44. The two inner lines of the doublets (2 and 3) have become more intense, and the two outer lines (1 and 4) have become less intense relative to the corresponding lines in an *AX* pattern. As the ratio J/δ increases, the distortion becomes more pronounced, and, in the limiting case of two identical protons (A_2), only a single line is observed.

The value of J is equal to the separation in Hz of lines 1 and 2 or of lines 3 and 4 in Fig. 15.44. The values of δ_A and δ_B can be calculated using equations 15.15 and 15.16.

$$(\nu_2 + \nu_3) = (\delta_A + \delta_B)_\nu \qquad (15.15)$$

$$\nu_2 - \nu_4 = \nu_1 - \nu_3 = \sqrt{J^2 + \nu^2(\delta_A - \delta_B)^2} \qquad (15.16)$$

where ν_1, ν_2, ν_3, and ν_4 represent the displacements of the respective peaks in Hz, ν represents the radiation frequency in MHz and δ_A and δ_B represent the chemical shifts in ppm.

341

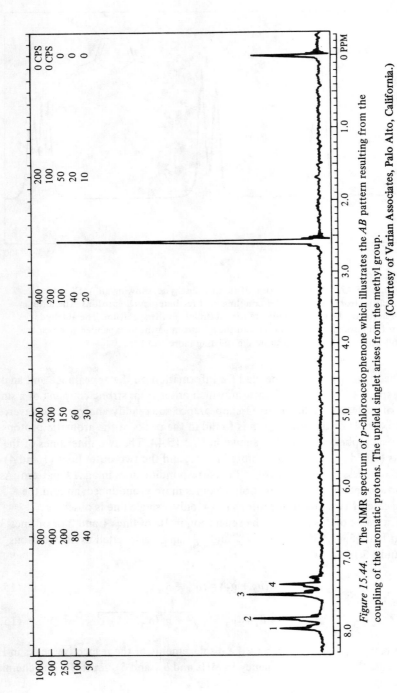

Figure 15.44. The NMR spectrum of *p*-chloroacetophenone which illustrates the *AB* pattern resulting from the coupling of the aromatic protons. The upfield singlet arises from the methyl group.

(Courtesy of **Varian Associates, Palo Alto, California**.)

15.7.4.3. Coupling Constants

In general, spin-spin coupling in saturated systems is important only between protons separated by no more than 3 bonds, but in rigid molecules, long range coupling between protons separated by 4 bonds can be significant. In unsaturated systems, such as aromatic molecules, long range coupling often can be detected between protons separated by four or five bonds. A short list of coupling constants for representative structures is given in Table 15.5. More extensive listings are given in most textbooks of proton NMR.[1-10] The magnitudes of coupling constants can provide valuable structural information and accurate measurements of coupling constants is a necessary step in the analysis of NMR spectra.

15.7.5. Experimental Techniques

To obtain well resolved NMR spectra, one must devote care both to the preparation of the sample and to the operation of the spectrometer. The following discussion describes important techniques and indicates the key aspects of these techniques. Because detailed operating instructions for NMR spectrometers are unique to each model, the discussion emphasizes the functions of the major controls and it gives general guidance concerning the adjustment of these controls to obtain optimal results, but it does not give detailed operating instructions.

Liquid samples of low viscosity are usually studied "neat," that is without the addition of solvent. Viscous liquids and solids must be dissolved in a solvent of low viscosity to obtain well resolved spectra, and usually non-protonic or deuterated solvents are used to avoid spectral interference by solvent peaks. Approximately $\frac{1}{2}-1$ ml of sample generally is required to obtain satisfactory spectra, and solutions should contain at least 10% of solute by weight. Smaller sample volumes or lower solute concentrations can be accommodated by using specialized techniques and equipment.

Commonly used NMR solvents include carbon tetrachloride, carbon disulfide **(Caution!)**, and the deuterated isomers of chloroform, acetone, benzene, dimethylsulfoxide, dioxane, methanol, methylene chloride, and water. If the solvent absorption peaks do not obscure important features of the sample spectrum, protonic solvents can be used.

Samples must be free of particulate matter, which broadens absorption peaks and reduces spectral resolution. Metallic particles are particularly troublesome, and particular care is necessary if the sample has been in contact with finely divided metal catalysts. Particulate matter usually can be removed by filtering the liquid through a small plug of glass wool which has previously been washed with the solvent.

Dissolved oxygen (a paramagnetic substance) tends to broaden NMR peaks, and occasionally this broadening can interfere with interpretation of the spectrum, but usually dissolved oxygen will not interfere with the use of NMR spectra for identification of organic compounds. Dissolved oxygen can usually be removed

TABLE 15.5

Representative Values of Proton Coupling Constants

Structure	J, Hz
$\underset{\diagdown}{\overset{\diagup}{C}}\underset{H}{\overset{H}{\diagdown}}$	11–15
$-\underset{\underset{H}{\mid}}{\overset{\mid}{C}}-\underset{\underset{H}{\mid}}{\overset{\mid}{C}}-$	2–10
$\underset{\underset{H}{\mid}}{H_2C}-\underset{\underset{H}{\mid}}{CH}\ X$	6–8
$(CH_3)_2CH\ X$	5–7
$C{=}C\overset{H}{\underset{H}{\diagdown}}$	0.4–3.5
$\underset{H}{\overset{}{\diagdown}}C{=}C\overset{H}{\diagup}$	12–18
$\underset{\diagup}{\overset{H}{\diagdown}}C{=}C\overset{H}{\diagdown}$	8–12
$\underset{H}{\overset{}{\diagdown}}C{=}C\overset{C-H}{\diagdown}$	0.4–3
$\underset{}{\overset{}{\diagdown}}C{=}C\overset{H}{\underset{C-H}{\diagdown}}$	4–9
$H-C{\equiv}C-C-H$	2.4
Aromatic ring	
Ortho	6–10
Meta	1–3
Para	0–1
$\underset{}{\overset{}{\diagdown}}CH-C\overset{H}{\underset{O}{\diagdown}}$	7–9

adequately by bubbling nitrogen through the sample in the sample tube and placing a plastic cap on the filled sample tube. Care is necessary to avoid loss of sample by entrainment. Occasionally, oxygen must be removed completely from samples, and this can be achieved by subjecting the sample to freeze-thaw cycles on a vacuum line as described in Chapter 14. The filled sample tube is attached to a vacuum line with a stopcock, the sample is frozen using a suitable cooling bath, and the stopcock to the vacuum line is opened, removing uncondensed gases. The stopcock is closed, the cooling bath is removed, and the sample is allowed to melt, releasing part of the dissolved gases. The cycle is repeated several times to complete the removal of oxygen, and the tube is sealed to prevent reentry of oxygen during the recording of the spectrum.

It is necessary to standardize the magnetic field in NMR spectroscopy by recording a peak of a reference substance which serves as a basis for calculating chemical shift values. The reference compound may be dissolved in the sample, the internal standard method, or it may be sealed in a capillary tube that is in turn placed in the sample, the external standard method. Tetramethylsilane, TMS, is the most widely used reference substance, and it is used almost always as an internal standard. Internal standard TMS samples are prepared by adding approximately 1% of TMS directly to the sample prior to recording the spectrum. This concentration range corresponds to addition of one or two drops of TMS using a drawn-out medicine dropper. The boiling point of TMS is 26°C, which facilitates removal of TMS from the sample, but also necessitates care to prevent undue losses from the stock bottle. The bottle should be capped tightly and stored in a refrigerator to prevent loss of TMS.

For solutions in water, dioxane and acetonitrile are suitable standards which give singlet peaks that are downfield from TMS by 3.68 ppm and 2.00 ppm respectively. Chemical shift values measured with respect to these standards thus are smaller by these magnitudes than values measured with respect to TMS.

Sample tubes used for NMR are glass tubes of thin walls approximately 0.5 cm o.d. and 15 cm long. Tubes made for use with different instruments vary significantly in o.d., and they are not interchangeable. For example, tubes for use with Perkin-Elmer spectrometers are 4.6 mm o.d., whereas those for use with Varian spectrometers are 5.0 mm o.d. Tubes must be cleaned of all foreign material, both inside and outside, to prevent spurious signals and possible distortion of the sample spectrum.

15.7.5.1. NMR Spectrometer Controls

Compared to a typical bench top infrared spectrometer, even a simple NMR spectrometer has a large number of controls which must be used for routine spectral studies. A working knowledge of these controls and the manner in which they influence NMR spectra is a prerequisite to obtaining spectra of high quality. In the following sections, the functions of these controls will be described and the manner in which they influence the NMR spectrum will be illustrated. The discussion will not give operating instructions for a particular type of NMR spectrometer, or detailed directions for obtaining spectra; rather it will provide general discussion and guidance which should be *generally* applicable to spectrometers used for routine measurements.

The controls of an NMR spectrometer can be divided into three functional groups; those that control the variation of the magnetic field, those that influence the quality of the peaks in the resultant NMR spectrum, and those that control the electronic peak integrator.

Sweep Controls. The variation of the magnetic field is controlled with the sweep controls, that determine the starting point of the sweep, the width of the sweep, and the time required to complete the sweep. The sweep controls allow selection of the optimum sweep width for a particular spectrum or expansion of a small portion of a spectrum for detailed study.

Sweep Zero. The sweep zero is used to position the TMS reference peak precisely at zero chemical shift on the recorder chart.

Sweep Offset. The sweep offset control permits continuous downfield displacement of the sweep by as much as 1000 Hz.

Sweep Time. The sweep time control permits selection of the optimum sweep time from among a number of pre-set values which are usually 50, 250, 500, and 1,000 sec. Normally, a sweep time of 250 sec is used to record spectra, and a sweep time of 50 sec is used to record integrals.

Sweep Width. The sweep width control permits stepwise adjustment of the range over which the magnetic field is scanned. The horizontal scale sensitivity of the recorder is adjusted automatically so that at each setting the sweep width corresponds to full scale horizontal deflection of the recorder. Standard values of sweep width are 25, 50, 100, 250, 500, and 1000 Hz. For routine spectral measurements a sweep width of 500 Hz is suitable unless it is desired to study protons with δ values greater than 8.33 ppm.

 A small portion of the spectrum may be expanded for study by using the sweep offset control in conjunction with the sweep width control. The sweep width is selected to cover the desired range in Hz and the sweep offset is adjusted so that the desired portion of the spectrum will be scanned. The position of each peak as read from the recorder chart must be corrected by adding the value of the sweep offset in consistent units to obtain the chemical shift value. As an example, suppose that it is desired to study in detail the peak of an aldehyde proton that has a δ value of 9.20 ppm. A number of combinations of sweep width and sweep offset can be used to expand a segment of the NMR spectrum in the range near 9.20 ppm, which corresponds to 552 Hz at an r.f. frequency of 60 MHz. For example, a sweep offset setting of 540 Hz and a sweep width setting of 25 Hz permits scanning the range from 565 Hz downfield with respect to TMS to 540 Hz downfield with respect to TMS, and places the peak of interest near the center of the sweep. Other combinations of sweep width and sweep offset can be chosen to vary the position of the peak on the recorder chart. The corresponding frequency relative to TMS at 30 MHz is 276 Hz, and a sweep offset of 260 Hz would be appropriate. Under these conditions and a sweep width of 25 Hz, the range from 260 Hz downfield to 285 Hz downfield is covered and again the peak is near the center of the sweep.

The amplitude and the shape of the absorption peaks are influenced by a number of controls, the functions of which are described below. The effect of certain control settings on the spectrum is complex and the figures showing peak shapes are meant to be representative.

R.F. Power Level. The r.f. power level control governs the intensity of the 60 MHz radiation received by the sample. The optimum value of the r.f. power level depends on the width of the absorption peak and the sweep rate and it must be determined by trial and error. As Fig. 15.45 shows, increasing the r.f. power level improves the signal to noise ratio, but excessive levels of r.f. power cause the peak to broaden and to diminish in height. The deterioration of the peak intensity above a certain r.f. power level results from a phenomenon known as saturation. Initially, there is a slightly larger population of protons in the ground state than in the excited state in accord with the Boltzmann equation. Exposing the sample to radiation of the proper frequency causes a net absorption of energy to occur, and the intensity of the absorption peak is directly proportional to the excess population of the protons in the ground state. As a consequence, the intensity of the absorption peak decreases. Fortunately, protons in the excited state can lose energy and return to the ground state by mechanisms known as relaxation processes. The downward transitions of protons undergoing relaxation oppose the upward transitions of protons absorbing energy and thus operate to restore the excess

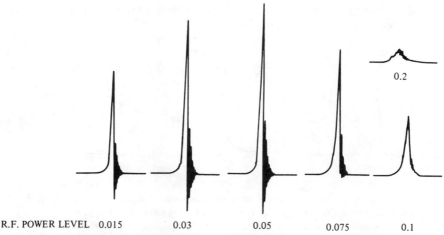

R.F. POWER LEVEL   0.015   0.03   0.05   0.075   0.1

Figure 15.45.   The effect of the r.f. power level on the intensity of a representative resonance peak. Optimum setting is represented by the third trace.
(Courtesy of Varian Associates, Palo Alto, California.)

population of the ground state and prevent saturation. Whether saturation is observed or not depends on the relative rates of the upward transitions and the downward transitions. The upward rate is determined by the r.f. power level and the downward rate is controlled by the characteristic relaxation rate of the proton in question. Each chemically distinct proton has its own net characteristic relaxation rate with the result that saturation may occur at different r.f. power levels for the different kinds of protons in the sample. This phenomenon, known as differential saturation, can interfere with accurate determinations of relative numbers of protons. Such problems are best avoided by using the lowest r.f. power level that gives an adequate signal to noise ratio. The optimum setting is usually between 0.015 and 0.050, but the best setting must be determined experimentally for each sample and sweep rate.

Detector Phase.   The detector phase control adjusts a reference signal in the detector to permit optimization of the shape of the absorption peaks. The detector phase is adjusted until the baseline immediately prior to a peak is at the same level as the baseline immediately after the peak. The influence of the detector phase control setting on the peak shape is illustrated in Fig. 15.46 that illustrates both proper and improper adjustments. Precise adjustment of the detector phase control is not necessary for recording the normal spectrum, but it is very important if the integral is to be recorded.

Filter.   The filter is a low-pass network which attenuates frequencies above a selected value and thus minimizes the effect of random noise on the spectrum. The proper setting is determined by the sweep rate and the peak widths but, in general, the setting should be as low as possible to minimize distortion of absorption peaks. For most spectra it should be possible to use the lowest setting (time constant).

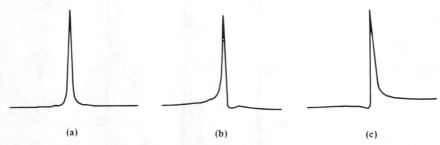

(a)   (b)   (c)

Figure 15.46.   The effect of the detector phase control setting on the shape of a representative absorption peak. The correct setting shown in curve *a*, is characterized by no change in the baseline level on passing through the peak. (a) correct adjustment, (b) phase control advanced too far, (c) phase control not advanced sufficiently.   (Courtesy of Varian Associates, Palo Alto, California.)

Sample Spinning Rate. The sample tube is rotated in the magnetic field to average minor spatial variations of the field strength across the diameter of the sample tube, and thus to improve the resolution. The sample tube is fitted with a plastic vane which is driven by a jet of air. The rotational velocity in revolutions per second (r/s) is controlled by a throttle valve and is read on a panel meter. Sample spinning rates of 30–50 r/s are normally used. One result of spinning the sample is the generation of spinning sidebands, that are beat signals arising from the interaction of the spinning frequency and the r.f. excitation frequency. As the sample spins, it encounters alternately regions of high and low magnetic field at a frequency equal to the spinning frequency or some multiple of the spinning frequency. In effect, the magnetic field is modulated at the spinning frequency and this modulated field couples with the applied r.f. field to generate spinning side bands corresponding to the sum and the difference of the two frequencies.

Spinning sidebands are rather weak relative to the absorption peak and they occur as pairs of peaks which are symmetrically displaced about the absorption peak by a frequency equal to the sample spinning rate. Sidebands can be identified readily by this dependence on the sample spinning rate. The amplitude of the spinning sidebands decrease sharply as the sample spinning rate is increased, and the sidebands should be sufficiently attenuated at normal spinning rates to prevent interference.

Spectrum Amplitude. The spectrum amplitude controls usually consist of a decade attenuator and a multiposition switch which permits attenuation in approximately 25% steps. The spectrum amplitude controls normally are adjusted so that the most intense peaks correspond to nearly full scale vertical deflection of the pen. The spectrum amplitude can be increased to increase the signal of a weak peak, but this change will also increase the noise proportionately, and may require readjustment of the sweep rate and filter settings.

Resolution. The resolution control (Y gradient) permits minor adjustment of the uniformity of the magnetic field to optimize the resolution. The optimum setting is indicated by maximum peak height and "ringing," a series of exponentially decaying oscillations immediately after the absorption peak. Ringing results from the beating of the magnetic field of the 60 MHz r.f. radiation with an r.f. magnetic field generated by the excess population of protons in the ground state. When the external magnetic field corresponds to the resonance condition, the protons begin to act in a cooperative manner to generate an r.f. magnetic field with a frequency of exactly 60 MHz. The frequency of the magnetic field generated by the protons is directly proportional to the external magnetic field so that as the external field increases above the value corresponding to resonance, the frequency of the field generated by the protons also increases above 60 MHz and the two r.f. fields interact to

349

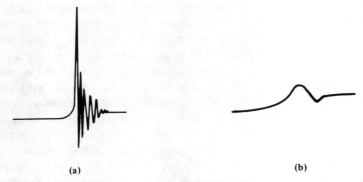

(a)   (b)

Figure 15.47.   The effect of the resolution control (*Y* gradient) on NMR peak shape. Optimum setting of the resolution control is indicated by maximum peak height and ringing. (a) correct setting, (b) incorrect setting.

give a beat signal. The frequency of the beat signal increases with time because the frequency difference between the 60 MHz radiation and the field generated by the protons increases as the external magnetic field increases. Simultaneously, the amplitude of the beat signal decreases with time as the protons relax from the cooperative state associated with the resonance condition to a random state. Resonance peaks illustrating ringing and proper and improper adjustment of the resolution controls are shown in Fig. 15.47.

Integrator Controls.   The operation of the electronic peak integrator is controlled with a set of controls that are usually grouped together on the front panel and set apart from the other controls. The functions of these controls and the manner in which they are used is described below.

Integrate-Hold Switch.   In the integrate position the signal from the NMR spectrometer is connected to the integrator input. This position is used to obtain integral traces. In the hold position, the integrator input is disconnected from the spectrometer output, and the integrator output remains equal to the integral accumulated to that point.

Reset.   Depressing the reset button deletes the accumulated integral and returns the integrator output to zero.

Integrator Zero.   The integrator zero adjusts the spectrometer output to precisely zero in the absence of a spectrum signal to prevent integrator error. The control is adjusted to minimize the drift of the recorder pen when the integrate switch is activated, so that the slope of the integral steps is minimized.

Amplitude. The amplitude control permits continuous adjustment of the amplitude of the integrated signal without changing the settings of the spectrum amplitude controls. The control is adjusted to obtain an output, the magnitude of which is consistent with accurate measurements of the relative height of the step associated with each peak in the normal spectrum.

Integral traces are obtained at high sweep rates to minimize errors caused by drift in the integrator circuitry. The appropriate sweep time for a 500 Hz sweep width is 50 sec. The integrator zero control and the detector phase control interact to some extent, and careful adjustment of both controls is necessary to obtain accurate integrals. The effects of these controls on integral traces are illustrated in Fig. 15.48. Correct adjustment of the two controls is shown in Fig. 15.48a, and is characterized by horizontal integral plateaus both before and after the absorption peak. Improper adjustment of the detector phase control is illustrated in Fig. 15.48b, and is indicated by the fact that the two plateaus have slopes of opposite sign. Improper adjustment of the integrator zero adjustment is illustrated in Fig. 15.48c, and results in the two plateaus having equal but non-zero slopes. The proper settings are best obtained by first adjusting the detector phase control using the normal spectrum as a guide (see Fig. 15.46), and then adjusting the integrator zero control using the integral trace as a guide. The height of each step is proportional to the number of hydrogen atoms associated with the corresponding peak in the normal spectrum. The height of each step is measured as illustrated in Fig. 15.36 and the relative heights are calculated. These can be reduced to a set of small whole numbers

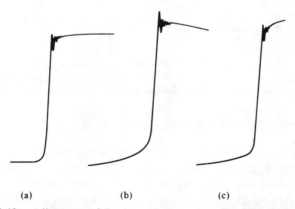

(a)   (b)   (c)

Figure 15.48. Adjustment of the detector phase control and the integrator zero control for integration. Correct adjustment, illustrated in curve *a*, results in horizontal plateaus in the integral trace. (a) phase and zero correct, (b) phase incorrect, (c) zero incorrect. (Courtesy of Varian Associates, Palo Alto, California.)

that indicate the relative numbers of each kind of proton in the molecule. It is best to record more than one value for the integrals and to average them for the most accurate determination.

General Sequence of Control Adjustments. The following order of control adjustments is recommended for obtaining spectra using a spectrometer that is well tuned. Initial tuning of NMR spectrometers requires considerable experience, and should not be attempted without the guidance of an experienced operator. If the following steps do not yield a spectrum of high quality, consult the instructor.

Representative settings of the key controls for survey spectra are given in Table 15.6. The spectrum amplitude may require adjustment to obtain appropriate peak heights. Appropriate changes in the sweep width and sweep offset settings may be made to expand or offset the sweep, as indicated earlier. The sweep zero is adjusted to bring the TMS line to the zero chart position. The detector phase is adjusted next, and the resolution control is set to maximize the peak amplitude and ringing. The r.f. power setting can now be optimized by trial and error adjustment. It is necessary to check the position of the TMS peak just before obtaining the sample spectrum to prevent error in chemical shift values caused by slow drift in the sweep circuitry.

TABLE 15.6

Nominal Settings of NMR Spectrometer Controls for Survey Spectra

Sweep width	500 Hz
Sweep time	250 s
Sweep offset	0 Hz
R.F. power	0.05
Spectrum amplitude	2
Filter	1
Spinning rate	40 r/s
Detector phase	obtain level baseline
Resolution	optimize peak height and ringing

15.7.6. Examination of Spectra

As a first step, an attempt should be made to recognize first order patterns, that may be slightly distorted from the ideal intensity ratios. It is well to bear in mind that first order multiplets may overlap, sometimes causing an apparently complex spectrum, as illustrated by the spectrum of 1-bromo-3-chloropropane in Fig. 15.49. The complex downfield multiplet in this spectrum is in fact two overlapping triplets which arise from the terminal methylene groups.

The integral trace should be analyzed to determine the relative numbers of protons of each chemical type. Evaluation of spectra should include determination

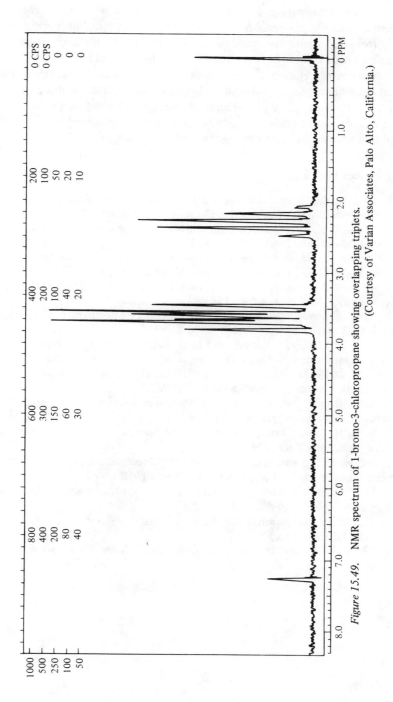

Figure 15.49. NMR spectrum of 1-bromo-3-chloropropane showing overlapping triplets. (Courtesy of Varian Associates, Palo Alto, California.)

of the coupling constant and chemical shift value for each multiplet. This information can be derived directly from first order spectra, but it is not obtainable easily from higher order patterns other than AB patterns. If the spectrum is clearly of higher order, one of the texts listed in Refs. 1–10 should be consulted for methods of interpretation. Once the coupling constants and the chemical shift values are determined, they should be compared with the values listed for proposed structures. Considerable trial and error often is necessary before a final identification can be made, and other information about the unknown may be helpful or necessary in interpreting the NMR spectrum.

References to 15.7

1. L. M. Jackman and S. Sternhell, "Applications of Nuclear Magnetic Resonance Spectroscopy in Organic Chemistry," Pergamon Press, Oxford, 1969.

2. D. W. Mathieson (ed.), "Nuclear Magnetic Resonance for Organic Chemists," Academic Press, New York, 1967.

3. J. A. Pople, W. G. Schneider, and H. J. Bernstein, "High-Resolution Nuclear Magnetic Resonance," McGraw-Hill, New York, 1959.

4. R. M. Lynden-Bell and R. K. Harris, "Nuclear Magnetic Resonance Spectroscopy," Nelson, London, 1969.

5. F. A. Bovey, "Nuclear Magnetic Resonance Spectroscopy," Academic Press, New York, 1969.

6. J. W. Emsley, J. Feeney, and L. H. Sutcliffe, "High Resolution Nuclear Magnetic Resonance Spectroscopy," Pergamon Press, London, 1965. Vols. 1 and 2.

7. E. D. Becker, "High Resolution NMR; Theory and Chemical Applications," Academic Press, New York, 1969.

8. W. Brügel, "Nuclear Magnetic Resonance Spectra and Chemical Structure," Academic Press, New York, 1967.

9. R. H. Bible, "Interpretation of NMR Spectra; An Empirical Approach," Plenum Press, New York, 1965.

10. R. M. Silverstein and G. C. Bassler, "Spectrometric Identification of Organic Compounds," 2d ed., Wiley, New York, 1967.

11. "Nuclear Magnetic Resonance Spectra," Sadtler Research Laboratories, Philadelphia. Vols. 1–34 plus indices.

12. "High Resolution NMR Spectra Catalog," Varian Associates, Palo Alto (1962, 1963). Vols. 1 and 2.

13. R. M. Silverstein and R. G. Silberman, *J. Chem. Educ.*, **50**, 484 (1973).

CHAPTER

16

X-Ray Diffraction

16.1. INTRODUCTION AND GENERAL PRINCIPLES

X-Ray diffraction studies comprise an important method of determining the absolute structures of molecules or crystals, lattice spacings, and atomic or molecular dimensions. Although much of the work on a research level is directed to single crystals, this discussion will be restricted largely to the simpler case of powder studies.

A crystal is made up of atoms, ions, or molecules arranged in a regular way, and this arrangement, or lattice, is capable of acting as a three-dimensional diffraction grating, with each structural unit reflecting incident X-rays. The basic relationship governing this reflection is the Bragg equation, which is derived in most physical chemistry texts.[1] The Bragg equation predicts that the rays reflected from successive planes of atoms in the crystal reinforce each other, and hence can be detected, only when

$$\lambda = \frac{2}{n} d \sin \theta \qquad (16.1)$$

where λ is the wavelength of the X-rays, d is the distance between successive reflecting planes, θ is the angle of reflection, and n is an integer corresponding to the order of reflection. Most often only first-order reflections, where $n = 1$ are considered. From this equation it is evident that if the angle of reflection can be measured for a given wavelength, the distance between successive planes of reflecting atoms can be found.

For any regular structure, an infinite number of planes of different slope can be drawn through some of the atoms, but, in general, only planes passing through an appreciable number of atoms are important. Different planes are distinguished by the reciprocal of their intercepts on a set of axes constructed for the crystal. These values are known as the Miller indices,[1] h, k, and l. A two-dimensional representation of a crystal lattice showing planes of atoms and the Miller indices is illustrated in Fig. 16.1.

If a single crystal is irradiated with a beam of monochromatic X-rays, the reinforced reflected beam consists of a series of rays that yield discrete spots on the detector film. These spots correspond to the planes of atoms for which θ and d satisfy the Bragg relationship. If a powder is used, the microcrystals are present in all possible orientations, so that instead of a series of spots, the rays are reflected in the form of concentric cones. Arcs of these cones can be detected as lines on a strip of film. The spacings of these lines are characteristic of the interatomic distances of the compound, and these along with the relative intensities of the lines can be used for identification purposes.

X-rays are produced by bombarding a metal electrode with electrons. When copper is used as the target, three intense X-ray lines, superimposed on a background of continuous radiation, are produced. These lines and their wavelengths are denoted as K_{α_1}, 1.54050 Å, K_{α_2}, 1.54434 Å and K_β 1.39217 Å. Passing this beam through a nickel foil results in absorption of most of the K_β component. The K_{α_1} and K_{α_2} lines are so close in wavelength that they are reflected

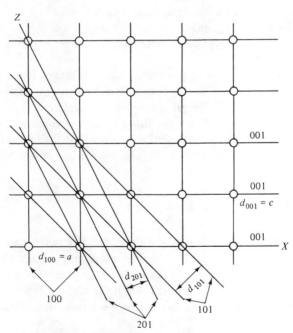

Figure 16.1. Two-dimensional representation of a crystal lattice, showing some planes of atoms. The indices are those for planes parallel to the Y axis.

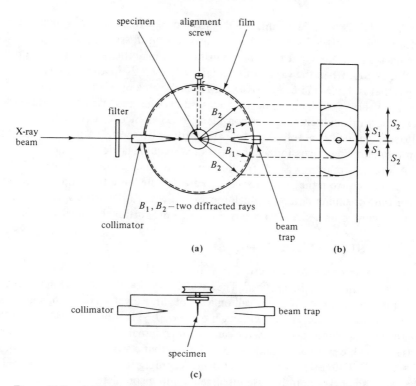

Figure 16.2. (a) X-ray powder diffraction camera, (b) the exposed X-ray film, (c) top view of the camera.

almost as if they were one. In high resolution work the lines of the diffraction pattern may appear as doublets, but normally this minor effect does not interfere with the structure determination.

A schematic diagram of an X-ray powder diffraction camera together with the appearance of the film is shown in Fig. 16.2. A cylindrical camera usually is used for this type of experiment, with the sample situated at the center of the camera and with the film held on the circumference. Various sizes of cameras are available, with the more common radii (R) being 57.3, 28.6 and 71.6 mm. These radii are used so that when the film is laid flat for measurement as in Fig. 1b, the distance from the center of the film to the arc S, is a simple multiple of the Bragg angle θ, that is

$$\frac{2\theta}{360} = \frac{S}{2\pi R} \tag{16.2a}$$

or

$$\theta = \frac{360S}{4\pi R} \tag{16.2b}$$

where $R = 57.3, 28.6$ or 71.6 mm, and $\theta = \frac{1}{2}S$, S or $\frac{4}{10}S$ respectively. The angles may therefore be determined by measuring the distances on the film S_1, S_2, S_3, etc., for all the lines on the X-ray pattern and the values of d may then be obtained from Bragg's equation. When the values of θ and d are known the pattern is then indexed; that is, Miller indices are assigned for each value of d. Unfortunately, except for a very special, but important, class of crystals known as cubic, this task is exceedingly difficult and requires considerable experience. The required equations[2-8] usually are solved by trial and error or by special techniques. For this reason the following discussion will be restricted to cubic lattices.

Another form of powder diffraction apparatus does not use a camera. The X-rays diffracted from the powder specimen are detected as a function of angle by a Geiger tube or similar detector which moves in a circle around the specimen. The output is recorded on a chart, with position being proportional to angle.

16.2 CRYSTAL STRUCTURE

The simplest repeating structural unit from which a crystal may be constructed is known as the *unit cell*. The crystal may therefore be characterized by the lengths intersected by the unit cell on a set of three crystal axes, a, b, and c and the three angles these axes make with each other α, β, γ. It should be noted that the axes a, b, and c are not necessarily the same as an external set of axes x, y and z. In the case of a cubic system the unit cell is a cube so that $a = b = c$ and $\alpha = \beta = \gamma = 90°$. Other structures are classified according to the restrictions placed on the axes and angles, and they are discussed in more complete treatments.[1,2]

A cubic crystal belongs to one of three sub-classes:

1. primitive or simple cubic (P) with one molecule per unit cell (one at each corner, shared by eight cells)
2. face-centered cubic (F) with four molecules per unit cell (one at each corner, shared by eight cells; one in each face, shared by two cells)
3. body-centered cubic (I) with two molecules per unit cell (one at each corner shared by eight cells, one at the center)

These lattices are illustrated in Fig. 16.3. Note that in each lattice, all "molecules" are identical; that is, the corner and center points in the I lattice, for example, are the same.

In the case of a cubic lattice the equation relating d and the Miller indices is

$$d = \left(\frac{a^2}{h^2 + k^2 + l^2} \right)^{1/2} \tag{16.3}$$

Substitution of this expression into the Bragg relationship yields:

$$\sin^2 \theta \frac{\lambda^2}{4a^2} (h^2 + k^2 + l^2) = \frac{\lambda^2 N}{4a^2} \tag{16.4}$$

where the integer N is $h^2 + k^2 + l^2$.

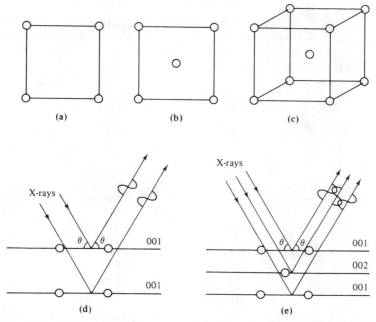

Figure 16.3. The unit cells for a cubic lattice: **(a)** primitive cubic unit cell (one face), **(b)** face-centered cubic unit cell (one face), **(c)** body-centered cubic unit cell, **(d)** reflection from **(a)**, rays in phase, and **(e)** reflection from **(b)**. Note that the 002 rays are now out of phase without those of 001.

Certain reflections will always be absent in the F and I lattices. In the case of an F lattice only those reflections will be present for which the values of the Miller indices are homogeneous, that is, either all odd or all even; while for an I lattice the only reflections which will be present are those for which the sum of the Miller indices is an even number, i.e., when $h + k + l = 2n$, where n is an integer. There are no restrictions on reflections from a P lattice. With these rules in mind Table 16.1 can be constructed. To appreciate these rules, consider a primitive cubic unit cell in two dimensions, as in Fig. 16.3a. The Miller indices (h, k, l) for a plane parallel to the x and y axes, and perpendicular to the z axis, are 001. (The intercepts on the x, y, and z axis are at ∞, ∞, and 1 units respectively. The Miller indices are the reciprocals of these intercepts, reduced to whole numbers, or 001 in this case). An incident beam of X-rays would be reflected when the Bragg relation is satisfied, as in Fig. 16.3d.

In the case of a face-centered unit cell, as in Fig. 16.3b, a second plane of atoms runs parallel to the 001 planes, mid-way between them. We can call this the 002 plane. When the 001 reflections reinforce one another, the 002 reflections are exactly out of phase with them and cancel them (Fig. 16.3e). Thus, reflections corresponding to 001 planes are absent. However, a line will be observed if sin θ is doubled, because then the 001 and 002 reflections are in phase. The result is that

TABLE 16.1

Values of $N = h^2 + k^2 + l^2$ for the Cubic Lattices (to $N = 50$)

N	P hkl	F hkl	I hkl	N	P hkl	F hkl	I hkl
1	100			28			
2	110		110	29	520		
3	111	111		29	432		
4	200	200	200	30	521		
5	210			31			
6	211		211	32	440	440	440
7				33	522		
8	220	220	220	33	441		
9	300			34	530		530
9	221			34	433		433
10	310		310	35	531	531	
11	311	311		36	600	600	600
12	222	222	222	36	442	442	442
13	320			37	610		
14	321		321	38	611		611
15				38	532		532
16	400	400	400	39			
17	410			40	620	620	620
17	322			41	621		
18	411		411	41	540		
18	330		330	41	443		
19	331	331		42	541		
20	420	420		43	533	533	
21	421			44	622	622	622
22	332		332	45	630		
23				45	542		
24	422	422	422	46	631		631
25	500			47			
25	430			48	444	444	444
26	510		510	49	700		
26	431		431	49	632		
27	511	511		50	710		710
27	333	333		50	550		550

reflections from planes with odd N are eliminated. Arguments of this sort yield the restrictions listed in Table 16.1.

The usual method for indexing the powder pattern of a cubic material is to compare the value of $\sin^2 \theta$ for all of the lines. That is:

$$\frac{\sin^2 \theta_2}{\sin^2 \theta_1} = \left(\frac{h_2^2 + k_2^2 + l_2^2}{h_1^2 + k_1^2 + l_1^2} \right) = \frac{N_2}{N_1} ; \qquad \frac{\sin^2 \theta_3}{\sin^2 \theta_1} = \frac{N_3}{N_1} \text{ etc.} \qquad (16.5)$$

These ratios $N_1/N_1, N_2/N_1, N_3/N_1$, etc. are reduced to whole numbers by using a common multiplier. These numbers are identical to the N given in Table

16.1 and if their values are 3, 4, 8, 11, etc. the substance is face-centered cubic. If N has the values 2, 4, 6, 8, etc. the lattice is body-centered cubic. When N has the values 1, 2, 3, 4, etc. the lattice is primitive or simple cubic.

After the pattern is indexed, values for a, the edge of the unit cube, can be calculated from Eq. (16.3). The volume of the unit cell of the compound is merely a^3 and combining this volume with Avogadro's number permits a calculation of the density of the compound. Alternatively, Avogadro's number can be calculated if the density is known, and this is, in fact, one of the methods used to obtain this fundamental constant.

16.3. EXAMPLE OF X-RAY ANALYSIS

When a sample of powdered silver was subjected to a narrow beam of X-rays from a copper target the developed film shows diffraction maxima at $S = 38.1$, 44.4, 64.4, 77.4, 81.5, 97.9, 110.5, 114.9, and 134.9 mm. The radius of the camera used was 57.3 mm.

The treatment of these experimental data is summarized in Table 16.2.

Because the values obtained for N in this case are 3, 4, 8, etc. the substance is arranged in a face-centered cubic lattice with four silver atoms per unit cell. The average of the last five values for a, the edge of the unit cell, is $a = 4.0863$ Å (4.0863×10^{-8} cm). Therefore the volume of this cube (a^3) is 68.232×10^{-24} cm^3. Because this volume contains four atoms of silver, each silver atom effectively occupies a volume of

$$\frac{a^3}{n} = \frac{68.232 \times 10^{-24}}{4} = 17.058 \times 10^{-24} \text{ cm}^3$$

Avogadro's principle would require that one gram-atomic weight of silver occupy a volume $V_M = N_A a^3/n$ where N_A is Avogadro's number and V_M is the molar volume. This volume is equal to the atomic weight of silver, A, divided by its density ρ; that is

$$V_M = \frac{N_A a^3}{n} = \frac{A}{\rho}$$

$$N_A = \frac{nA}{a^3 \rho}$$

In this case $A = 107.880$ g and $\rho = 10.550$ g/cm^3, so that

$$N_A = \frac{(107.880)(4)}{(10.500)(68.232 \times 10^{-24})}$$

$$N_A = 6.0231 \times 10^{23}$$

TABLE 16.2

Parameters obtained in the X-ray analysis of powdered silver

See note:		1		2	3	4	5	6	7
Line	S, mm.	$\theta°$	$\sin^2\theta$	$\dfrac{\sin^2\theta}{0.1064}$	$\dfrac{3\sin^2\theta}{0.1064}$	N	hkl	d_{hkl} Å	a Å
1	38.1	19.05	0.1064	1.000	3.000	3	111	2.3611	4.0895
2	44.4	22.20	0.1423	1.337	4.011	4	200	2.0420	4.0840
3	64.4	32.20	0.2840	2.669	8.007	8	220	1.4455	4.0885
4	77.4	38.72	0.3912	3.677	11.031	11	311	1.2314	4.0920
5	81.5	40.75	0.4263	4.006	12.018	12	222	1.1798	4.0869
6	97.9	48.95	0.5689	5.347	16.031	16	400	1.0212	4.0848
7	110.5	55.25	0.6750	6.344	19.032	19	331	0.93756	4.0867
8	114.9	57.45	0.7103	6.676	20.028	20	420	0.91389	4.0870
9	134.9	67.45	0.8528	8.015	24.045	24	422	0.83408	4.0861

Note 1. $\theta° = 0.5$ times the displacement of the line on the film from the undiffracted ray.

2. The values for $\sin^2\theta$ are divided by the value of $\sin^2\theta$ for the first line. In this case the value for the line is 0.1064.

3. Because the values for the preceding column are not whole numbers inspection shows that the smallest multiplier that will convert these to whole numbers is 3.

4. The result of rounding the numbers in the preceding column. See Table 16.1.

5. See Table 16.1.

6. $\lambda = 2d_{hkl}\sin\theta$ (Bragg's equation).

7. $a = d_{hkl}(h^2 + k^2 + l^2)^{1/2} = d_{hkl}N^{1/2}$.

Alternatively the atomic (molecular) weight or the density of the solid could have been determined as

$$A = \frac{N_A a^3 \rho}{n}$$

and

$$\rho = \frac{nA}{N_A a^3}$$

16.4. PROCEDURES FOR OBTAINING POWDER PATTERNS

16.4.1. Sample Preparation

The sample should be ground into a very fine powder using an agate mortar and pestle. Some of the powder then should be placed in a very fine glass X-ray capillary tube. To get some powder to the lower end of the capillary tube a file can be rubbed gently across the top of the tube; alternatively allowing the capillary tube to fall through a larger diameter tube and to strike a hard surface can be of assistance. The tube normally is filled to a depth of about 1 cm, with care being taken to avoid air spaces. The tube can be sealed with a lighted match. Another good procedure for mounting the sample, if it is not air sensitive, is to cause a *thin* coating of the powder to adhere to a very fine glass fiber lightly covered with vaseline or stopcock grease.

16.4.2. Loading the Capillary Tube into the Specimen Holder of the Camera

The cover, collimator, and beam trap should be removed from the camera first. The specimen holder is normally situated on a metal plate that can be rotated with a pulley system. The holder often is filled with clay and a hole to secure the capillary should be made in the center of the holder using a pin. The capillary tube then should be placed in the hole and carefully secured. **It is critical that the capillary be mounted in a position perpendicular to the cover of the camera.** After the collimator is replaced, a light source is placed opposite the remaining opening and the adjusting screw is used to ensure complete alignment of the sample tube. The specimen is completely centered when it remains visible through the collimator and remains stationary for a complete revolution. After this adjustment is accomplished the beam trap and cover should be replaced.

16.4.3. Loading the Camera

Preparation of the Film

The operations described below all are performed in a darkroom so it is important to allow some time to adjust to the darkroom conditions before proceeding. Also it is wise to perform a dry run with some used film before carrying out the actual experiment.

The camera, film, and film cutter must be placed so that they may be found easily. To dismantle the camera, the front cover and the two collimators should be removed and it is recommended that the collimators be placed behind the camera with the right hand one on the right hand side of the camera and the left hand one on the left hand side. It is very important to avoid knocking the collimators over or dropping them, because they are easily damaged.

The film (X-ray film, 35 mm wide in rolls 25 or 50 ft long) then should be removed from its container and inserted into the film cutter, until the end of the film is against the stop, making sure that the knife edge is raised. Then the film is cut with the knife edge, holes are punched in it using a hand punch, and it is removed from the cutter. Unused film should be returned to its container at once.

Loading Film into the Camera

This operation is difficult at first, and it is strongly suggested that a dry run using a piece of developed film be conducted.

First, locate the holder clips at the top of the camera to gain the correct orientation for inserting the film and make sure that the front knob at the top of the camera is loose. One end of the film is then placed in the leftmost clip and the film is fitted snugly against the walls of the camera with the other end of the film being attached to the right hand clip. Then the film should be tightened using the appropriate knob. The cover and collimators are replaced, and the camera is ready to be mounted on the diffractometer.

16.4.4. Operation of the X-ray Generator

The detailed operating procedure will vary for different instruments and specific information normally is posted near the instrument which will be supervised by a responsible person. **Take care to avoid any exposure to X-rays!** The general procedures are as follows:

1. Place the camera on the track and secure it in position.
2. Make sure that the voltage and current controls are *off*.
3. Set the desired exposure time on the appropriate dial.
4. Depress the *Start* button.
5. When water flow is established, depress the X-ray *on* switch.
6. Adjust the appropriate switches to select the correct voltage (e.g. 35 kV) and current (e.g. 20 mA).
7. A red light normally is lighted if the generator is functioning correctly. The most common problems are an inadequate flow of water, or an incorrect setting of the timer.

16.4.5. Developing the Film

In the dark-room (a safe light can be used) remove the collimator, beam trap and cover of the camera, and take out the film. Development can be performed with a reel and developing tank, but open tray development is just as suitable. Place the film in the developer, making sure that it is completely immersed. Leave it there for the time specified (e.g., 3 min) with occasional agitation. Rinse the film with water, and place it in the fixer for the time specified (e.g., 2 min). Finally, wash the film in running water for 20 min, and dry it by hanging it from a line.

Return the developer and fixer to their proper containers, and clean up the dark room.

16.5. X-RAY POWDER DATA FILE

X-ray diffraction analysis is a useful method for the identification of unknown solid compounds. An X-ray pattern is measured for the unknown in the normal way, the d values are calculated for each line, and the relative intensities for each line are estimated visually. The identification can be based on comparison with known spectra or on matching d values and relative intensities with tabulated values. Several tabulations exist and one of the more comprehensive compilations is the Powder Diffraction File published by the Joint Committee on Powder Diffraction Standards, which is a continuation of the file initiated by The American Society for Testing & Materials (ASTM).[9] The cards are arranged both alphabetically and numerically, based on intensities. The former index is useful if the unknown has been narrowed to several substances.

The upper left hand corner of each card lists the three most intense reflections with the corresponding d values, d_1, d_2, and d_3. The cards are arranged on a relative scale based on 100 for the most intense reflection observed. To use the file, search first for the d value for the most intense (darkest) line of the unknown in the index. Then match the next two darkest lines against the several possibilities that result from the initial search. All the lines on the film and the data card must agree to identify the unknown.

For unknown mixtures, each component is treated as a single substance. Further instruction on the use of these tables is given in the original tabulations.

REFERENCES TO CHAPTER 16

1. W. J. Moore, "Physical Chemistry," 4th ed., Prentice-Hall, Englewood Cliffs, New Jersey, 1972, chap. 18.

2. L. V. Azaroff and M. J. Buerger, "The Powder Method, in X-ray Crystallography," McGraw-Hill, New York, 1958.

3. S. Nyburg, "X-ray Analysis of Organic Structures," Academic Press, New York, 1961.

4. H. Lipson and H. Steeple, "Interpretation of X-ray Powder Diffraction Patterns," MacMillan, London, 1970.

5. L. V. Azaroff, "Elements of X-ray Crystallography," McGraw-Hill, New York, 1968.

6. N. F. Henry, H. Lipson, and W. A. Wooster, "The Interpretation of X-ray Diffraction Photographs," MacMillan, London, 1961.

7. W. R. D'Eye and E. Wait, "X-ray Powder Photography in Inorganic Chemistry," Academic Press, New York, 1960.

8. G. H. Stout and L. H. Jensen, "X-ray Structure Determination — A Practical Guide," MacMillan, New York, 1968.

9. "Alphabetical and Grouped Numerical Index of X-ray Diffraction Data," American Society for Testing & Materials, Philadelphia, Spec. Pub. 48E, 1955.

17

Electrochemical Techniques

17.1. INTRODUCTION

Electrochemical techniques can be used in many ways; e.g., for analytical purposes, in synthetic methods, to study equilibria and kinetics, to provide thermodynamic data, and as a means of obtaining information about the structures and properties of liquids and solutions. Flow of electric current may occur as a flow of electrons (metals and other electronic conductors) or as a flow of ions (electrolytic conductors). Normally, electronic conduction is associated with solids and ionic conduction with liquids, but this distinction is not always the case. Some solids will have ionic conduction (solid electrolytes) and some liquids are electronic conductors.

Electrochemical studies may be concerned with the transport of ions (conductance) or with the processes of electron transfer between the electrode and the ion. Electron transfer processes include the study of the equilibrium potentials of oxidation-reduction reactions (potentiometry), the relation between the current which flows and the chemical reactions produced (electrolysis, coulometry), and a variety of aspects of the mechanisms of ion-electrode interactions (voltammetry).

Electrochemical cells consist of two electrodes dipping into a solution of electrolyte. A simple example (Fig. 17.1) of Zn and Cd electrodes immersed in $1\,M$ solutions of $ZnSO_4$ and $CdSO_4$, respectively, can be considered as an illustration. A potential difference can be measured between the electrodes, and if they are connected, electrons will flow from Zn to Cd through the external circuit. At the electrodes the reactions

$$Zn \longrightarrow Zn^{2+} + 2e^- \tag{17.1}$$

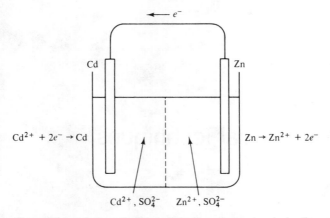

Figure 17.1. Schematic illustration of an electrochemical cell.

and

$$Cd^{2+} + 2e^- \longrightarrow Cd \tag{17.2}$$

take place, and the spontaneous overall cell reaction is

$$Zn + Cd^{2+} \longrightarrow Zn^{2+} + Cd \tag{17.3}$$

If a potential larger in magnitude than that developed by the cell is applied so that the Zn electrode is made negative, this reaction would be reversed. In constructing such a cell, it is necessary to prevent mixing of the $CdSO_4$ and $ZnSO_4$ solutions, because if mixing occurred, the reaction would take place directly at the Zn electrode and no electrons would be transferred through the external circuit. This requirement to avoid mixing of the solutions in the electrode compartment arises often in electrochemical work. A direct junction between two dissimilar solutions may be formed in special cases but this technique is not practical in general. In addition, such a junction between dissimilar solutions (liquid junction) generates a potential difference called a junction potential, the magnitude of which is generally non-reproducible. Junction potentials are of importance in any attempt to measure electrode potentials in cells with junctions, and must be corrected for or minimized if possible.[1]

A salt bridge is often used to connect the two dissimilar electrode compartments of a cell such as described above, and some examples are shown in Fig. 17.2. A salt bridge consists of a concentrated solution of a salt, the ions of which will not interfere with the desired electrode reactions. These ions are present in the bridge in high concentrations, and virtually the entire ionic current between the cell compartments is carried by this salt. A salt bridge also can be used to reduce the junction potential to small (for many purposes, negligible) values. In this case, salts in which both ions have very nearly the same mobilities must be used; that is, each carries half of the current. Saturated KCl or saturated NH_4NO_3 solutions are chosen most frequently. A junction potential arises because different ions tend to move at different rates. The effects of the

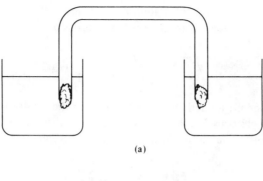

(a)

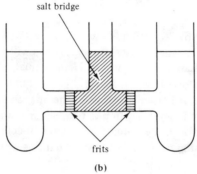

(b)

Figure 17.2. A salt bridge: **(a)** A U-tube filled with salt solution, with the ends plugged with cotton, **(b)** the bridge compartment of a cell with fritted glass dividers.

two liquid junctions produced when a salt bridge is formed will cancel if the cations and anions of the bridge electrolyte move at the same rate and if they make up the majority of the ions moving at each junction.

To reduce convective mixing, the ends of the salt bridge may be plugged with cotton, (Fig. 17.2a) or closed with fritted glass discs (Fig. 17.2b). Still less diffusion will occur if the salt bridge solution is an agar gel, made by heating a mixture of approximately 3% agar with the concentrated salt solution, pouring the liquid into the tube, and allowing it to cool. Agar gels may be dissolved with hot water. For less critical applications, separation of the electrode compartments with a porous plug or a glass frit may be satisfactory, although some mixing will occur, and the junction potential may remain significant.

17.2. POTENTIOMETRY

A redox reaction can be characterized by a cell potential that is a direct measure of the tendency for the reaction to take place. The potential-concentration relationship of an electrode is given by the Nernst equation.

For a metal, M, in equilibrium with a solution of its ion, M^{n+}, $(M^{n+}_{(soln)} + ne^- \rightleftharpoons M_{(s)})$, the electrode potential is given by

$$E = E^\circ - \frac{RT}{nF} \ln \frac{1}{[M^{n+}]} \qquad (17.4)$$

where E° is the standard reduction potential (that for the ideal system at unit activities), R is the gas constant in Joules/mol, T is the absolute temperature and F is the Faraday, 96,496 coulombs. This equation can be rewritten at 25°C as

$$E = E^\circ + \frac{0.0592}{n} \log[M^{n+}] \qquad (17.5)$$

Strictly speaking, the above equation is valid only if activity is employed instead of concentration, although in many cases concentrations will give acceptable results. Within the limits of this approximation, the Nernst equation predicts a straight-line relationship when E is plotted against $\log[M^{n+}]$. The number of electrons transferred can be evaluated from the slope of the line, and the intercept when $\log[M^{n+}] = 0$ ($[M^{n+}] = 1$) gives E°. In practice, one measures the potential of a cell, that is, the difference of the potentials of two half-cells. The Nernst equation for the cell will be the sum of the two Nernst equations which apply to the half-cell reactions.

If activities are considered instead of concentrations, the equation for the potential at 25°C is written

$$E = E^\circ + \frac{0.0592}{n} \log a_M^{n+} \qquad (17.6)$$

where a_M^{n+} is the activity of M^{n+}. For ionic species, the activity of species i is given by $a_i = \gamma_i m_i$, where γ_i is the activity coefficient, and m_i is the molal concentration. In dilute solution, molal and molar concentrations are approximately equal (see Sec. 3.4). The activity coefficients of individual ions cannot be evaluated, but the mean ionic activity coefficient for the salt used, $\gamma_\pm$, may be used in its place. Hence,

$$E = E^\circ + \frac{0.0592}{n} \log m_M^{n+} + \frac{0.0592}{n} \log \gamma_\pm \qquad (17.7)$$

The Debye-Hückel theory predicts that $\gamma_\pm$ should be proportional to the square root of the ionic strength. The ionic strength $I = \frac{1}{2} \Sigma C_i Z_i^2$ where Z_i is the charge and C_i the concentration of species i. For uni-univalent electrolytes, the ionic strength equals the concentration, and in this case, where $n = 1$

$$E = E^\circ + \frac{0.0592}{1} \log m_{M+} + BT\sqrt{C_{M+}} \qquad (17.8)$$

where B is a constant. A plot of $E - (0.0592/n)\log m_M^+$ vs. $\sqrt{C}$, should yield a straight line, which on extrapolation to $C = 0$, gives E°. Potential measurements often are used to evaluate activities, when E° values are already known.

One cannot measure the potential of a single electrode, only the potential difference between two electrodes. It is necessary therefore to report electrode potentials relative to the potential of some reference electrode. Conventionally, the potential of the standard hydrogen electrode is defined as zero at all temperatures and all other potentials are referred to this reference. In practice, it often is more convenient to use some other electrode as the reference. Measured potentials can be corrected to the hydrogen scale if the potential of the reference electrode relative to the hydrogen electrode is known. A very common reference electrode is the calomel electrode, consisting of a paste of mercury, insoluble Hg_2Cl_2 (calomel) and KCl solution. If the KCl concentration is fixed (conveniently it is kept saturated) the potential of this electrode is constant at a constant temperature. These electrodes are described more fully in Sec. 17.4.4. In general, any electrode can serve as a reference electrode if its potential is constant while the measurements are being made. That is, its potential must not depend upon concentrations that vary during the experiment, or on the effect of any current that may be drawn by the measuring instrument. If standard potential data are required, then the reference electrode potential must be known on the hydrogen scale, and considerations of reproducibility and stability are involved. For many purposes, however, only relative potentials are of interest.

Some care is necessary in dealing with the signs of the potentials. It is common practice to tabulate E° values as reduction potentials; that is for reactions written as

$$M^{n+} + ne^- \longrightarrow M \tag{17.9}$$

If the reaction is written in reversed form, the sign of the potential must be changed. When measuring the potential of a cell, it is necessary to note the polarity of the electrodes if the overall potential is to be used to determine a particular electrode potential.

17.2.1. Potentiometers

Potentiometers are standard devices for the precise measurement of small voltages such as those obtained from an electrochemical cell or from a thermocouple. A potentiometer operates by comparing the unknown potential with a known voltage by impressing the two voltages in opposition across a resistance wire, and finding the point on the wire at which the potentials are balanced. The ratio of the potentials is then given by the ratio of the lengths of the resistance wire (Fig. 17.3), and the instrument dials normally are calibrated to read potential directly. A galvanometer is used as the detector; at balance, no current flows and the galvanometer needle is not deflected. The ultimate known voltage source is a standard cell, but to protect this cell it is not used for the routine measurements

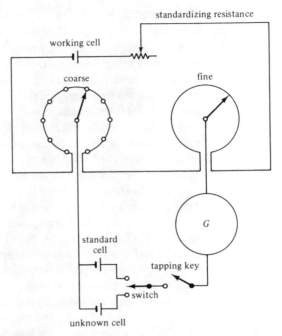

Figure 17.3. Schematic circuit of a potentiometer. The total measuring "slidewire" is constructed of a set of resistors for coarse adjustment, and a continuous slidewire for fine adjustment. The selected resistance setting is read as a voltage.

but only to standardize the output of the working cells which usually are ordinary 1.5 V dry cells. The standard cell is usually a Weston cell (Fig. 17.4). One should use care never to draw significant current from a standard cell, or to shake it.

Various styles of potentiometer may be encountered. Some are self-contained portable models, but high precision instruments have separate galvanometers and cells. With all types, however, readings are made by tapping the potentiometer key, noting the galvanometer deflection, and adjusting the controls to bring the deflection to zero. This adjustment is done in successive steps, starting with the coarse adjustment controls. The key should not be depressed continuously while the instrument is unbalanced, because current will be drawn from the cell under test and this may upset the equilibria upon which the potential depends. If keys of different sensitivity are present, that of lowest sensitivity is used as long as the deflection is reasonably large; then the next key, etc., as the balance point is approached.

When using a potentiometer, correct polarity of all connections is essential. To check connections quickly, set the dial reading at 0.000 V, and tap the key. Note the direction of galvanometer deflection. Now change to a high setting, e.g., 1.5 V, and tap the key again. If the galvanometer deflects in the reverse direction,

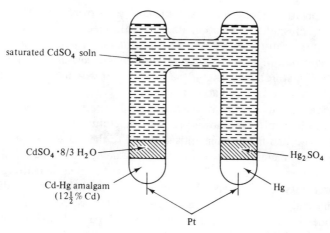

saturated CdSO$_4$ soln

CdSO$_4 \cdot 8/3$ H$_2$O

Cd-Hg amalgam
(12$\frac{1}{2}$% Cd)

Hg$_2$SO$_4$

Hg

Pt

Figure 17.4. A Weston standard cell.

the polarity is correct. If it deflects in the same direction, reverse the polarity of the leads to the potential to be measured. If there is no deflection, check for loose connections, for an electrode not dipping into the solution, or for weak batteries. Potentiometers of the type described here will not function if the resistance of the cell to be measured is too large. In this case, electronic instruments (electrometers or pH meters) are used. Further discussion of the principles of the potentiometer may be found in the references.[1,2] These references will also discuss some of the requirements necessary for high precision potentiometry.

17.2.2. The pH-meter

17.2.2.1 General Operation

A pH meter is used to measure potential and is calibrated to be read in terms of pH values. Most pH meters can be read directly in millivolts also, so that a variety of electrodes in addition to the glass-calomel pair used for pH measurements can be employed. The concept of pH is discussed further in Sec. 17.3.

The details of operation of different pH meters vary, but certain features common to most are discussed below. The names given by various manufacturers to the controls are not the same, but the purpose is usually clear once the essential functions are understood. The common controls of direct reading pH meters are:

1. A function selector which selects measurement of potential or pH. Often two ranges of differing sensitivity are provided. In connection with pH measurements, provision often is included for setting to "manual" or "automatic" temperature operation. Because the potential of a glass and calomel electrode pair varies appreciably with temperature, some means of compensating for this behavior is necessary to convert potential to pH. To use

the "automatic" setting, a temperature compensator probe is attached to the meter and dips into the solution along with the electrodes. More commonly, the "manual" setting is used, and *a temperature control dial is set to the value of the temperature of the solution or buffer being measured* (not necessarily room temperature). Note that if the solution temperature changes, this control should be changed accordingly.

2. A "standardization" control is necessary for pH measurements. The electrodes are placed in a buffer of known pH, the temperature control is set to the proper value, and the instrument is switched to the "measure" or "use" position. The meter is then set to read the value of the standard buffer with the standardization control. This control is not touched thereafter in taking measurements. Use of a second standard buffer to check the operation of the instrument is recommended; this procedure is discussed further below. Before electrodes are removed from a solution, the pH meter should be switched to the "standby" or "zero" position.

   Note that the pH of a buffer solution varies with temperature. The temperature control dial does *not* compensate for this behavior. The instrument must be standardized to the value of pH which the buffer has at the temperature in question.

   Millivolt measurements are made with the function switch in the appropriate position. In some instruments, the meter must be set to read zero with the standardization control when the electrode leads are shorted if absolute values are required. For relative values, any convenient setting can be used. If the initial value is set near one end of the meter scale, consideration must be given as to the direction in which the potential will change so that the meter will not go off-scale.

3. Expanded scale operation. Some pH meters permit the sensitivity to be increased from the normal value (typically ±0.01–0.10 pH unit) to a more sensitive value (±0.001 pH unit). Normally, pH ranges of only a few pH units can be examined at one time in this mode (e.g., 1–3 pH units may cover the whole meter scale), and generally the desired range must be brought on scale with the "standardize" control. The meter reading then will not be the true pH, and it is important to remember that the meter reading, X, is always Y pH units higher (or lower) than the true pH. Y is the difference between the meter reading and the actual pH of a known buffer solution used for standardization (X = true pH + Y). It is important to realize that a pH-meter reading is only as accurate as the standardization procedure. Hence, if the expanded scale mode is used, the meter must be standardized in the expanded scale mode, if absolute values are required.

The behavior of a glass electrode generally is not ideal so that calibration at a single value of pH usually is not adequate for work over a large pH range. It is always best to standardize with a buffer of a pH near those to be measured. If a large range of pH is to be measured, precise work requires that the meter be standardized at least twice; once with a buffer of low pH and once with a buffer of high pH. In practice, the meter is set to the pH of the first buffer solution with the calibration control, and the apparent pH of the second is measured. A calibration curve of apparent pH vs. true pH (actual value of the buffer) can then be plotted and used to correct other readings. A linear curve based on two points generally is sufficiently accurate. Some meters have controls which permit the meter to be adjusted to give correct readings directly over the full range. It should be clear that when only *changes* in pH, but not actual values, are of interest (as in a titration), careful calibration is *not* necessary.

Although a pH meter may malfunction because of electronic faults, the most common causes of trouble stem from the electrodes. If drifting or erratic behavior is observed, check:

1. that the glass electrode has not been allowed to become dry
2. that the electrode connections to the meter are good, especially that the glass electrode connector is fully in its socket, and that there is no break in the cable shielding
3. that there is adequate KCl solution in the calomel electrode to make contact with the inner element
4. that both electrodes are dipping into the solution, and that the solution is well stirred
5. that the tip of the glass electrode is not broken or cracked
6. that the fiber tip of the calomel electrode is not clogged with solid

Systematic substitution of electrodes which function properly on another instrument can be used to locate the source of the problem. An instructor's assistance should be requested if difficulties arise.

17.2.2.2. Electrode Connections to the pH Meter

Various pH meters require different types of electrode connections. Most meters are now built with American Standard connections for the glass electrode (Fig. 17.5), but many older models have a different connection. Some examples are illustrated in Fig. 17.5. Other types of coaxial connectors such as BNC connectors are encountered occasionally. A good contact between the plug and socket is essential, and poor or dirty connections are a frequent source of pH-meter malfunction. There must be good connection between the central conductor of the plug and socket, and also between the outer metal sleeve on the plug and the socket. Adapters are available which permit plugs of one type to be used with different sockets. The reference electrode usually is attached through a pin connector, but some non-standard types may use a "banana" plug. Many silver and platinum electrodes also have pin type connectors. A pin-to-glass electrode adapter permits such electrodes to be used in the glass electrode socket, if necessary.

Electrodes, especially the glass electrode, are fragile and expensive. They must not touch the vessel. Particular care must be taken to see that the tip is not pushed

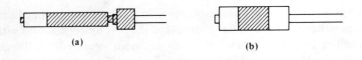

the cross-hatched regions are metal

Figure 17.5.   Typical connectors on electrodes intended for use with pH meters: (a), (b), (c) glass electrode connectors of American Standard, Leeds and Northrup, and phono-jack types respectively, (d) pin type connector used for most other electrodes.

against the bottom of the vessel, and that they are not struck by stirring bars. Only the electrode tips need be immersed in the solution to be measured.

17.3. pH

One of the most frequent measurements made in the laboratory is that of pH.[2] Commonly, pH is defined as

$$pH = -\log C_H \tag{17.10}$$

although an alternate definition,

$$pH = -\log a_{H^+} \tag{17.11}$$

is also used, with the implication that Eq. (17.11) is a thermodynamically exact relation, and that Eq. (17.10) is an approximation. In fact, the pH scale normally employed is based on an operational definition. The definition is

$$pH = pH_s + \frac{E - E_s}{2.303RT/F} \tag{17.12}$$

where pH_s is the pH of a standard solution, the pH of which is assigned from the potential of the cell

$$Pt; H_2(1 \text{ atm}), H^+, Cl^- \| \text{reference electrode} \tag{17.13}$$

The Nernst equation is used to calculate a_{H^+}. E and E_s are the potentials of the cell

electrode reversible to H^+	unknown or standard (s)	salt bridge	reference electrode	(17.14)

Thus, the pH of the reference solution is defined in terms of the hydrogen ion activity, with the limitations of activity measurements as discussed in Sec. 17.2. The pH of the unknown solution is measured with respect to the pH of the standard, but is subject to uncertainties such as changes in junction potentials. Absolute pH values of accuracy greater than 0.01 pH units are difficult to obtain. The estimated uncertainties in pH_s values themselves at room temperature are 0.005 pH units.

Buffer solutions used to standardize pH meters, and also to maintain a fixed pH in solution for many other purposes, can be prepared from a weak acid or base and its salt. Some representative compositions that have defined values of pH_s are given in Table 17.1. Others can be found in Refs. 2–5, and in standard handbooks. Commercial buffer solutions, or buffer tablets which give a specified pH when dissolved to make a specified volume of solution, are used commonly. The pH of a buffer solution varies with temperature, and for precise work it is necessary to control the temperature.

TABLE 17.1

Composition of Some Standard Buffer Solutions

pH at 25°C	Buffer composition
3.0	22.3 ml 0.1M HCl and 50 ml 0.1M KHPhthalate, diluted to 100 ml
4.0	0.1 ml 0.1M HCl and 50 ml 0.1M KHPhthalate, diluted to 100 ml
5.0	22.0 ml 0.1M NaOH and 50 ml 0.1M KHPhthalate, diluted to 100 ml
6.0	5.6 ml 0.1M NaOH and 50 ml 0.1M KH_2PO_4, diluted to 100 ml
7.0	29.1 ml 0.1M NaOH and 50 ml 0.1M KH_2PO_4, diluted to 100 ml
8.0	49.1 ml 0.1M NaOH and 50 ml 0.1M KH_2PO_4, diluted to 100 ml
9.0	4.0 ml 0.1M HCl and 50 ml 0.025M borax, diluted to 100 ml
10.0	18.3 ml 0.1M NaOH and 50 ml 0.025M borax, diluted to 100 ml
11.0	22.7 ml 0.1M NaOH and 50 ml 0.05M Na_2HPO_4 diluted to 100 ml

17.4. TYPES OF ELECTRODES

17.4.1. Introduction

Many metals in the form of wires, gauzes, foils, or rods can be used as electrodes for measurement of the potential of the metal-metal ion couple. Not all metals will reach equilibrium potentials of the simple oxidation-reduction reaction in aqueous solution, however. Very reactive metals, e.g., sodium, may be utilized as electrodes as dilute amalgams in liquid mercury under some conditions. It sometimes is possible to have the reaction between the element and its ions take place on the surface of an inert metal that serves as the conductor. The hydrogen electrode described below is such a case. No attempt will be made to describe the many possible electrodes that may be encountered in specialized work, but rather this section will describe some of those electrodes in common use. Many of these examples are commercial electrodes intended especially for use with pH-meters. Frequently commercial electrodes are not labeled as to type, and the descriptions given below should help in recognition of the various types.

17.4.2. The Hydrogen Electrode

The ultimate reference electrode is the standard hydrogen electrode. The electrode reaction is

$$H^+_{(aq)} + e^- \quad \rightleftharpoons \quad \tfrac{1}{2}H_{2(g)}. \qquad (17.15)$$

Other potentials are referred to this electrode, which is defined as having a potential of zero volts at all temperatures. This electrode under standard conditions consists of a piece of platinum foil dipping into an acid at unit activity of hydrogen ion with H_2 gas at one atmosphere pressure bubbling over it (Fig. 17.6). In fact, a hydrogen electrode generally is not operated under standard conditions.[6] A hydrogen electrode is not difficult to prepare, although some care is required during its use to avoid "poisoning" the metal surface with impurities, and to exclude oxygen. The hydrogen electrode is unwieldy when compared to many other types of electrode, and the flammable and explosive nature of mixtures of hydrogen gas and air is a hazard. **Safe venting of the hydrogen gas exhaust is essential.**

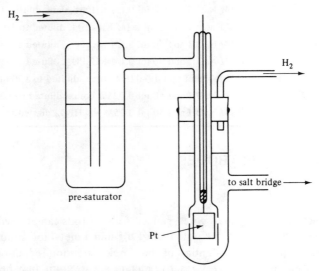

Figure 17.6. A hydrogen electrode with pre-saturator containing HCl of the same concentration as that in the electrode compartment. This precaution is necessary to prevent concentration changes in the electrode caused by evaporation. Another trap in the outlet gas stream is useful to prevent back-diffusion of air.

If the platinum electrode is to function properly, it must be "platinized", that is, its surface must be covered with a layer of finely divided platinum (platinum black). To platinize the electrode it is thoroughly cleaned in conc. HNO_3, followed by thorough rinsing with distilled water. The electrode and a platinum counter-electrode are suspended in a solution of chloroplatinic acid, (about 5% platinic chloride in $2N$ HCl) and a dc voltage source (e.g., a battery charger) is attached so that the electrode to be platinized is negative. A voltage which produces a current density of 30–50 mA/cm^2 for 3–5 min is applied to cover the bright surface of the platinum with a black film of finely divided platinum. The layer should be quite thin; heavy coatings are apt to produce slow response from the electrode. After thorough rinsing in distilled water, the electrode is ready for use. It should be stored in distilled water. Two electrodes may be prepared simultaneously by alternating the polarity of the leads every 15–30 s. Other detailed directions are given in Ref. 7, p. 106.

In following this procedure, use care not to allow the electrodes to touch and short-circuit the voltage source. Retain the platinizing solution for further use.

17.4.3. Glass Electrodes

Glass electrodes are commonly used in the measurement of pH, and occasionally as reference electrodes in potentiometric titrations. The tip of a glass electrode is usually one of two shapes, a conical type (typified by the Leeds and Northrup types) and a bulb type which is typical of most other manufacturers. (Fig. 17.7a and b, respectively). Other shapes may be encountered (Fig. 17.7c), but in all of them there will be no obvious fiber or porous plug connection through the glass tip. There may be a plastic shield over the electrode. Glass microelectrodes for pH measurements in very small volumes and other special designs are available. Glass electrodes are equipped with a coaxial cable, usually with a large plastic and metal plug for attachment to the pH meter (see Fig. 17.5). If this plug is absent, the electrode probably will not be a glass electrode. The measuring part of a glass electrode consists of a thin and fragile glass bulb with a high electrical resistance. The inside of the measuring bulb contains an HCl solution of fixed strength, and an internal reference electrode reversible to chloride ions. When the measuring bulb is immersed in a solution, a difference in hydrogen ion activity on the two sides of the glass causes a potential difference which can be calibrated in terms of the pH of the external solution. The potential produced across the glass membrane is transmitted to the meter by means of the internal electrode dipping into the fixed internal HCl solution. Normally, an Ag/AgCl electrode, or a calomel electrode (mercury-mercurous chloride paste with an inert lead wire) is used. The internal construction is illustrated in Fig. 17.7d. Some glass electrodes are labeled as to this internal element; do not be confused into thinking that this label refers to the electrode as a whole. Glass electrodes are designed for particular operating conditions, in terms of both temperature and pH range. General purpose electrodes are useful over the temperature range of $-5°C$ to $60°C$, and at pH values up to 11.

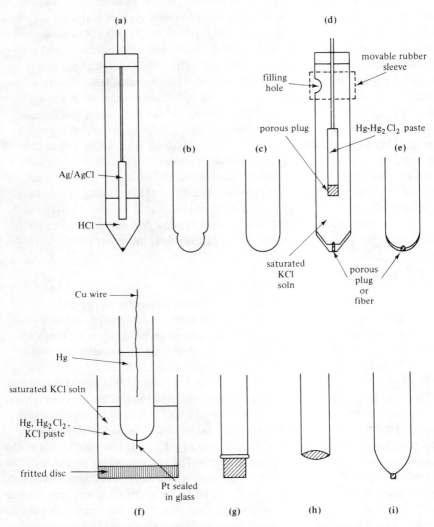

Figure 17.7. Some common commercial electrodes for use with pH meters:

(a) a glass electrode showing internal construction.

(b), (c) other possible shapes of glass electrode tips.

(d) a calomel electrode showing internal construction.

(e) another shape for a calomel electrode tip.

(f) a home-made calomel electrode of low resistance. The glass tube with a short piece of Pt sealed in the bottom makes electrical contact with the calomel electrode; the mercury and copper wire form a convenient way of making contact to the measuring equipment.

(g) a silver billet electrode (also used as an Ag/AgCl electrode).

(h), (i) platinum electrodes.

Above this pH limit, alkali metal ions begin to affect the potential and result in pH readings which are below the true value. Electrodes accurate at high pH values and at high temperatures are available. The use of any glass electrode in a medium that will attack glass significantly (e.g., HF solutions or very concentrated base) is to be avoided.

Before initial use, all glass electrodes must be conditioned by soaking in water or a buffer solution for several hours, and they should be stored in water or a buffer when not in use. The response time of the electrode, that is, the time required to reach a steady potential, normally is a few seconds. In poorly buffered solutions, the layer of water on the surface of the glass may reach a pH value determined by the glass or by material absorbed from a prior solution. Such solutions may cause very slow response, and good stirring is essential.

The electrode should always be rinsed thoroughly when transferred from one solution to another. This precaution is especially important with plastic-shielded designs which can carry over considerable liquid. Excess liquid can be removed from the electrode tip with a piece of filter paper but normally rinsing with small portions of the solution to be measured is preferable.

17.4.4. Calomel Reference Electrode (Saturated Calomel Electrode; SCE)

The reference electrode for pH measurements usually is a calomel electrode, which can be identified by the porous connector in the bottom of the bulb. This connector usually is very small. Figure 17.7d and (e) show two common shapes. Another type has a ground glass sleeve at the tip, with electrical contact being made through a film of KCl solution between the ground glass surfaces. Calomel electrodes almost always have a simple pin type connector for attachment to the pH-meter (Fig. 17.5).

The calomel reference electrode consists of a tube of mercury-mercurous chloride paste surrounded with a KCl solution. The calomel paste makes contact with the KCl solution through a porous plug, while the KCl makes contact with the solution to be measured through the porous fiber channel in the glass jacket which acts as a salt bridge. The potential of the calomel electrode essentially is independent of the solution in which it is immersed. Ions can migrate through the fiber, but little material is actually transferred. However, because of the high concentration of Cl^- ions, calomel electrodes should not be used in solutions that can react with this ion, such as those containing Ag^+. Also, a significant concentration of Hg(II) is dissolved in the KCl solution. Most calomel electrodes must be refilled periodically because the KCl solution tends to leak out. In some designs the outer envelope may be unscrewed, while in others there is a filling hole in the side of the envelope. These electrodes may be filled with saturated KCl solution and solid KCl crystals are often present in the envelope. Commercial calomel electrodes of this design have a high resistance and are polarized easily by a flow of current. Calomel electrodes of lower resistance which can tolerate the flow of small currents are easily prepared by placing a paste of Hg and Hg_2Cl_2, along with KCl solution containing excess solid KCl, in a tube with a fritted glass bottom (Fig. 17.7f). Connection to the external circuit is made with a platinum wire dipping into the paste.

17.4.5. Combination Electrodes

Glass and calomel electrodes often are combined in the same envelope for convenience in pH measurements. Such combination electrodes can be recognized because of the double lead with a large glass electrode type plug and a small calomel electrode type plug.

17.4.6. Silver or Silver-Silver Chloride Electrodes

The silver electrodes generally available are the silver billet type, Fig. 17.7g, although silver wire or foil can be used. If the silver is coated with AgCl, these serve as Ag/AgCl electrodes; that is, they develop a potential dependent on the chloride ion concentration. These electrodes can be used to determine the ions to which other anions forming insoluble silver salts. Ag/AgCl electrodes for routine use can be made by anodizing the silver electrode (i.e., making it the positive electrode) in 1 M HCl solution against a platinum cathode using a 1.5 V battery for 10 min. Directions for preparing Ag/AgCl electrodes for critical work are given in Ref. 7, p. 203.

17.4.7. Platinum Electrodes

Platinum electrodes are useful as inert electrodes, and commonly are used in potentiometric redox titrations where they serve to measure the potential of the system, even though they take no part in the actual redox reaction. Common types have a flat bottom formed from a platinum disc, or a conical point with a platinum metal pin at the end (Fig. 17.7h and i). Platinum wire or foil also can be used.

17.4.8. Ion Selective Electrodes

In recent years, many electrodes have been developed that respond to the concentrations or activities of a single type of ion (or to a limited number of ions with similar properties). These electrodes can be used to determine the ions to which they are sensitive just as a glass electrode is used to determine pH. (Indeed, a glass electrode is an example of a specific-ion electrode). Other special electrodes will respond to gases, e.g., the ammonia electrode which consists of a plastic membrane surrounding a glass electrode and permeable to NH_3. The details of the construction of such electrodes depend upon the specific type. Details are given in Ref. 8. In most cases, these electrodes are used with expanded-scale pH meters. The electrodes are calibrated with solutions of known concentration or activity, and then can be used to measure solutions of unknown concentration.

17.5. ELECTROLYSIS

17.5.1. Introduction

Passage of an electric current through a solution can be used to bring about chemical reactions that can be applied for synthetic purposes and for quantitative analysis. If a potential is applied between two inert electrodes (e.g., platinum, gold, graphite) immersed in a solution, a current will flow as electroactive components lose or gain electrons at the electrodes. If the electrode is not inert, it may itself react. For example, a metal such as Zn may be oxidized to Zn^{2+} ions if it is made the anode. The reactions that can take place depend on the applied potential. If this potential is high enough, several processes may take place simultaneously. The extent of chemical reaction that occurs is related directly to the amount of electricity passed. One Faraday, 96,486 coulombs, results in the transfer of Avogadro's number of electrons and the electrochemical reaction of 1 equiv of material. The number of coulombs passed, at a fixed current, is given by $Q = It$ where I = current (A) and t = time (s). (More generally, $Q = \int I dt$, if I is not constant). Consequently, if the potential applied permits only a single reaction to take place, then measurement of the amount of electricity passed is a quantitative measure of the amount of material reacted. Methods based on the amount of current passed often are referred to as coulometric methods.

17.5.2. Coulometric Syntheses

Many synthetic procedures, both inorganic and organic, can be based on electrochemical reactions. Generally, these are qualitative procedures because measurement of current passed is not important. In some cases, the potential must be controlled to prevent unwanted secondary reactions, but in many cases this precaution is not necessary. A dc power supply such as a variable output battery charger often is an adequate current source for this application. Voltages used in simple cases typically are 3–6 V, and the main concern is to avoid passing a current so large that it causes overheating. A storage battery with a rheostat for voltage control also may be used.

17.5.3. Coulometric Analyses

Quantitative coulometric methods find many applications in analysis. The procedures are sometimes called titrations, but the reagent added (titrant) normally is generated coulometrically in the reaction cell, and the amount added is determined from current-time measurements. Two general procedures are used.

(a) *Controlled Potential Coulometry* includes processes in which the potential is controlled at a value that permits only the desired reaction to take place. The current is measured until it falls essentially to zero, at which time all of the electroactive material will have been consumed. Measurement of the total number of coulombs passed gives the amount of material reacted, because 96,486 coulombs = 1 equiv (i.e., corresponds to 1 mol of electrons).

The amount of current passed in this case may be measured by plotting current readings vs. time (or recording them on a strip-chart recorder) and determining the area under the curve. Alternatively, a coulometer may be used. This device may take the form of an electronic instrument, but more often a chemical coulometer is used in instructional laboratories. This system consists of a simple electrochemical reaction that occurs at 100% current efficiency. The coulometer is connected in series with the experimental cell so that the same current passes through both. The quantity of material reacted in the coulometer is determined by chemical methods or by weighing a deposit and the number of coulombs that have passed is calculated from this information. The best known example is the silver coulometer in which silver metal is deposited on an inert platinum electrode from a silver nitrate solution. The amount of silver ion that has been reduced to metal is determined by weighing the electrode before and after current is passed.

A constant potential is established with a potentiostat, an electronic device which controls the potential difference between a reference electrode (e.g., a standard calomel electrode) and the working electrode (the one at which the desired reaction occurs) at a desired value. A third electrode (auxiliary electrode) usually is employed to carry the current, because the current between the working and the reference electrode must be essentially zero. The experimental arrangement is shown schematically in Fig. 17.8.

In the absence of a potentiostat, potentials may be held roughly constant by manual adjustment of the voltage source. However, this procedure is tedious and of limited accuracy.

(b) *Controlled current coulometry* is performed with the current fixed at a known value. Completion of the reaction must be signaled by some change in property as in a classical volumetric titration procedure, and a visual indicator or potentiometric end-point detection may be employed (see Sec. 9.4). In this case, the potentiometric electrodes are not pertinent to the electrochemical titration reaction as such; they merely serve as indicators of concentration.

A constant current is generated with an electronic supply that automatically adjusts its output potential to the value necessary to achieve the desired current. The potential is determined both by the electrode potential of the reaction taking place and by the ohmic (iR) drop in the cell. If the cell resistance is high or infinite from a poor connection or other cause of an open circuit, the output voltage of the constant current source may assume a high value. The voltage limit depends on the apparatus used, but some common devices can generate several hundred volts. **Caution is essential in using these instruments because the potential and current levels can cause fatal shocks**. If the voltage required to pass the desired current

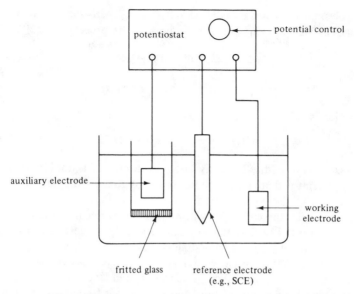

Figure 17.8. Schematic illustration of a cell used for controlled potential coulometry. The potential of the working electrode is set with respect to the reference electrode, but the current flows between the working and the auxilliary electrodes. The auxilliary electrode is separated from the working electrode compartment to prevent mixing of the anode and cathode products. The auxilliary electrode may be a tube with a fritted glass bottom (a layer of agar saturated with a salt can be above the frit if necessary), with an inert Pt or graphite electrode inside.

exceeds the capability of the source, the current will fall below the desired constant value.

Although the electrochemical reaction of the substance being determined may take place directly at the electrode, this situation usually is not practical because, as the concentration of the substance being determined decreases, the current that can be maintained by the direct reaction of that substance at the electrode becomes smaller. When the current that can be maintained by the direct electrolysis of the substance drops below the level of the applied current, a second electrode reaction must begin to dissipate it. To be useful quantitatively, the desired reaction must be the only one to occur; that is, for every Faraday of electricity passed 1 equiv of the sample must react. However, the sample may react directly at the electrode or it may react indirectly, i.e., with a reagent generated at the electrode. The usual procedure is to add a large excess of a second substance that will react electrochemically to yield a product that reacts immediately and completely with the substance to be determined.

This process is illustrated by the coulometric titration of As(III) to As(V). Although the oxidation of As(III) could occur directly at the anode, the actual titration involves mainly oxidation of added I^- to I_3^- with subsequent reaction of I_3^- with As(III). The following equations represent the process:

$$3I^- \xrightarrow{\text{coulometric}} I_3^- + 2e^- \tag{17.16}$$

$$I_3^- + \text{As(III)} \xrightarrow{\text{chemical}} \text{As(V)} + 3I^- \tag{17.17}$$

The net reaction is

$$\text{As(III)} \longrightarrow \text{As(V)} + 2e^- \tag{17.18}$$

Whether the As(III) is oxidized directly or by reaction in I_3^- is not important. The important constraint is that all the anodic coulombs go toward oxidation of As(III).

In this example, the end-point will be marked by accumulation of I_3^- that will not be consumed. The first trace of this can easily be detected visually by the blue color formed by the starch-iodine complex.

The number of coulombs passed in a constant current process is the product of current and time. Because small currents can be used, and current and time can be determined with high accuracy, coulometric titrations frequently are useful for the analysis of small amounts of material.

A schematic diagram for an apparatus which may be used for constant current electrolysis is given in Fig. 17.9.

17.5.4. Electrodeposition or Electrogravimetric Analysis

Electrolysis which deposits an insoluble product on an electrode can be used for quantitative analysis. Usually the product employed is a metal produced by electrochemical reduction, although very occasionally a compound is formed. The apparatus ordinarily used consists of a platinum gauze cathode of large area on which the metal is deposited from solution. The electrode is cleaned, dried, and weighed before electrolysis, and carefully rinsed, dried, and weighed again after all of the metal in solution has been deposited. Solution composition, current density, and other factors must be selected to ensure that an adherent coating is deposited. These conditions often are specific for the metal being analyzed, and the appropriate reference sources must be consulted for details. The counter-electrode is usually a smaller piece of platinum wire. Many commercial units are built so that the counter electrode can be rotated to accomplish the necessary stirring. In simple systems, i.e., those with only a single reducible metal ion present, the potential can be set at any reasonable value. Completeness of the deposition can be checked by weighing the electrode, continuing electrolysis, and re-weighing the electrode to be sure that no more material is deposited. More conveniently, the top of the electrode is left uncovered by the solution until the electrolysis is expected to be complete. The electrode is lowered (or the liquid level raised) and electrolysis continued. Absence of a deposit on the fresh surface indicates that deposition was complete.

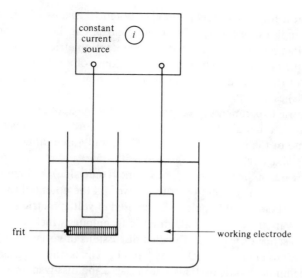

Figure 17.9. Schematic illustration of a cell used for constant current coulometry.

Mixtures of metal ions may be analyzed with controlled potential electrolysis, using a potentiostat and a three-electrode system with the potential set so that only the ion of interest will be discharged. When the current drops to zero at this fixed potential virtually all of the element in question will have been reduced. After completion of the analysis, the electrode is cleaned with an appropriate reagent.

17.6 VOLTAMMETRY

Voltammetry, the study of the behavior of an electrode at which an electrode reaction is proceeding, is concerned with the relationships between the electrical variables (e.g., current, potential, charge, time) and the chemical properties of the system undergoing the electrode reaction (e.g., thermodynamic or kinetic aspects of the electrode reaction, concentrations, and diffusion coefficients of substances involved in the electrode reaction). The electrode of interest, usually denoted as the indicator electrode, is coupled to a suitable reference electrode, usually through a salt bridge, to form a complete electrochemical cell. Because the area of the indicator electrode usually is of the order of a square millimeter or less, the indicator electrode often is referred to as a "micro-electrode." The unique feature of the indicator electrode is that it is polarizable; i.e., its potential is very sensitive to the current passing through it, with the result that the potential of the indicator electrode can be varied easily by application of an external current or potential source. The reference electrode, on the other hand, must be non-polarizable under the experimental conditions. Its potential is not influenced significantly by the flow of current so that it provides a stable reference potential against which to define the potential of the indicator electrode. The flow of current at the indicator electrode distinguishes voltammetric measurements from potentiometric measurements, in which no current flows, but under suitable conditions voltammetric measurements can yield much of the same thermodynamic information (e.g., formation constants, formal potentials) as potentiometric measurements.

Voltammetric experiments are characterized by the application of an electrical impulse to the indicator electrode and the recording and analysis of the electrical response of the indicator electrode. Commonly used impulses include the ramp voltage (a voltage that increases linearly with time) and the current step (a current that increases stepwise from zero to some fixed value), and among the more common responses are the current-potential curve and the potential-time curve. A number of different electrical impulses can be used in voltammetric studies, and discussion of the various resulting voltammetric techniques which are presently in use is beyond the scope of this text. Detailed discussions of voltammetric methods and their applications are given by Delahay[9] and by Thirsk and Harrison.[10] Each voltammetric technique has its particular advantages and disadvantages, both theoretical and practical, but in general, among the applications of each technique are the following:

quantitative analysis
qualitative and quantitative studies of electrode reaction mechanisms
determination of diffusion coefficients
study of the thermodynamics of electrode reactions

The substance which exchanges electrons with the electrode is known as the electroactive species, and the mechanisms by which the electroactive species appears at the electrode surface and by which it undergoes electron exchange are two topics of major interest in voltammetry. The electroactive species may arrive at the electrode surface by mass transport from the bulk of the solution or it may be generated at the electrode surface by a chemical reaction. If the electroactive species is generated by a chemical reaction, the chemical reaction is said to be coupled to the electrode reaction, and under favorable conditions, it is possible to study the kinetics of the coupled chemical reaction by suitable voltammetric measurements. Coupled chemical reactions are discussed in detail in the monographs by Delahay[9] and by Thirsk and Harrison.[10] We shall consider only those cases in which the electroactive species arrives at the electrode surface by mass transfer from the bulk of the solution, but we first need to consider briefly the electron exchange step. In some electrode reactions, the electron exchange process is the slowest (rate determining) step and the kinetics of the electron exchange process determine the current. Such electrode reactions, which are referred to as activation controlled, are often, but less accurately, denoted as "irreversible." Excellent discussions of activation controlled electrode reactions are given by Delahay[9] and by Thirsk and Harrison,[10] but consideration of this topic is beyond the scope of this text. We shall discuss only those cases in which the electrode reaction rate is controlled by the mass transfer of reactants and products. Such electrode reactions which are referred to as mass transfer controlled, are often denoted as "diffusion controlled," or reversible."

The mass transfer of the electroactive species to the electrode surface constitutes a step of major importance in the overall electrode reaction and it is necessary that the mass transfer process be well defined and reproducible. The two mass transfer processes most often employed in voltammetry are forced convection at a rotated solid electrode, and diffusion at a stationary electrode. If mass transfer by other than the desired mode becomes significant, serious problems often arise in the interpretation of the results, and it is necessary to suppress the contribution of undesired modes of mass transfer through careful experimental design. Two undesired modes of mass transfer merit special comment: electrical transport (migration) of ionic reactants and unintended convection. The latter may arise from density gradients in the solution caused by compositional changes induced by the electrode reaction, or from vibrations accidently transmitted to the electrochemical cell from the surroundings.

Electrical transport of reactants in voltammetric measurements is suppressed by the presence of a supporting electrolyte (background electrolyte, indifferent electrolyte) at a concentration that typically ranges from 100 fold to 1,000 fold higher than the concentration of the electroactive species. Because it is present in such high concentrations, the supporting electrolyte carries essentially all of the current in the solution and thus reduces the transport number (see Sec. 17.7.3) of the electroactive species to virtually zero. Hence the fraction of electroactive species that reaches the electrode by migration is vanishingly small when a supporting electrolyte is used. In addition to suppressing migration, the supporting electrolyte also minimizes the potential drop in the solution that arises from the solution resistance, and that reduces the effective potential of the indicator electrode.

In selecting a supporting electrolyte for voltammetric studies, one should consider two factors. First, the supporting electrolyte must not undergo electrode reactions in the potential range of interest, because the current that would result would prevent study of the electroactive species. Second, certain supporting electrolytes may complex the electroactive species and thus affect the electrode reaction by altering the nature of the electroactive species. For many voltammetric studies, KNO_3, KCl, or $NaClO_4$ at 0.1 M or 1.0 M concentration are suitable supporting electrolytes.

Convection increases the rate of supply of the electroactive species to the electrode surface and thus increases the current. Convective mass transfer arising from density gradients in the solution can interfere with electrode reactions where the desired mode of mass transfer is diffusion in quiescent solution, if the duration of the experiment is longer than $ca.$ 20 s. Vibrations transmitted to the indicator electrode from nearby motors or other mechanical equipment can also cause convective mass transfer that interferes seriously with interpretation of voltammetric measurements.

Voltammetric measurements fall into one of two general categories with regard to the nature of the output signal, steady state and non-steady state. In steady state measurements, the response of the indicator electrode is independent of time, whereas in non-steady state measurements, the time dependence of the response of the indicator electrode is of importance.

Steady State Voltammetry (Polarography)

Polarography is useful both for measuring concentrations of electroactive species and for studying the thermodynamics of electrode reactions, e.g., the formation constant of a complex ion. In polarographic experiments, the potential of the indicator electrode is increased slowly and continuously, and the current is recorded to yield a current-potential plot which is sometimes called a polarogram. Two types of indicator electrodes are used in polarographic measurements; solid noble metal electrodes in the form of a disc or a wire, rotated to provide steady state currents, and the dropping mercury electrode (DME), a continuous stream of mercury droplets that form at the end of a capillary, grow to a fixed size, and fall periodically (3–8 s). Although the current at the DME varies periodically as the drop grows, the average current or the peak current behaves as a steady state current.

There are several advantages to using the dropping mercury electrode (DME) in polarography including:

1. the large overvoltage* for reduction of hydrogen ions, that extends the useful range of potential to more negative values than the H_2 evolution potential; e.g., to about -2 V vs. SCE. In comparison, the limit for platinum is about -0.45 V vs. SCE. (These values depend on pH.)
2. the drop is replaced continually providing a clean, reproducible surface.
3. the repetitive formation of new drops provides fresh solution at the electrode surface.

The chief disadvantage to the use of the DME is the ease with which mercury is oxidized, which limits its useful positive potential to approximately 0.25 V vs. SCE, whereas platinum is useful at potentials close to the oxygen evolution potential, about 0.65 V vs. SCE. A polarographic cell and electrodes are shown in Fig. 17.10.

Analytically, polarography is applicable to low concentrations $(10^{-3}-10^{-6} M)$ and it can be applied to both organic and inorganic materials that react electrochemically at an accessible potential.

An idealized polarogram of a reducible substance is shown in Fig. 17.11. Very little current flows until the applied potential is sufficient to cause an electrochemical reaction to occur. The small current which does flow is known as the residual current and is caused by the reaction of impurities and a so-called

---

* Overvoltage refers to the situation when the reaction, $H^+ + e^- \rightarrow \frac{1}{2}H_2$ in this case, does not take place at a significant rate at the potential predicted from the Nernst equation, but only at some larger potential as a result of kinetic limitations.

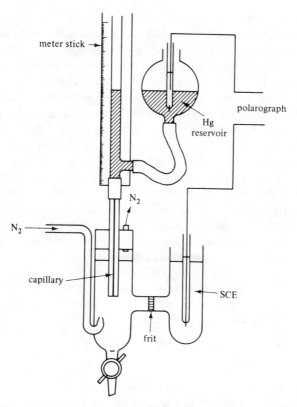

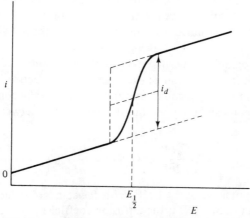

Figure 17.10. A dropping mercury electrode and polarographic cell. The Hg reservoir is attached to the capillary with flexible (Tygon) tubing. The height can be adjusted to control the drop rate. The meter stick provides a guide for the application of reproducible pressure heads.

Figure 17.11. A polarographic wave. The diffusion current is i_d.

charging current that arises because the mercury drop and its surrounding sheath of ions act as a small capacitor of continually changing area. When the potential becomes large enough to cause the reduction of the electroactive material, the current increases sharply above the residual current as the substance reacts at the working electrode. The electrode reaction depletes the concentration of the electroactive material in the vicinity of the working electrode relative to that in the bulk of the solution, and material diffuses to the working electrode. The value that the current can reach is limited by the rate at which material can diffuse from the bulk of the solution to the electrode surface. This in turn depends upon the difference between the bulk concentration and the surface concentration of the electroactive material. At sufficiently negative potentials, the surface concentration is essentially zero and no further increase in current is possible. The resulting current plateau is known as a diffusion limited current, i_d. In practice, the plateau obtained from a plot of current vs. potential may have a non-zero slope because of the increase in the residual current. The current increases sharply again at the potential where reduction of a second electroactive species becomes possible. To ensure that mass transfer of the species of interest takes place solely by diffusion, an excess of an inert supporting electrolyte such as KCl must be present.

A fairly accurate relationship for the average diffusion current, i_d, flowing during the life of the drop is given by the Ilkovic equation

$$i_d = 607nD^{1/2}Cm^{2/3}t^{1/6} \tag{17.19}$$

where

n = number of electrons transferred in the electrode reaction
D = diffusion coefficient of the electroactive ion, cm^2/s
C = concentration of electroactive ion, mmol/ℓ
m = flow rate of mercury, mg/s
t = lifetime for one drop, s

If the flow rate and drop lifetime are held constant, a plot of i_d vs. C is a straight line, and a calibration curve can therefore be constructed and concentrations can be determined without knowledge of the individual parameters of the DME.

In many polarographic experiments a simple metal ion is reduced at the DME to form an amalgam. If the reaction is reversible, i.e., the electron transfer step is sufficiently fast that the Nernst equation can be applied to the concentrations at the electrode surface, it can be shown that the relationship between the current and potential is[11-13]

$$E_{DME} = E^\circ - \frac{RT}{2nF} \ln(D/D') - \frac{RT}{nF} \ln\left(\frac{i}{i_d - i}\right) \tag{17.20}$$

where E_{DME} is the potential of the DME, E° is the standard reduction potential for the half cell reaction (vs. SCE), D is the diffusion coefficient of the metal ion in solution, D' is the diffusion coefficient of the reduced metal in mercury, i is the current observed, R is the gas constant, T is the absolute temperature, and F is the Faraday. If the electrode reaction is such that the reduced species remains in solution, e.g., $Fe^{3+} \rightarrow Fe^{2+}$, then D' is the diffusion coefficient of the reduced species in the solution. This equation predicts that for a reversible reaction at $25°C$, with soluble reactants and products a plot of E_{DME} vs. $\log (i/i_d - i)$ will be a straight line with a slope of $-0.0592/n$.

Polarograms are often characterized by the half-wave potential, $E_{1/2}$, which is the potential at which $i = i_d/2$ (see Fig. 17.11). In this case the last term in Eq. (17.20) becomes zero so that the half-wave potential is given by:

$$E_{1/2} = E° - \frac{RT}{2nF} \ln(D/D') \qquad (17.21)$$

The diffusion coefficients usually are not very different, so that the half-wave potential is very nearly equal to the standard reduction potential for the half-cell. In the rest of this discussion it will be assumed that the correction term involving the diffusion coefficients is sufficiently small that it can be ignored.

Because different substances react at different potentials, a mixture of materials may be analyzed. A series of waves will then be obtained, one after another, if the half-wave potentials of the respective reactions are sufficiently separated. The height of the plateau above the background current produced by the previous reduction is proportional to the concentration of the species being reduced. Furthermore, the $E_{1/2}$ value may be used to identify the species present in the solution, although polarography is more useful as a quantitative technique than as a qualitative technique.

It should be noted that electrode reactions do not always behave ideally. Non-reversibility is one form of non-ideal behavior. Another type of non-ideality is the appearance of a peak or "maximum" rather that a simple plateau. The maximum usually can be eliminated by adding a maximum suppressor to the solution. Gelatin often was used in the past, but its solutions are unstable and they must be prepared every day or two. The detergent Triton X-100, solutions of which are stable, is much more widely used. When using a suppressor, however, care must be taken to add the minimum quantity necessary, because an excess can distort the wave. In general, the concentration of the suppressor in the solution should not exceed 0.004%.

The above discussion treated the case in which no complexing agents were present. When the metal ion undergoes complex formation the half-wave potential is shifted and the dependence of this shift on the concentration of complexing agent yields information about both the formula and the stability of the complex formed. The discussion below is applicable when a large excess of complexing agent of known total concentration is present, the metal complex is reduced to the metallic state in a reversible manner, and only one complex is formed.

The equation for the reduction of the metallic ion, Ox, to the metallic state, Red, is

$$Ox + ne \longrightarrow Red \qquad (17.22)$$

and the Nernst equation for this reaction is

$$E = E° + \frac{0.0592}{n} \log \frac{[Ox]}{[Red]} \qquad (17.23)$$

If a complexing agent, L is present which forms a complex with the ion, the equation for complex formation may be written as:

$$Ox + pL \rightleftharpoons OxL_p \qquad (17.24)$$

The equilibrium constant (i.e., the formation or stability constant) for this reaction is

$$K = \frac{[OxL_p]}{[Ox][L]^P} \qquad (17.25)$$

Solving this equation for [Ox] and substituting the result into the Nernst equation yields

$$E = E° + \frac{0.0592}{n} \log \frac{[OxL_p]}{K[L]^P[Red]} \qquad (17.26)$$

Because, by definition, the half-wave potential is the value when one half of the oxidized form which reaches the electrode is reduced, at the half-wave potential $[OxL_p] = [Red]$. Therefore,

$$E_{1/2} = E° + \frac{0.0592}{n} \log \frac{1}{K[L]^P} \qquad (17.27)$$

which can be rewritten as

$$E_{1/2} = E° - \frac{0.0592}{n} \log K - \frac{0.0592p}{n} \log[L] \qquad (17.27a)$$

Recall that for the free (i.e., uncomplexed) metal ion $E_{1/2} = E°$ (when the diffusion coefficients are equal, as has been assumed). Thus, the difference between the half-wave potentials of the free and complexed ion, $\Delta E_{1/2}$ is

$$\Delta E_{1/2} = E_{1/2} \text{ (free)} - E_{1/2} \text{ (complex)} \qquad (17.28)$$

$$= \frac{0.0592}{n} \log K + \frac{0.0592p}{n} \log[L] \qquad (17.28a)$$

and a plot of $\Delta E_{1/2}$ vs. $\log[L]$ should be a straight line of slope $0.0592p/n$ and intercept $(0.0592/n)\log K$. Therefore, the formula and stability constant of the complex may be determined by measuring the half-wave potentials in the presence of a non-complexing electrolyte and in several solutions containing a known excess of complexing agent along with the non-complexing electrolyte. It is necessary to employ a large excess of complexing agent so that its concentration will remain essentially constant, even though it is being liberated at the electrode.

The above derivation was for the case in which the metal ion was reduced to the metal at the DME. A derivation for the situation in which the reduced species is still in solution and complexed with q mol of complexing agent is similar and is given by Gayer et al.[14]

Eq. (17.28a) was derived assuming that the reduction was reversible, because only then can the Nernst equation be applied. Therefore, it is necessary to verify the reversibility of the reduction. The most rigorous test of reversibility is to measure $E_{1/2}$ for the cathodic wave obtained when only the oxidized form is present, for the anodic wave obtained when only the reduced form is present, and for the composite wave obtained with a mixture of the two. If the half reaction is reversible then $E_{1/2}$ for all three cases will be equal. This is rarely done, however, and usually only the analysis suggested by Eq. (17.20) is performed. The danger here is that waves corresponding to irreversible processes often seem to yield straight lines, but with erroneous low values for n (often non-integers). Data must be secured using a cell with sufficiently low resistance that the iR drop is reduced to no more than a few millivolts, and the potentials must be corrected for iR drop to avoid obtaining an n value that is too small. In particular the commercial calomel electrodes employed for pH measurements have such a high resistance that they may not be used as reference electrodes, unless a three-electrode potentiostat is used as the source of potential. For a more complete discussion of reversibility and some of these complications consult a text such as Ref. 11.

Non-Steady State Voltammetry

When an electrochemical reaction occurs at a stationary indicator electrode in an unstirred solution, the output signal of the indicator electrode varies with time as the reactants are depleted and products accumulate at the electrode surface. A number of such non-steady state voltammetric experiments can be performed, but two of these, linear sweep voltammetry and chronopotentiometry, are sufficient for illustration. Further details can be found in the reference books by Delahay[9] and by Thirsk and Harrison.[10]

Linear sweep voltammetry is analogous to polarography both in respect to the nature of the experiment, and the general type of information which can be obtained. In linear sweep voltammetry the potential of the indicator electrode is increased linearly with time over the potential range of interest. Sweep rates, which range from 0.01 V/s to 10 V/s, are much higher than in steady state voltammetry in which the sweep rate is typically of the order of 0.001 V/s. The current, which initially is low, rises when the potential is sufficient to cause appreciable electrochemical reaction, reaches a maximum and then drops gradually as the reactant becomes depleted in the vicinity of the electrode. An important variation of linear sweep voltammetry is cyclic voltammetry, in which the initial potential sweep is followed by a second potential sweep which returns the potential to the initial value. Cyclic voltammetry often is used to determine whether an electrode reaction is reversible; that is whether the electrode reaction can be reversed and if so, how readily.

In chronopotentiometry, a constant current sufficiently large to deplete the electroactive species at the electrode surface within a few seconds is applied to the stationary indicator electrode, and the potential of the indicator electrode is monitored as a function of time. Initially the potential is relatively constant, but as the reactant is depleted and the product accumulates, the potential shifts more rapidly. Ultimately the flow of reactant is insufficient to maintain the applied current and the potential of the indicator electrode changes sharply. The current may be reversed to study the reversibility of the electrode reaction in the same manner as the potential sweep is reversed in cyclic voltammetry.

17.7. TRANSPORT PROPERTIES OF ELECTROLYTE SOLUTIONS

17.7.1. Introduction

There are several areas, for example, battery technology and the study of electrolyte solutions, where it is important to have information on the *electrical conductance* of a particular solution or the *transference number* or *mobility* of a particular ion to discuss deviations from ideality caused by long-range, ion-ion forces and ion-solvent interactions. Conductance, transference number, and mobility are all related. The conductance of a solution also is widely used in practice to monitor water purity, for analytical use in conductometric titrations, and to follow the course of chemical reactions. Properties of electrolyte solutions are discussed by Robinson and Stokes.[15]

17.7.2. Conductance

Under certain conditions electrolyte solutions obey Ohm's law

$$I = \frac{\Delta V}{R} \qquad (17.29)$$

where I (amp) is the current passing through the electrolyte caused by an applied potential of ΔV (volts) across the two electrodes and R (ohms) is the resistance of the solution. The resistance of the solution may be written as

$$R = \rho \frac{l}{A} \qquad (17.30)$$

where l is the distance between the electrodes, A is the cross-sectional area of the electrodes, and the proportionality constant ρ is known as the specific resistance. Because conductance, L, is defined as the reciprocal of the resistance, the specific conductance, κ, is the reciprocal of the specific resistance, i.e.,

$$\kappa = \frac{1}{\rho} = \frac{1}{R}\left(\frac{l}{A}\right) = L\left(\frac{l}{A}\right) = LK \qquad (17.31)$$

17.7. Transport Properties of Electrolyte Solutions

The quantity (l/A) depends on the measuring cell is referred to as the cell constant (K).

For comparisons among different types of electrolyte solutions it is convenient to refer the measured conductances to some reference concentration and for this purpose the molar conductance Λ is used:

$$\Lambda = \frac{\kappa}{C} \tag{17.32}$$

where C is the concentration in mol cm^{-3}. The units of Λ are cm^2 Ω^{-1} mol^{-1}.

The molar conductance defined by Eq. (17.32) depends on the type of electrolyte (i.e., strong, weak, or whether 1 : 1 or 2 : 1 type, etc.) and the temperature, but in principle Eq. (17.32) eliminates the effects of variation of concentration. In practice, however, it is found that Λ increases with dilution. This trend is attributed to an increase in the mobility of the ions as the concentration is lowered, which occurs because the electric fields of surrounding ions becomes less important as the concentration decreases. Theoretical treatments of the variation of molar conductance with concentration proposed by Debye, Hückel, and Onsager may be found in standard physical chemistry texts.

The value of Λ obtained by extrapolating a plot of Λ vs. $C^{1/2}$ to zero concentration is referred to as Λ_0, the molar conductance at infinite dilution, and represents the conductance of the solution when there is no interaction between the ions. Consequently, the ions can be regarded as moving independently of each other. Kohlrausch's law

$$\Lambda_0 = \Lambda_0^+ + \Lambda_0^- \tag{17.33}$$

is often useful in computing the Λ_0 of a weak electrolyte. Here Λ_0^+ and Λ_0^- are single ion conductances. To a good approximation the degree of dissociation of a weak electrolyte is given by:

$$\alpha = \frac{\Lambda}{\Lambda_0} \tag{17.34}$$

The molar conductance-concentration dependence of a strong electrolyte was expressed on a more quantitative basis by Debye, Hückel and Onsager (see Ref. 16) as:

$$\Lambda = \Lambda_0 - (a\Lambda_0 + b)C^{1/2} \tag{17.35}$$

Eq. (17.35) was derived using the concept of an "ionic atmosphere" and invoking the ideas of a "relaxation effect" and an "electrophoretic effect." The theory was refined by Onsager, who derived expressions for a and b in Eq. (17.35). The equation fits experimental data very well for 1 : 1 electrolytes at low concentrations.

For practical purposes the most convenient form of the Debye-Hückel-Onsager equation for electrical conductance data in any solvent is:

$$\Lambda = \Lambda_0 - \left[\frac{1.981 \times 10^6}{(\epsilon T)^{3/2}} w\Lambda_0 + \frac{29.14(|z_+| + |z_-|)}{\mu(\epsilon T)^{1/2}}\right](|z_+| + |z_-|)^{1/2}C^{1/2}$$

(17.36)

where ϵ is the dielectric constant of the solvent, μ is the viscosity of the solvent and z_+ and z_- are the charges on the cation and anion respectively and

$$w = \frac{q|z_+ z_-|}{1 + q^{1/2}}$$

where

$$q = \frac{|z_+ z_-|\Lambda_0}{[|z_+| + |z_-|]\,[|z_+|(\Lambda_-)_0 + |z_-|(\Lambda_+)_0]}$$

One can therefore check whether the measured slope of a plot of Λ vs. $C^{1/2}$ agrees with the slope predicted by theory.

Fuoss and Onsager[17] have extended the approach on a semi-empirical basis for data at higher concentrations and different types of electrolytes, *viz*:

$$\Lambda = \Lambda_0 + AC^{1/2} + BC + DC \log C$$

(17.37)

where A is the slope predicted by Debye-Hückel-Onsager and B and D are empirical constants.

Typical conductivity cells are illustrated in Fig. 17.12. The simple dip-type cell is useful for approximate measurements, while the more elaborate cell of Fig. 17.12b would be used for precise work. Effects of capacitance of the leads and

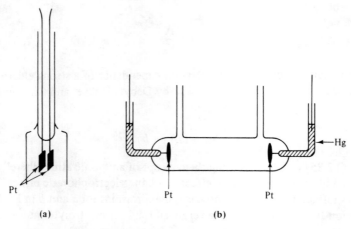

(a)   (b)

Figure 17.12. Typical conductance cells: **(a)** simple dip-type cell for routine measurements, **(b)** cell for precise work.

electrodes are minimized in designs such as this and all such effects must be carefully reduced. In precise work, capacitance can be quite significant. The cell constant of a cell employed for accurate measurements should provide a total resistance that can be measured accurately by the bridge used. Thus, different cells might be needed with dilute and with concentrated solutions if highest accuracy is to be achieved. The electrodes usually are platinum, and for most purposes must be platinized (see Sec. 17.4.2).

Figure 17.13 shows a schematic circuit of a bridge used to measure conductance. Note that for a good null the capacitance as well as the resistance must be balanced. The detector is most commonly an oscilloscope, although ear-phones or other null detectors can be used. For approximate measurements, simple self-contained conductance meters are available, but for precise measurements very elaborate equipment is encountered; details are beyond the scope of this chapter. Alternating current is usually used (often 60 or 1,000 Hz) to avoid the effects of the concentration polarization. Good temperature control is required for accurate measurements, because the temperature coefficient of conductance is about 2% per degree C.

The cell constant must be determined experimentally for each cell. (Cell constants are provided for some commercial cells, and these values are adequate for routine measurements.) Calibration normally is done with a solution of known conductance, most commonly solutions of potassium chloride, values for which are given in Table 17.2. Current practices employed in the calibration of conductance cells have been reviewed by Janz and Tomkins.[18]

Conductometric titrations are useful when plots of conductance vs. added titrant change slope at the end point. This is the case if ions are removed from solution in the titration reaction, as in the titration of sodium chloride ion with silver nitrate. The original conductance is that of a NaCl solution. As $AgNO_3$ is added, AgCl precipitates. The net effect is a replacement of Cl^- ions by NO_3^- ions and the conductance does not change appreciably. Past the end point, however, the conductance increases because Ag^+ and NO_3^- ions are being added to the cell. The plot of conductance vs. volume of $AgNO_3$ added is illustrated in Fig. 17.14a. The

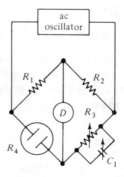

Figure 17.13. Schematic diagram of a conductance bridge circuit.

TABLE 17.2

Conductance Data for Aqueous KCl Solutions for Conductance Cell Calibration[a]

	Concentration by *Volume*		
Concentration (*M*)	KCl (g/1,000 g of solution) (*in vacuo*)	$\kappa(\Omega^{-1}cm^{-1})$	
		18°C	25°C
1	71.3828	0.09822	0.11180
0.1	7.43344	0.01119	0.01288
0.01	0.746558	0.001225	0.001413
	Concentration by *Weight*		
Concentration (Demality[b])	KCl (g/1,000 g of solution) (*in vacuo*)	$\kappa(\Omega^{-1}cm^{-1})$	
		18°C	25°C
1	71.1352	0.09783_8	0.11134_2
0.1	7.41913	0.011166_7	0.012856
0.01	0.745263	0.0012205_2	0.0014087

[a] G. Jones and C. Bradshaw; *J. Am. Chem. Soc.*, **55**, 1780 (1933).

[b] A demal solution (1D) is one containing one gram mole of salt in one cubic decimeter of solution at 0°C.

Table 17.2 is taken with permission from "Conductance Cell Calibrations," by G. J. Janz and R. P. T. Tomkins, J. Electrochem. Soc. **124**, 55C (1977).

end point is indicated by the discontinuity in the plot and is located by extrapolation of the two segments to their intersection.

Such titrations can be used for the titration of weak acids and bases, and are sometimes more sensitive than normal potentiometric acid-base titrations. Figure 17.14b illustrates the conductometric titration curve of boric acid ($K_a = 6 \times 10^{-10}$) with NaOH. Before the end point, the reaction is

$$Na^+ + OH^- + B(OH)_3 \longrightarrow Na^+ + B(OH)_4^- \qquad (17.38)$$

and the conductance increases slightly as titrant is added because of the addition of Na^+ ions. Past the end-point, the highly conducting OH^- ions accumulate, and the conductance increases rapidly as excess titrant is added. These and other examples are discussed in more detail by Skoog and West.[13]

17.7.3. Transference Numbers

In electrolysis, the various ions generally do not all carry equal amounts of current, because they move with different velocities. The velocity of an ion often is

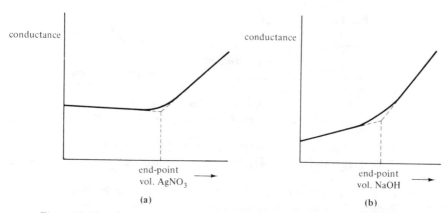

Figure 17.14. Conductometric titration curves; (a) NaCl titrated with $AgNO_3$, (b) $B(OH)_3$ titrated with NaOH.

defined in terms of the mobility, the velocity under a unit potential gradient. The fraction of the total current carried by a particular ion defines the transference number, or transport number, of that ion in the solution in question. Theoretical and experimental aspects of transport number measurements are considered in many physical chemistry texts.

From Kohlrausch's law of the independent migration of ions (Eq. (17.33)), the transference numbers of cations or anions at infinite dilution may be expressed as

$$t_0^+ = \frac{\Lambda_0^+}{\Lambda_0} ; \qquad t_0^- = \frac{\Lambda_0^-}{\Lambda_0} \tag{17.39}$$

The mobility μ of an ion is defined as the velocity of the ion under an electric field of unit strength. An important equation can be derived to show the relation of the transference number and the ionic mobility, viz:

$$\Lambda_i = F\mu_i = t_i\Lambda \tag{17.40}$$

where Λ_i is the molar conductance of the ion i, μ_i is its mobility, t_i the transference number, and F is the Faraday.

The variation of transference number with electrolyte concentration can be obtained by application of the Onsager limiting law (see conductance). The resulting equation is

$$t^+ = t_0^+ + \frac{t^+ - z^+/(z^+ + z^-) \Lambda_e}{\Lambda_0} \tag{17.41}$$

Here Λ_e is the electrophoretic contribution to the conductance, given by

$$\Lambda_e = \beta(z^+ + z^-)\sqrt{I}$$

where $\beta = 40.25/\eta(\epsilon T)^{1/2}$, η, ϵ and T are the solvent viscosity, dielectric constant and absolute temperature, respectively and I is the ionic strength.

401

REFERENCES TO CHAPTER 17

1. D. P. Shoemaker, C. W. Garland, and J. I. Steinfield, "Experiments in Physical Chemistry," 3d ed., McGraw Hill, New York, 1974.

2. R. G. Bates, "Determination of pH," Wiley, New York, 1964.

3. E. V. Bower and R. G. Bates, *J. Res. Natl Bur. Stds* (U.S.), **55**, 197 (1955).

4. R. G. Bates, *J. Res. Natl Bur. Stds* (U.S.), **60A**, 179 (1962).

5. B. R. Staples and R. G. Bates, *J. Res. Natl Bur. Stds* (U.S.), **73A**, 37 (1969).

6. T. Biegler and R. Woods, *J. Chem. Educ.*, **50**, 604 (1973).

7. D. G. J. Ives and G. J. Janz, "Reference Electrodes," Academic Press, New York, 1961.

8. R. A. Durst (ed.), "Ion-Selective Electrodes," Special Publication 314, National Bureau of Standards, Washington, D.C., 1969.

9. P. Delahay, "New Instrumental Methods in Electrochemistry," Wiley-Interscience, New York, 1954.

10. H. R. Thirsk and J. A. Harrison, "A Guide to the Study of Electrode Kinetics," Academic Press, London, 1972.

11. L. Meites, (ed.), "Polarographic Techniques," 2d ed., Interscience, New York, 1965.

12. H. H. Willard, L. L. Merritt, Jr., and J. A. Dean, "Instrumental Methods of Analysis," 5th ed., Van Nostrand, New York, 1974.

13. D. A. Skoog and D. M. West, "Fundamentals of Analytical Chemistry," 3rd ed., Holt, Rinehart and Winston, New York, 1976.

14. K. H. Gayer, A. Demmler, and M. J. Elkind, *J. Chem. Educ.*, **30**, 557 (1953).

15. R. A. Robinson and R. H. Stokes, "Electrolyte Solutions," 2d ed., Butterworths, London, 1959.

16. H. S. Harned and B. B. Owen, "Physical Chemistry of Electrolytic Solutions," Reinhold, New York, 1958.

17. R. M. Fuoss and L. Onsager, *J. Phys. Chem.*, **61**, 668 (1957).

18. G. J. Janz and R. P. T. Tomkins, *J. Electrochem. Soc.*, **124**, 556 (1977).

Identification of Unknowns

18.1. INTRODUCTION

It often is necessary to identify a substance or mixture of substances encountered in the course of laboratory work. Common cations found in inorganic materials can be separated and identified according to rigorous and systematically controlled schemes.[1] Many procedures, based primarily on the differential solubilities of the chlorides, hydroxides, sulfides, and carbonates, are useful for the separation and identification of cations. These techniques are most successful when coupled with a significant level of manipulative and interpretive skill. In some cases spot-tests[2] can be used, particularly if certain elements are suspected, thus avoiding the rigorous schemes mentioned above.

Similar schemes exist for common anions. Qualitative schemes for both anions and cations have been replaced to a large degree by separation schemes based on chromatography and solvent extraction,[3] and by emission spectroscopy[4,5] for metals (see Sec. 15.33).

The apparent magnitude of the task encountered when working with organic compounds can be reduced to manageable proportions by taking advantage of the characteristic physical and chemical properties of the materials in question. The process of identification relies on a systematic approach to the determination of physical and chemical properties of the unknown substance or of products derived from these compounds by well defined chemical transformations. A number of textbooks[6,7,8] are available which delineate the procedures to be used, and one of these should be consulted for details. In this section, the general philosophy of the identification of unknown substances will be treated.

Much can be learned from an examination of the physical properties of a compound, but before anything else is done, some measure of the purity of the compound should be obtained. Thin layer chromatography (see Sec. 10.6.3, 10.6.4) is the most efficient way to assay the purity of solids and non-volatile liquids. Gas chromatography (see Sec. 10.6.5) similarly may be applied to the analysis of thermally stable liquids and some solids. It must be remembered that this approach assumes that optimum conditions have been used so that a mixture could be detected if it were present. The breadth of the melting range of solids will give an indication of the purity of a compound but it cannot demonstrate how many components are present, as TLC can. Similarly, the boiling point of a liquid cannot indicate the number of components present on the scale at which most analyses are conducted. Further, boiling point is less sensitive to the presence of small amounts of impurities than is the melting point. The chromatographic methods not only will indicate the level of purity of the sample, but, in the case of mixtures, will also give some relative measure of the amounts of the various components that are present. The presence of solids (or non-volatile liquids) dissolved in volatile solvents can be detected simply by allowing a few drops of the presumed solution to evaporate on a watch glass and observing if a residue remains.

18.2. GUIDELINES FOR IDENTIFICATION OF UNKNOWN ORGANIC SUBSTANCES

The identification of unknowns can be very time consuming and frustrating if a systematic approach is not used. Guides for the analysis of unknowns are given in the references. Note the following points and compare the behavior of the unknown sample with that of the known substances.

Make a preliminary examination of the sample. Most pure organic compounds are colorless. Those which are colored generally contain conjugated multiple bonds such as aromatic nitro or azo compounds (orange to red) or quinones (yellow to red). Brown or black materials generally are contaminated with oxidized impurities. Similarly, pale yellow samples may be contaminated slightly with some decomposition product.

The odor of volatile organic materials can be quite characteristic ranging from unpleasant (mercaptans) to pleasing fragrances (esters). The range of difference in odors for compounds of a given type is relatively slight so that with a little experience one can narrow further the identification process. Caution must be exercised when smelling organic compounds because of possible toxicity.

Ignition of an organic substance on a spatula or in a crucible cover can be quite informative. Do not put the sample immediately in the hottest portion of the burner flame, but rather raise the temperature of the sample slowly by gradually moving the sample into the flame. Observe if the sample melts or evaporates. Aliphatic materials generally burn with a clear blue or yellow flame, while aromatics burn with sooty, yellow flames. The formation of a non-volatile residue usually indicates the presence of a metal ion in the sample.

Check the physical constants (see Sec. 12) of the sample, such as boiling or melting ranges. Broad ranges ($>1-2°C$) generally indicate impure samples but a sharp melting point can be the result of the presence of a eutectic mixture. Be careful to observe any changes that occur on heating. Be sure that the values determined are reproducible because some compounds undergo irreversible chemical and/or physical change on heating. Phenomena such as dehydration or phase changes often are reversible and can be quite characteristic of a particular compound. Liquids also can be characterized by their refractive indices and densities.

Establish the solubility properties of the compound and tabulate the results. This approach usually will provide a good deal of information about the general chemical class to which a compound belongs, but the assignment of a formal classification is somewhat arbitrary (see Fig. 18.1).

Solubility tests in water are made semi-quantitatively; that is, approximately known amounts of material are used, but with a little experience an "eye-ball" estimate of amounts may be close enough. Note that in water, "solubility" is defined *for this purpose* as 3 g per 100 g of water at room temperature. Tests should be made on a test-tube scale using only about 0.1 g, or less, of sample. In assessing the solubility in acids or bases, it is important to determine whether the solubility is increased over that in pure water. Materials that dissolve in acid or base but not in water should precipitate on neutralization.

Care should be taken not to be confused by chemical reactions. For example, if a material (e.g., an ester) is readily hydrolyzed by NaOH to water soluble products, the observed solubility will be misleading. For this reason HCl and NaOH solutions should not be warmed when performing simple solubility tests. Water may be warmed but the solution should then be cooled and seeded before interpreting the results. To speed the attainment of equilibrium, solid samples should be finely ground. A more detailed description of solubility tests is provided in the next section.

The analysis of an unknown sample for the presence of N, S, and halogens (X) is accomplished by fusing the sample with sodium. In this process the elements are converted to anions (S^{2-}, CN^-, and X^-) which may be detected by conventional inorganic analysis.

Caution! Always wear safety glasses when doing a fusion.

The fusion is conducted in a small test tube held in place by pushing it through a tight-fitting hole in a wire gauze supported on an iron ring or tripod. Carefully free a piece of sodium of the adhering sodium hydroxide so that the remaining cube is about 0.5 cm on a side. (Use tongs or tweezers for this procedure.) Alternatively sodium pellets can be used. Melt the metal in the test tube, described above, with the flame of a microburner. Remove the flame and add about 5–10 mg of the sample directly onto the molten sodium. Reheat the tube and add a second increment of compound. After any reaction subsides, heat the tube until it glows red and then let it cool. Excess sodium should be destroyed by the addition of a little ethanol. Cautiously boil the ethanol until it completely evaporates. Add enough water (10–15 ml) to a small beaker so that, when the beaker is raised around the fusion tube, the bottom half of the tube will be below the water level. Heat the bottom of the tube until it is red-hot and raise the beaker around the tube. The tube should break under these conditions and the glass fragments should be well stirred with a glass stirring rod. Heat the solution to boiling, filter it and use the filtrate to conduct the following tests.

1. *Sulfur.* Acidify a 1-ml portion of the filtrate with acetic acid and add a few drops of 5% lead acetate solution. A black precipitate (PbS) indicates the present of sulfur.

2. *Nitrogen.* Add sufficient concentrated NaOH solution to change the pH of 1 ml of the filtrate to 13 (pH paper). Add two drops each of saturated ferrous sulfate solution and 30% potassium fluoride. Boil the mixture briefly (30 sec) and then acidify it by adding 30% H_2SO_4 *dropwise* until the precipitate (iron hydroxide) just dissolves. The appearance of a deep blue precipitate (Prussian Blue) indicates the presence of nitrogen. Nitrogen can also be detected by using a few drops of ammonium polysulfide which is added to about 2 ml of the filtrate. Evaporate the solution to dryness on a steam bath and add 5 ml of dilute HCl, warm and filter. Add a few drops of $FeCl_3$ solution to the filtrate. A red color indicates the presence of nitrogen.

3. *Halogens.* Acidify (litmus paper) 2 ml of the filtrate by dropwise addition of nitric acid. Boil the solution for a few minutes to expel any hydrogen sulfide or hydrogen cyanide present (**Hood!**). Cool the solution and add a few drops of 5% silver nitrate solution. The formation of a thick precipitate of silver halide indicates the presence of chlorine (AgCl is white), bromine (AgBr is pale yellow), or iodine (AgI is yellow). Turbidity may be the result of an impurity, and the fusion should be repeated. This procedure generally is effective for all halogen-containing compounds. Addition of silver nitrate solution to an alcoholic solution of the starting compound *before* the fusion process resulting in the formation of a precipitate indicates the presence of a *labile* halide ion (e.g., an acid chloride or a tetraalkyl ammonium salt, among others.)

To confirm the identity of the halide ion the following procedure should be adopted: Acidify a portion of the stock solution (about 10 ml) with dilute sulfuric acid and add 1 ml of carbon tetrachloride to 1 ml of the solution followed by *one* drop of 5% sodium hypochlorite solution and shake. The presence of iodine will be indicated by a violet color in the CCl_4 layer. If iodine is confirmed, add $NaNO_2$ to the remainder of the acidified solution and extract the iodine with CCl_4. To test for bromide boil the solution for a few minutes and then allow it to cool. Add 0.5 ml of CCl_4 to 1 ml of this solution followed by two drops of chlorine water. A brown color indicates the presence of bromine. Dilute the remaining solution to about 60 ml and add 2 ml of concentrated H_2SO_4, followed by 0.5 g of potassium persulfate and boil the solution for 5 min. Cool the mixture and add silver nitrate solution. A white precipitate indicates chlorine.

Classification tests for functional groups should now be applied. The solubility and elemental analysis results, coupled with the visual observations and ignition tests should make only two or three tests necessary. Many of these tests are performed in connection with particular experiments, but if a test has not been done it should be conducted on known compounds so that it can be identified as positive or negative. Good notebook records are helpful in developing and maintaining experience in identification.

Spectroscopic results (IR, NMR) will be very helpful. Sometimes a compound can be identified from these alone, but chemical confirmation should always be included. It is good practice to keep all spectra taken and gradually accumulate a library of information which can be helpful in subsequent work.

Physical constants can now be used to identify, tentatively, the particular compound. In addition to melting or boiling points, data such as refractive indices, neutralization equivalents, and molecular weights can be useful. Note that the Rast freezing point depression method, described in Sec. 12.2, is a rapid and easy way to obtain a molecular weight. The method of mixed melting points is a good confirmatory test if the unknown compound is a solid and an authentic sample of it can be obtained.

If possible, a solid derivative should be made and its melting point compared with the literature value to confirm the identity of the unknown compound. Many derivatives are described in the literature (see Chap. 2). A derivative is chosen because of its ease and rapidity of preparation and purification. The derivative of choice should be a solid with a melting point which differs appreciably from that of the starting material but which is also different from the derivatives of possible alternatives for the unknown being identified. Thus the derivative should be the final demonstration of the identity of a compound. Consult the references for details of preparation and choices of suitable derivatives.

18.3. PROCEDURE FOR THE APPLICATION OF SOLUBILITY TESTS TO ORGANIC COMPOUNDS

Either the following general procedure or one given in the reference texts may be followed. The procedure is summarized in Fig. 18.1. A fresh sample of unknown may be used for each solubility test, but the procedure below does not always require a fresh sample. However, judgement is necessary to determine when to omit unnecessary tests or those of doubtful value.

1. *Solubility in Water.* Place 0.1 g of a finely powdered solid or 0.2 ml of a liquid in a small test tube. A solid may be weighed (to within 0.01 g) and a liquid sample should be measured roughly with a calibrated dropper kept for this purpose. A dropper may be calibrated by counting the number of drops required to add some volume, e.g., 1 ml, to a graduated cylinder. Add cold water dropwise with thorough shaking (do *not* use your fingers as stoppers) until the substance dissolves, or until 3 ml of water has been added. If the substance dissolves in less than 3 ml of water it is classified as *soluble*. A

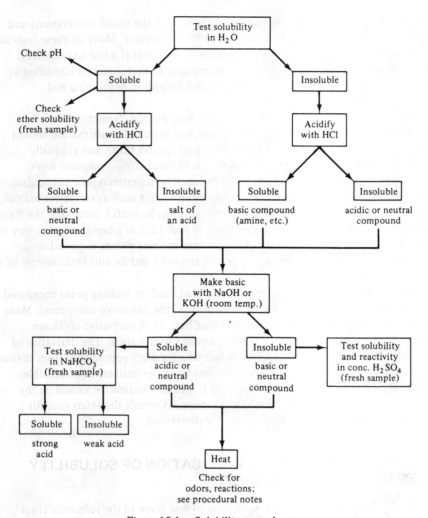

Figure 18.1. Solubility test scheme.

record of the amount of water used may be of value for the subsequent identification.

If a solid fails to dissolve in 3 ml of water, warm and shake the mixture. If it still fails to dissolve, it is *insoluble.* If it dissolves in the warm water, cool the tube with tap water and note if any solid separates. It is advisable to shake the solution well and to seed the cooled solution with a crystal of the original solid to minimize the possibility of the solution remaining supersaturated. If no precipitate appears, the substance is soluble. *Save the solution or mixture for Part 2.*

In the preceding experiment note any indication of a reaction (e.g., evolution of heat); compounds such as acyl halides may react vigorously with water. In any case, test the solution with litmus paper or indicator paper (see Table 18.1). Keep in mind that the litmus test is only an indication and not a conclusive test for the presence of an acid or base, because chemical reaction may transform a neutral group to an acid or a base.

TABLE 18.1

Ether Solubility-pH Classification for Water Soluble Compounds

pH of water solution	Ether soluble	Ether insoluble
<4	Low molecular weight mono-carboxylic acids and their anhydrides, acyl halides	Sulfonic acids, amine salts, polyfunctional carboxylic acids
4–8	Low molecular weight neutral oxygen and nitrogen compounds	Amine salts, polyfunctional neutral nitrogen and oxygen compounds, low molecular weight amides, amine carboxylic acids
8–10	Low molecular weight amines	Alkali metal carboxylates, polyfunctional amines
>10	Low molecular weight amines	Polyfunctional amines

2. *Solubility in Dilute Hydrochloric Acid.* Add dropwise, with shaking, 1–2 ml of 20% hydrochloric acid to the aqueous solution or mixture which remains from 1 above. Note whether this leads to dissolution of an insoluble substance or precipitation if the material was water soluble. *Save the solution or mixture for Part 3.*

    If the addition of hydrochloric acid to water-insoluble material leads to dissolution of the compound, the unknown is most likely to be a basic nitrogen compound i.e., an amine. Conversely, if the addition of hydrochloric acid to an aqueous solution causes a precipitate, the unknown is very probably a water-soluble salt of a water-insoluble acid. A soluble salt of some metals, e.g., silver or lead, could also give a precipitate of an insoluble chloride at this point, but these compounds would be indicated by the presence of an ash residue from the ignition test.

3. *Solubility in Base.*

    a. *Sodium or Potassium Hydroxide.* To the acid solution or mixture from Part 2 above add dropwise, with shaking, 30% potassium or sodium hydroxide solution until the mixture is neutral to litmus. Cool the tube as necessary to prevent excessive heating and add 0.5–1 ml more of the 30% hydroxide solution. Note the effect. Now warm the cold alkaline solution and cautiously note the odor. Ammonia will be liberated from ammonium salts, and many amides and imides, among other groups. Free amines that may be present, e.g., trimethylamine, aniline, usually will be more readily detectable. Do not confuse the smell of some of these materials with ammonia. Hot hydroxide solutions will decompose acyl halides, esters, alkyl halides, acid anhydrides, amides, nitriles, and many other compounds.

b. *5% Sodium Bicarbonate*. It is helpful to investigate the behavior of a dilute sodium bicarbonate solution on a new 0.1 g sample of the unknown, if it is soluble in NaOH.

Of the compounds which dissolve in sodium or potassium hydroxide, only carboxylic acids, sulfonic acids, sulfinic acids and some phenols substituted with strong electron withdrawing groups are soluble in bicarbonate. Other phenols are not soluble in this reagent. If the substance is not soluble in the hydroxide solution, the solubility test with the bicarbonate solution is unnecessary.

4. *Cold and Hot, Concentrated Sulfuric Acid*. Treat about 0.2 g of the substance with 1 ml of cold, concentrated sulfuric acid. Saturated and aromatic hydrocarbons and their halogen derivatives are insoluble. Most other compounds dissolve.

After checking for solution, carefully warm the test tube and look for the evolution of gas. Concentrated sulfuric acid is a strong oxident and there may be liberation of toxic or noxious gases such as carbon monoxide, hydrocyanic acid, cyanogen, and other toxic compounds.

The evolution of a gas without blackening of the sample may indicate the presence of formic or oxalic acid or their salts. A pungent vapor alone may be caused by certain acids, phenols, salts or lachrymators. Blackening with gas evolution may result from carbohydrates, glycosides, alkaloids, salts, and similar materials.

5. *Solubility in 85% Phosphoric Acid*. Because phosphoric acid is not a strong oxidizing agent as is H_2SO_4, it usually does not develop heat and color when a compound dissolves in it. Alcohols, aldehydes, methyl ketones, cyclic ketones, and esters having fewer than nine carbon atoms dissolve in 85% phosphoric acid, while the size limit for ethers is somewhat less. Some compounds form products with the phosphoric acid. This test is not very widely used.

To apply the phosphoric acid solubility test treat about 0.2 g of the finely powdered solid, or the liquid, with 3 ml of phosphoric acid.

6. *Solubility in Ether*. Check the solubility in ether in the same way as in water. The pH in water and the ether solubility properties can give useful information for water soluble compounds; this is summarized in Table 18.1. A compound soluble in both water and ether will probably be of low molecular weight.

18.4. DISSOLUTION OF INORGANIC SAMPLES FOR ANALYSIS

Most inorganic analyses require that the sample be dissolved prior to determination of the components of interest. The dissolution of organic materials, which has been discussed earlier in this chapter, and the preliminary treatment of inorganic samples, such as sampling, crushing, and drying, which is discussed in comprehensive quantitative analysis texts such as Skoog and West[9] will not be treated here.

Many inorganic materials can be dissolved in water or in dilute acidic or basic solutions, and these solvents should be tried first. Approximately 50 mg of sample is added to a few milliliters of the solvent in a small test tube and is shaken for a few moments. Gentle heating may be useful to increase the rate and/or extent of dissolution.

Many metals and metallic compounds that are insoluble in dilute acid or base can be dissolved in concentrated mineral acids. Dissolution often is accompanied by vigorous gas evolution and care must be taken to prevent loss of sample by spattering. Solutions that must be diluted to a specified volume can be prepared in volumetric flasks, while samples that need not be diluted precisely are dissolved in beakers or Erlenmeyer flasks. A watch glass should be used to prevent loss of sample when the solution is boiled or gases are evolved. Heating should be done slowly to prevent uncontrollable reactions and loss of sample. The dissolution technique must be chosen to prevent accidental loss of a desired constituent. For example, H_2S is evolved rapidly when sulfide solutions are acidified, and some metal chlorides are volatilized from hot, concentrated HCl.

Reagents that will interfere in later steps of the analysis can be used to effect dissolution, if they are removed prior to the point of interference. For example, HNO_3 used to dissolve a copper sample which is to be analyzed by electrodeposition can be removed by heating the solution in the hood with H_2SO_4 until white fumes of SO_3 appear. Some useful mineral acids and their applications include:

HCl. Concentrated (12 M) hydrochloric acid dissolves most metals which are as easily oxidized as hydrogen, and it is a good solvent for oxides if they are not refractory. HCl gas is evolved when hydrochloric acid is boiled to yield a constant-boiling solution that is approximately 6 M.

HNO₃. Hot concentrated (16 M) nitric acid is a strong oxidizing agent that will dissolve most common metals except aluminum and chromium. Tin reacts but forms insoluble SnO_2. For many applications it is preferable to use 1 : 1 HNO_3/H_2O to reduce the reactivity of this solvent.

H₂SO₄. Hot concentrated sulfuric acid (18 M) is a relatively strong oxidant, and it can be heated to over 300°C to accelerate its action. It attacks most metals, although some, such as barium and lead, form insoluble sulfates.

HF. Hydrofluoric acid rapidly dissolves and volatilizes silica, especially when used with concentrated H_2SO_4. **Caution! Solutions of HF can cause extremely painful burns which heal slowly, and HF vapors are extremely irritating. Solutions of HF must be vaporized only in a hood, and protective gloves must be worn.**

HClO₄. **Caution! Hot concentrated perchloric acid is a powerful and somewhat unpredictable oxidizing agent, that has been the cause of numerous**

explosions. It can be particularly hazardous in contact with organic matter. *The proper use of HClO$_4$ requires both experience and care. Dolezal* et al.[10] *have described the precautions necessary when inorganic samples are treated with HClO$_4$.*

Aqua regia. A mixture of three volumes of concentrated HCl and one volume of concentrated HNO$_3$ combines the oxidizing power of HNO$_3$ and the complexing power of HCl. It is a powerful solvent that readily dissolves noble metals except iridium.

Some samples which are not attacked significantly by aqueous solutions can be rendered soluble by fusion, i.e., strong heating of an intimate mixture of the sample with a large excess of a dry reagent (flux) to produce a melt which can be dissolved after cooling. Fluxes may be classified as basic, acidic, or oxidative.

Sodium carbonate is an alkaline flux that attacks siliceous minerals.

Potassium pyrosulfate and potassium bisulfate are strongly acidic fluxes that solubilize many metal oxides, including refractory materials which resist aqueous solutions.

Sodium peroxide is a strongly alkaline oxidative flux that raises most elements to their highest oxidation states. **Caution! It reacts violently with readily oxidized materials**.

Fusions are performed in crucibles, and the proper choice of the crucible material is necessary to prevent destruction of the crucible and contamination of the sample. Sodium carbonate and potassium pyrosulfate fusions may be performed in a platinum crucible. Fusions with sodium peroxide are performed in nickel crucibles, which are slightly attacked by the reagent, thus introducing variable quantities of nickel into the sample. Details concerning fluxes and fusion techniques are given by Skoog and West.[9]

18.5. DESTRUCTION OF ORGANIC MATTER

Organic materials which are to be analyzed for inorganic components are almost always first treated to destroy the organic material. Two general approaches are used: dry ashing or ignition, and wet ashing or chemical oxidation.

Dry ashing is performed by heating the material in air, leaving most metals behind as oxides or other compounds; these are then dissolved as described earlier. This technique is not applicable if the inorganic components to be determined are appreciably volatile. In dry ashing, it is necessary to avoid rapid combustion of the organic material that could result in mechanical loss of the sample. The sample is first heated gently, usually in a closed crucible, to convert the organic material largely to carbon. The crucible is then opened and heated strongly in an oxidizing flame to complete the oxidation of carbon. The full flame of a Meker burner or blast lamp is necessary for complete oxidation of carbon. The usual error that is made is the failure to heat the sample long enough, or at a sufficiently high temperature, to effect complete removal of the organic components.

Wet ashing is based on the use of liquid oxidants to destroy organic material, and offers the advantage that loss of volatile inorganic compounds can be minimized or eliminated in many cases. Concentrated H_2SO_4 commonly is used, often with added HNO_3. **Perchloric acid can be very effective, but it should be used only under carefully controlled conditions and only after the hazards associated with its use are thoroughly understood. The misuse of perchloric acid for destruction of organic material has been the cause of many explosions. Under *no* circumstances should perchloric acid be used to destroy organic matter without clear documentation that the technique is free of hazard. Gorsuck[11] gives an excellent discussion of the proper use of perchloric acid for this purpose.**

REFERENCES TO CHAPTER 18

1. L. Meites (ed.), "Handbook of Analytical Chemistry," McGraw-Hill, New York, 1962.
2. See for example, F. Feigl, "Qualitative Analysis by Spot Tests," 4th completely revised English ed., Elsevier, Amsterdam, 1954.
3. G. H. Morrison and H. Freiser, "Solvent Extraction in Analytical Chemistry," John Wiley & Sons, Inc., New York, 1957.
4. L. H. Ahrens and S. R. Taylor, "Spectrochemical Analysis," 2d ed., Addison-Wesley, Reading, Massachusetts, 1961.
5. N. H. Nachtrieb, "Principles and Practice of Spectrochemical Analysis," McGraw-Hill, New York, 1950.
6. R. L. Shriner, R. C. Fuson, and D. Y. Curtin, "The Systematic Identification of Organic Compounds, 5th ed., Wiley, New York, 1961.
7. N. D. Cheronis, J. G. Entrikin, and E. M. Hodnett, "Semimicro Qualitative Organic Analysis," 3d ed., Interscience, New York, 1965.
8. T. R. Hogness, W. C. Johnson, and A. R. Armstrong, "Qualitative Analysis and Chemical Equilibrium," 5th ed., Holt, Rinehart and Winston, Inc., New York, 1966.
9. D. A. Skoog and D. W. West, "Fundamentals of Analytical Chemistry," 3rd ed., Holt, Rinehart and Winston, New York, 1976.
10. J. Dolezal, P. Povondra, and Z. Sulcek, "Decomposition Techniques in Inorganic Analyses," Elsevier, Amsterdam, 1968.
11. T. T. Gorsuck, "The Destruction of Organic Matter," Pergamon Press, Oxford, 1970.

Index